AN INTRODUCTION TO PROBABILISTIC NUMBER THEORY

Despite its seemingly deterministic nature, the study of whole numbers, especially prime numbers, has many interactions with probability theory, the theory of random processes and events. This surprising connection was first discovered around 1920, but in recent years, the links have become much deeper and better understood.

Aimed at beginning graduate students, this textbook is the first to explain some of the most modern parts of the story. Such topics include the Chebychev bias, universality of the Riemann zeta function, exponential sums, and the bewitching shapes known as Kloosterman paths. Emphasis is given throughout to probabilistic ideas in the arguments, not just the final statements, and the focus is on key examples over technicalities. The book develops probabilistic number theory from scratch, with short appendices summarizing the most important background results from number theory, analysis, and probability, making it a readable and incisive introduction to this beautiful area of mathematics.

Emmanuel Kowalski is Professor in the Mathematics Department of the Swiss Federal Institute of Technology, Zurich. He is the author of five previous books, including the widely cited *Analytic Number Theory* (2004) with H. Iwaniec, which is considered to be the standard graduate textbook for analytic number theory.

CAMBRIDGE STUDIES IN ADVANCED MATHEMATICS

All the titles listed below can be obtained from good booksellers or from Cambridge University Press. For a complete series listing, visit www.cambridge.org/mathematics.

Already Published
152 V. I. Paulsen & M. Raghupathi *An Introduction to the Theory of Reproducing Kernel Hilbert Spaces*
153 R. Beals & R. Wong *Special Functions and Orthogonal Polynomials*
154 V. Jurdjevic *Optimal Control and Geometry: Integrable Systems*
155 G. Pisier *Martingales in Banach Spaces*
156 C. T. C. Wall *Differential Topology*
157 J. C. Robinson, J. L. Rodrigo & W. Sadowski *The Three-Dimensional Navier-Stokes Equations*
158 D. Huybrechts *Lectures on K3 Surfaces*
159 H. Matsumoto & S. Taniguchi *Stochastic Analysis*
160 A. Borodin & G. Olshanski *Representations of the Infinite Symmetric Group*
161 P. Webb *Finite Group Representations for the Pure Mathematician*
162 C. J. Bishop & Y. Peres *Fractals in Probability and Analysis*
163 A. Bovier *Gaussian Processes on Trees*
164 P. Schneider *Galois Representations and* (φ, Γ)*-Modules*
165 P. Gille & T. Szamuely *Central Simple Algebras and Galois Cohomology (2nd Edition)*
166 D. Li & H. Queffelec *Introduction to Banach Spaces, I*
167 D. Li & H. Queffelec *Introduction to Banach Spaces, II*
168 J. Carlson, S. Müller-Stach & C. Peters *Period Mappings and Period Domains (2nd Edition)*
169 J. M. Landsberg *Geometry and Complexity Theory*
170 J. S. Milne *Algebraic Groups*
171 J. Gough & J. Kupsch *Quantum Fields and Processes*
172 T. Ceccherini-Silberstein, F. Scarabotti & F. Tolli *Discrete Harmonic Analysis*
173 P. Garrett *Modern Analysis of Automorphic Forms by Example, I*
174 P. Garrett *Modern Analysis of Automorphic Forms by Example, II*
175 G. Navarro *Character Theory and the McKay Conjecture*
176 P. Fleig, H. P. A. Gustafsson, A. Kleinschmidt & D. Persson *Eisenstein Series and Automorphic Representations*
177 E. Peterson *Formal Geometry and Bordism Operators*
178 A. Ogus *Lectures on Logarithmic Algebraic Geometry*
179 N. Nikolski *Hardy Spaces*
180 D.-C. Cisinski *Higher Categories and Homotopical Algebra*
181 A. Agrachev, D. Barilari & U. Boscain *A Comprehensive Introduction to Sub-Riemannian Geometry*
182 N. Nikolski *Toeplitz Matrices and Operators*
183 A. Yekutieli *Derived Categories*
184 C. Demeter *Fourier Restriction, Decoupling and Applications*
185 D. Barnes & C. Roitzheim *Foundations of Stable Homotopy Theory*
186 V. Vasyunin & A. Volberg *The Bellman Function Technique in Harmonic Analysis*
187 M. Geck & G. Malle *The Character Theory of Finite Groups of Lie Type*
188 B. Richter *Category Theory for Homotopy Theory*
189 R. Willett & G. Yu *Higher Index Theory*
190 A. Bobrowski *Generators of Markov Chains*
191 D. Cao, S. Peng & S. Yan *Singularly Perturbed Methods for Nonlinear Elliptic Problems*

An Introduction to Probabilistic Number Theory

EMMANUEL KOWALSKI
Swiss Federal Institute of Technology, Zürich

CAMBRIDGE
UNIVERSITY PRESS

CAMBRIDGE
UNIVERSITY PRESS

University Printing House, Cambridge CB2 8BS, United Kingdom

One Liberty Plaza, 20th Floor, New York, NY 10006, USA

477 Williamstown Road, Port Melbourne, VIC 3207, Australia

314–321, 3rd Floor, Plot 3, Splendor Forum, Jasola District Centre, New Delhi – 110025, India

79 Anson Road, #06–04/06, Singapore 079906

Cambridge University Press is part of the University of Cambridge.

It furthers the University's mission by disseminating knowledge in the pursuit of education, learning, and research at the highest international levels of excellence.

www.cambridge.org
Information on this title: www.cambridge.org/9781108840965
DOI: 10.1017/9781108888226

First published 2021

A catalogue record for this publication is available from the British Library.

ISBN 978-1-108-84096-5 Hardback

Les probabilités et la théorie analytique des nombres, c'est la même chose.
paraphrase of Y. Guivarc'h, Rennes, July 2017

Contents

Preface

The style of this book is a bit idiosyncratic. The results that interest us belong to number theory, but the emphasis in the proofs will be on the probabilistic aspects and on the interaction between number theory and probability theory. In fact, we attempt to write the proofs so that they use *as little arithmetic as possible*, in order to clearly isolate the crucial number-theoretic ingredients that are involved.

This book is quite short. We attempt to foster an interest in the topic by focusing on a few key results that are accessible and at the same time particularly appealing, in the author's opinion, without targeting an encyclopedic treatment of any. We also try to emphasize connections to other areas of mathematics – first, to a wide array of arithmetic topics, but also to some aspects of ergodic theory, expander graphs, and so on.

In some sense, the ideal reader of this book is a student who has attended at least one introductory advanced undergraduate or beginning graduate-level probability course, including especially the Central Limit Theorem, and maybe some aspects of Brownian motion, and who is interested in seeing how probability interacts with number theory. For this reason, there are almost no number-theoretic prerequisites, although it is helpful to have some knowledge of the distribution of primes.

Probabilistic number theory is currently evolving very rapidly, and uses more and more refined probabilistic tools and results. For many number theorists, we hope that the detailed and motivated discussion of basic probabilistic facts and tools in this book will be useful as a basic "toolbox".

Acknowledgments. The first draft of this book was prepared for a course Introduction to Probabilistic Number Theory that I taught at ETH Zürich during the fall semester of 2015. Thanks to the students of the course for their interest, in particular to M. Gerspach, A. Steiger, and P. Zenz for sending corrections and to B. Löffel for organizing and preparing the exercise sessions.

Thanks to M. Burger for showing me Cauchy's proof of the Euler formula,

$$\sum_{n=1}^{+\infty} \frac{1}{n^2} = \frac{\pi^2}{6},$$

in Exercise 1.3.4. Thanks to V. Tassion for help with the proof of Proposition B.11.11 and to G. Ricotta and E. Royer for pointing out a small mistake in [79].

Thanks to M. Radziwiłł and K. Soundararajan for sharing their proof [95] of Selberg's Central Limit Theorem for $\log|\zeta(\frac{1}{2} + it)|$, which was then unpublished.

This work was partially supported by the DFG-SNF lead agency program grant 200020L_175755.

Prerequisites and Notation

The basic requirements for most of this text are standard introductory graduate courses in algebra, analysis (including Lebesgue integration and complex analysis), and probability. Of course, knowledge and familiarity with basic number theory (for instance, the distribution of primes up to the Bombieri–Vinogradov Theorem) are helpful, but we review in Appendix C all the results that we use. Similarly, Appendix B summarizes the notation and facts from probability theory that are the most important for us.

We will use the following notation:

(1) For subsets Y_1 and Y_2 of an arbitrary set X, we denote by $Y_1 - Y_2$ the difference set, that is, the set of elements $x \in Y_1$ such that $x \notin Y_2$.

(2) A locally compact topological space is always assumed to be separated (i.e., Hausdorff), as in Bourbaki [15].

(3) For a set X, $|X| \in [0, +\infty]$ denotes its cardinal, with $|X| = \infty$ if X is infinite. There is no distinction in this text between the various infinite cardinals.

(4) If X is a set and f, g two complex-valued functions on X, then we write synonymously $f = O(g)$ or $f \ll g$ to say that there exists a constant $C \geqslant 0$ (sometimes called an "implied constant") such that $|f(x)| \leqslant Cg(x)$ for all $x \in X$. Note that this implies that in fact $g \geqslant 0$. We also write $f \asymp g$ to indicate that $f \ll g$ and $g \ll f$.

(5) If X is a topological space, $x_0 \in X$ and f and g are functions defined on a neighborhood of x_0, with $g(x) \neq 0$ for x in a neighborhood of x_0, then we say that $f(x) = o(g(x))$ as $x \to x_0$ if $f(x)/g(x) \to 0$ as $x \to x_0$, and that $f(x) \sim g(x)$ as $x \to x_0$ if $f(x)/g(x) \to 1$.

(6) We write $a \mid b$ for the divisibility relation "a divides b"; we denote by (a, b) the gcd of two integers a and b, and by $[a, b]$ their lcm.

(7) Usually, the variable p will always refer to prime numbers. In particular, a series $\sum_p(\cdots)$ refers to a series over primes (summed in increasing order, in case it is not known to be absolutely convergent), and similarly for a product over primes.
(8) We denote by $\mathbf{F}_p$ the finite field $\mathbf{Z}/p\mathbf{Z}$, for p prime, and more generally by $\mathbf{F}_q$ a finite field with q elements, where $q = p^n$, $n \geqslant 1$, is a power of p. We will recall the properties of finite fields when we require them.
(9) For a complex number z, we write $e(z) = e^{2i\pi z}$. If $q \geqslant 1$ and $x \in \mathbf{Z}/q\mathbf{Z}$, then $e(x/q)$ is well defined by taking any representative of x in $\mathbf{Z}$ to compute the exponential.
(10) If $q \geqslant 1$ and $x \in \mathbf{Z}$ (or $x \in \mathbf{Z}/q\mathbf{Z}$) is an integer that is coprime to q (or a residue class invertible modulo q), we sometimes denote by $\bar{q}$ the inverse class such that $x\bar{x} = 1$ in $\mathbf{Z}/q\mathbf{Z}$. This will always be done in such a way that the modulus q is clear from context, in the case where x is an integer.
(11) Given a probability space $(\Omega, \Sigma, \mathbf{P})$, we denote by $\mathbf{E}(\cdot)$ (resp. $\mathbf{V}(\cdot)$) the expectation (resp. the variance) computed with respect to $\mathbf{P}$. It will often happen that we have a sequence $(\Omega_{\mathrm{N}}, \Sigma_{\mathrm{N}}, \mathbf{P}_{\mathrm{N}})$ of probability spaces; we will then denote by $\mathbf{E}_{\mathrm{N}}$ or $\mathbf{V}_{\mathrm{N}}$ the respective expectation and variance with respect to $\mathbf{P}_{\mathrm{N}}$.
(12) Given a measure space (Ω, Σ, μ) (not necessarily a probability space), a set Y with a σ-algebra Σ' and a measurable map $f : \Omega \longrightarrow \mathrm{Y}$, we denote by $f_*(\mu)$ (or sometimes $f(\mu)$) the image measure on Y; in the case of a probability space, so that f is seen as a random variable on Ω, this is the probability law of f seen as a "random Y-valued element." If the set Y is given without specifying a σ-algebra, we will view it usually as given with the σ-algebra generated by sets $\mathrm{Z} \subset \mathrm{Y}$ such that $f^{-1}(\mathrm{Z})$ belongs to Σ.
(13) As a typographical convention, we will use sans-serif fonts like X to denote an *arithmetic* random variable and more standard fonts (like X) for "abstract" random variables. When using the same letter, this will usually mean that somehow the "purely random" X is the "model" of the arithmetic quantity X.

1

Introduction

1.1 Presentation

Different authors might define "probabilistic number theory" in different ways. Our point of view will be to see it as *the study of the asymptotic behavior of arithmetically defined sequences of probability measures, or random variables*. Thus the content of this book is based on examples of situations where we can say interesting things concerning such sequences. However, in Chapter 7, we will quickly survey some topics that might quite legitimately be seen as part of probabilistic number theory in a broader sense.

To illustrate what we have in mind, the most natural starting point is a famous result of Erdős and Kac.

Theorem 1.1.1 (the Erdős–Kac Theorem) *For any positive integer $n \geqslant 1$, let $\omega(n)$ denote the number of prime divisors of n, counted without multiplicity. Then, for any real numbers $a < b$, we have*

$$\lim_{N\to+\infty} \frac{1}{N}\left|\left\{1 \leqslant n \leqslant N \mid a \leqslant \frac{\omega(n) - \log\log N}{\sqrt{\log\log N}} \leqslant b\right\}\right| = \frac{1}{\sqrt{2\pi}}\int_a^b e^{-x^2/2}dx.$$

To spell out the connection between this statement and our slogan, one sequence of probability measures involved here is the sequence $(\mu_N)_{N\geqslant 1}$, defined as the uniform probability measure supported on the finite set $\Omega_N = \{1, \ldots, N\}$. This sequence is defined arithmetically, because the study of integers is part of arithmetic. The *asymptotic behavior* is revealed by the statement. Namely, consider the sequence of random variables

$$X_N(n) = \frac{\omega(n) - \log\log N}{\sqrt{\log\log N}}$$

defined on Ω_N for $N \geqslant 3$,[1] and the sequence (ν_N) of their probability distributions, which are (Borel) probability measures on $\mathbf{R}$ defined by

$$\nu_N(A) = \mu_N(X_N \in A) = \frac{1}{N}\left|\left\{1 \leqslant n \leqslant N \mid \frac{\omega(n) - \log\log N}{\sqrt{\log\log N}} \in A\right\}\right|$$

for any measurable set $A \subset \mathbf{R}$. These form another *arithmetically defined sequence of probability measures*, since primes are definitely arithmetic objects. Theorem 1.1.1 is, by basic probability theory, equivalent to the fact that the sequence (ν_N) converges in law to a standard Gaussian random variable as $N \to +\infty$. (We recall here that a sequence of real-valued random variables (X_N) converges in law to a random variable X if

$$\mathbf{E}(f(X_N)) \to \mathbf{E}(f(X))$$

for all bounded continuous functions $f : \mathbf{R} \to \mathbf{C}$, and that one can show that it is equivalent to

$$\mathbf{P}(a < X_N < b) \to \mathbf{P}(a < X < b)$$

for all $a < b$ such that $\mathbf{P}(X = a) = \mathbf{P}(X = b) = 0$; for the standard Gaussian, this means for all a and b; see Section B.3 for reminders about this.)

The Erdős–Kac Theorem is probably the simplest case where a natural deterministic arithmetic quantity (the number of prime factors of an integer), which is individually very hard to grasp, nevertheless exhibits a statistical or probabilistic behavior which fits a very common probability distribution. This is the prototype of the kinds of statements we will discuss (although sometimes the limiting measures will be far from standard!).

We will prove Theorem 1.1.1 in the next chapter. Before we do this, we will begin with a few results that are much more elementary but which may, with hindsight, be considered as the simplest cases of the type of results we want to describe.

1.2 How Does Probability Link with Number Theory Really?

Before embarking on this, however, it might be useful to give a rough idea of the way probability theory and arithmetic will combine to give interesting limit theorems like the Erdős–Kac Theorem. The strategy that we outline here

[1] Simply so that $\log\log N > 0$.

will be, in different guises, at the core of the strategy of the proofs of many theorems in this book.

We typically will be working with a sequence (X_n) of arithmetically interesting random variables, and we wish to prove that it converges in law. In many cases, we do this with a two-step process.

(1) We begin by approximating (X_n) by another sequence (Y_n), in such a way that convergence in law of these approximations implies that of (X_n), with the same limit. In other words, we see Y_n as a kind of perturbation of X_n, which is small enough to preserve convergence in law. Notably, the approximation might be of different sorts: the difference $\mathsf{X}_n - \mathsf{Y}_n$ might, for instance, converge to 0 in probability, or in some L^p-space; in fact, we will sometimes encounter a process of successive approximations, where the successive perturbations are small in different senses, before reaching a convenient approximation Y_n (this is the case in the proof of Theorem 4.1.2).

(2) Having found a good approximation Y_n, we prove that it converges in law using a probabilistic criterion that is sufficiently robust to apply; typical examples are the method of moments, and the convergence theorem of P. Lévy based on characteristic functions (i.e., Fourier transforms), because analytic number theory often gives tools to compute approximately such invariants of arithmetically defined random variables.

Both steps are sometimes quite easy to motivate using some heuristic arguments (for instance, when X_n or Y_n are represented as a sum of various terms, we might guess that these are "approximately independent," to lead to a limit similar to that of sums of independent random variables), but they may also involve quite subtle ideas.

We will not dwell further on this overarching strategy, but the reader will be able to recognize how it fits into this skeleton when we discuss the steps of the proof of some of the main theorems.

In many papers written by (or for) analytic number theorists, the approximations of Step 1, as well as (say) the moment computations of Step 2, are performed using notation, terminology and normalizations coming from the customs and standards of analytic number theory. In this book, we will try to express them instead, as much as possible, in good probabilistic style (e.g., we attempt to mention as little as possible the "elementary events" of the underlying probability space). This is usually simply a matter of cosmetic transformations, but sometimes it leads to slightly different emphasis, in particular concerning the nature of the approximations in Step 1. We suggest that the reader compare our presentation with that of some of the original source papers, in order to assess whether this style is enlightening (as we often find it to be), or not.

1.3 A Prototype: Integers in Arithmetic Progressions

As mentioned above, we begin with a result that is so elementary that it is usually not presented as a separate statement (let alone as a theorem!). Nevertheless, as we will see, it is the basic ingredient (and explanation) for the Erdős–Kac Theorem, and generalizations of it become quite quickly very deep.

Theorem 1.3.1 *For* $\mathrm{N} \geqslant 1$, *let* $\Omega_{\mathrm{N}} = \{1, \ldots, \mathrm{N}\}$ *with the uniform probability measure* $\mathbf{P}_{\mathrm{N}}$. *Fix an integer* $q \geqslant 1$, *and denote by* $\pi_q \colon \mathbf{Z} \longrightarrow \mathbf{Z}/q\mathbf{Z}$ *the reduction modulo* q *map. Let* X_{N} *be the random variables given by* $\mathrm{X}_{\mathrm{N}}(n) = \pi_q(n)$ *for* $n \in \Omega_{\mathrm{N}}$.

As $\mathrm{N} \to +\infty$, *the random variables* X_{N} *converge in law to the uniform probability measure* μ_q *on* $\mathbf{Z}/q\mathbf{Z}$. *In fact, for any function*

$$f \colon \mathbf{Z}/q\mathbf{Z} \longrightarrow \mathbf{C},$$

we have

$$\left|\mathbf{E}(f(\mathrm{X}_{\mathrm{N}})) - \mathbf{E}(f)\right| \leqslant \frac{2}{\mathrm{N}} \|f\|_1, \tag{1.1}$$

where

$$\|f\|_1 = \sum_{a \in \mathbf{Z}/q\mathbf{Z}} |f(a)|.$$

Proof It is enough to prove (1.1), which gives the convergence in law by letting $\mathrm{N} \to +\infty$. This is quite simple. By definition, we have

$$\mathbf{E}(f(\mathrm{X}_{\mathrm{N}})) = \frac{1}{\mathrm{N}} \sum_{1 \leqslant n \leqslant \mathrm{N}} f(\pi_q(n))$$

and

$$\mathbf{E}(f) = \frac{1}{q} \sum_{a \in \mathbf{Z}/q\mathbf{Z}} f(a).$$

The idea is then clear: among the integers $1 \leqslant n \leqslant \mathrm{N}$, roughly N/q are in any given residue class $a \pmod q$, and if we use this approximation in the first formula, we obtain precisely the second.

To do this in detail, we gather the integers in the sum according to their residue class a modulo q. This gives

$$\frac{1}{N}\sum_{1\leqslant n\leqslant N} f(\pi_q(n)) = \sum_{a\in \mathbf{Z}/q\mathbf{Z}} f(a) \times \frac{1}{N}\sum_{\substack{1\leqslant n\leqslant N\\ n\equiv a \,(\mathrm{mod}\, q)}} 1.$$

The inner sum, for each a, counts the number of integers n in the interval $1 \leqslant n \leqslant N$ such that the remainder under division by q is a. These integers n can be written $n = mq + a$ for some $m \in \mathbf{Z}$, if we view a as an actual integer, and therefore it is enough to count those integers $m \in \mathbf{Z}$ for which $1 \leqslant mq + a \leqslant N$. The condition translates to

$$\frac{1-a}{q} \leqslant m \leqslant \frac{N-a}{q},$$

and therefore we are reduced to counting integers *in an interval*. This is not difficult, although we have to be careful with boundary terms, since the bounds of the interval are not necessarily integers. The length of the interval is $(N-a)/q-(1-a)/q = (N-1)/q$. In general, in an interval $[\alpha,\beta]$ with $\alpha \leqslant \beta$, the number $N_{\alpha,\beta}$ of integers satisfies

$$\beta - \alpha - 1 \leqslant N_{\alpha,\beta} \leqslant \beta - \alpha + 1$$

(and the boundary contributions should not be forgotten, although they are typically negligible when the interval is long enough).

Hence the number N_a of values of m satisfies

$$\frac{N-1}{q} - 1 \leqslant N_a \leqslant \frac{N-1}{q} + 1, \tag{1.2}$$

and therefore

$$\left| N_a - \frac{N}{q} \right| \leqslant 1 + \frac{1}{q}.$$

By summing over a in $\mathbf{Z}/q\mathbf{Z}$, we deduce now that

$$\left|\frac{1}{N}\sum_{1\leqslant n\leqslant N} f(\pi_q(n)) - \frac{1}{q}\sum_{a\in\mathbf{Z}/q\mathbf{Z}} f(a)\right| = \left|\sum_{a\in\mathbf{Z}/q\mathbf{Z}} f(a)\left(\frac{N_a}{N}-\frac{1}{q}\right)\right|$$
$$\leqslant \frac{1+q^{-1}}{N}\sum_{a\in\mathbf{Z}/q\mathbf{Z}} |f(a)| \leqslant \frac{2}{N}\|f\|_1.$$

□

Remark 1.3.2 As a matter of notation, we will sometimes remove the variable N from the notation of random variables, since the value of N is usually made clear by the context, frequently because of its appearance in an expression involving $\mathbf{P}_N(\cdot)$ or $\mathbf{E}_N(\cdot)$, which refers to the probability and expectation on Ω_N.

Despite its simplicity, this result already brings up a number of important features that will occur extensively in later chapters.

A first remark is that we actually proved something much stronger than the statement of convergence in law: the bound (1.1) gives a rather precise estimate of the speed of convergence of expectations (or probabilities) computed using the law of X_N to those computed using the limit uniform distribution μ_q. Most importantly, as we will see shortly, these estimates are uniform in terms of q, and give us information on convergence, or more properly speaking on the "distance" between the law of X_N and μ_q, even if q depends on N in some way.

To be more precise, take f to be the characteristic function of a residue class $a \in \mathbf{Z}/q\mathbf{Z}$. Then since $\mathbf{E}(f) = 1/q$, we get

$$\left|\mathbf{P}(\pi_q(n) = a) - \frac{1}{q}\right| \leqslant \frac{2}{N}.$$

This is nontrivial information as long as q is a bit smaller than N. Thus, this states that the probability that $n \leqslant N$ is congruent to a modulo q is close to the intuitive probability $1/q$, uniformly for all q just a bit smaller than N, and also uniformly for all residue classes. We will see, both below and in many similar situations, that uniformity aspects are essential in applications.

The second remark concerns the interpretation of the result. Theorem 1.3.1 can explain what is meant by such intuitive statements as *the probability that an integer is divisible by* 2 *is* 1/2. Namely, this is the probability, according to the uniform measure on $\mathbf{Z}/2\mathbf{Z}$, of the set $\{0\}$, and this is simply the limit given by the convergence in law of the variables $\pi_2(n)$ defined on Ω_N to the uniform measure μ_2.

This idea applies to many other similar-sounding problems. The most elementary among these can often be solved using Theorem 1.3.1. We present one famous example: what is the "probability" that an integer $n \geqslant 1$ is squarefree, which means that n is *not* divisible by a square m^2 for some integer $m \geqslant 2$ (or, equivalently, by the square of some prime number)? Here the interpretation is that this probability should be

$$\lim_{N \to +\infty} \frac{1}{N} |\{1 \leqslant n \leqslant N \mid n \text{ is squarefree}\}|.$$

If we prefer (as we do) to speak of sequences of random variables, we can take the sequence of Bernoulli random variables B_N indicators of the event that $n \in \Omega_N$ is squarefree, so that

$$\mathbf{P}(\mathsf{B}_N = 1) = \frac{1}{N}|\{1 \leqslant n \leqslant N \mid n \text{ is squarefree}\}|.$$

We then ask about the limit in law of (B_N). The answer is as follows:

Proposition 1.3.3 *The sequence* (B_N) *converges in law to a Bernoulli random variable* B *with* $\mathbf{P}(\mathrm{B} = 1) = \frac{6}{\pi^2}$. *In other words, the "probability" that an integer n is squarefree, in the interpretation discussed above, is* $6/\pi^2$.

Proof The idea is to use inclusion-exclusion: to say that n is squarefree means that it is not divisible by the square p^2 of any prime number. Thus, if we denote by $\mathbf{P}_N$ the probability measure on Ω_N, we have

$$\mathbf{P}_N(n \text{ is squarefree}) = \mathbf{P}_N\Big(\bigcap_{p \text{ prime}} \{p^2 \text{ does not divide } n\}\Big).$$

There is one key step now that is both obvious and crucial: because of the nature of Ω_N, the infinite intersection may be replaced by the intersection over primes $p \leqslant \sqrt{N}$, since all integers in Ω_N are $\leqslant N$. Applying the inclusion-exclusion formula, we obtain

$$\mathbf{P}_N\Big(\bigcap_{p \leqslant N^{1/2}} \{p^2 \text{ does not divide } n\}\Big) = \sum_{I} (-1)^{|I|}\, \mathbf{P}_N\Big(\bigcap_{p \in I} \{p^2 \text{ divides } n\}\Big), \tag{1.3}$$

where I runs over the set of subsets of the set $\{p \leqslant N^{1/2}\}$ of primes $\leqslant N^{1/2}$, and $|I|$ is the cardinality of I. But, by the Chinese Remainder Theorem, we have

$$\bigcap_{p \in I} \{p^2 \text{ divides } n\} = \{d_I^2 \text{ divides } n\},$$

where d_I is the product of the primes in I. Once more, note that this set is empty if $d_I^2 > N$. Moreover, the fundamental theorem of arithmetic shows that $I \mapsto d_I$ is injective, and we can recover $|I|$ also from d_I as the number of prime factors of d_I. Therefore, we get

$$\mathbf{P}_N(n \text{ is squarefree}) = \sum_{d \leqslant N^{1/2}} \mu(d)\, \mathbf{P}_N(d^2 \text{ divides } n),$$

where $\mu(d)$ is the Möbius function, defined for integers $d \geqslant 1$ by

$$\mu(d) = \begin{cases} 0 & \text{if } d \text{ is not squarefree,} \\ (-1)^k & \text{if } d = p_1 \cdots p_k \text{ with } p_i \text{ distinct primes} \end{cases}$$

(see Definition C.1.3).

But d^2 divides n if and only if the image of n by reduction modulo d^2 is 0. By Theorem 1.3.1 applied with $q = d^2$ for all $d \leqslant \mathrm{N}^{1/2}$, with f the indicator function of the residue class of 0, we get

$$\mathbf{P}_{\mathrm{N}}(d^2 \text{ divides } n) = \frac{1}{d^2} + \mathrm{O}(\mathrm{N}^{-1})$$

for all d, where the implied constant in the $\mathrm{O}(\cdot)$ symbol is independent of d (in fact, it is at most 2). Note in passing how we use crucially here the fact that Theorem 1.3.1 was uniform and explicit with respect to the parameter q.

Summing the last formula over $d \leqslant \mathrm{N}^{1/2}$, we deduce

$$\mathbf{P}_{\mathrm{N}}(n \text{ is squarefree}) = \sum_{d \leqslant n^{1/2}} \frac{\mu(d)}{d^2} + \mathrm{O}\Big(\frac{1}{\sqrt{\mathrm{N}}}\Big).$$

Since the series with terms $1/d^2$ converges, this shows the existence of the limit, and that $(\mathsf{B}_{\mathrm{N}})$ converges in law as $\mathrm{N} \to +\infty$ to a Bernoulli random variable B with success probability

$$\mathbf{P}(\mathrm{B} = 1) = \sum_{d \geqslant 1} \frac{\mu(d)}{d^2}, \qquad \mathbf{P}(\mathrm{B} = 0) = 1 - \sum_{d \geqslant 1} \frac{\mu(d)}{d^2}.$$

It is a well-known fact (the "Basel problem," first solved by Euler; see Exercise 1.3.4 for a proof) that

$$\sum_{d \geqslant 1} \frac{1}{d^2} = \frac{\pi^2}{6}.$$

Moreover, a basic property of the Möbius function states that

$$\sum_{d \geqslant 1} \frac{\mu(d)}{d^s} = \frac{1}{\zeta(s)}$$

for any complex number s with $\mathrm{Re}(s) > 1$, where

$$\zeta(s) = \sum_{d \geqslant 1} \frac{1}{d^s}$$

is the Riemann zeta function (Corollary C.1.5), and hence we get

$$\sum_{d\geqslant 1}\frac{\mu(d)}{d^2}=\frac{1}{\zeta(2)}=\frac{6}{\pi^2}. \qquad \square$$

Exercise 1.3.4 In this exercise, we explain a proof of Euler's formula $\zeta(2)=\pi^2/6$.

(1) Assuming that

$$\frac{\sin(\pi x)}{\pi x}=\prod_{n\geqslant 1}\left(1-\frac{x^2}{n^2}\right)$$

(another formula of Euler), find a heuristic proof of $\zeta(2)=\pi^2/6$. [**Hint**: First, express the sum of the inverses of the roots of a polynomial (with nonzero constant term) in terms of its coefficients.]

The following argument, due to Cauchy, can be seen as a way to make rigorous the previous idea.

(2) Show that for $n\geqslant 1$ and $x\in \mathbf{R}-\pi\mathbf{Z}$, we have

$$\frac{\sin(nx)}{(\sin x)^n}=\sum_{0\leqslant m\leqslant n/2}(-1)^m\binom{n}{2m+1}\operatorname{cotan}(x)^{n-(2m+1)}.$$

(3) Let $m\geqslant 1$ be an integer, and let $n=2m+1$. Show that

$$\sum_{r=1}^{m}\operatorname{cotan}\left(\frac{r\pi}{n}\right)^2=\frac{2m(2m-1)}{6}$$

and

$$\sum_{r=1}^{m}\frac{1}{\sin\left(\frac{r\pi}{n}\right)^2}=\frac{2m(2m+2)}{6}.$$

[**Hint**: Using (1), view the numbers $\operatorname{cotan}(r\pi/n)^2$ as the roots of a polynomial of degree m, and use the formula for the sum of the roots of a polynomial.]

(4) Deduce that

$$\frac{2m(2m-1)}{6}<\sum_{k=1}^{m}\left(\frac{2m+1}{k\pi}\right)^2<\frac{2m(2m+2)}{6},$$

and conclude.

The proof of Proposition 1.3.3 above was written in probabilistic style, emphasizing the connection with Theorem 1.3.1. It can be expressed more straightforwardly as a sequence of manipulation with finite sums, using the formula

$$\sum_{d^2 \mid n} \mu(d) = \begin{cases} 1 & \text{if } n \text{ is squarefree,} \\ 0 & \text{otherwise} \end{cases} \tag{1.4}$$

for $n \geqslant 1$ (which is implicit in our discussion) and the approximation

$$\sum_{\substack{1 \leqslant n \leqslant \mathrm{N} \\ d \mid n}} 1 = \frac{\mathrm{N}}{d} + \mathrm{O}(1)$$

for the number of integers in an interval which are divisible by some $d \geqslant 1$. This goes as follows:

$$\begin{aligned}\sum_{\substack{n \leqslant \mathrm{N} \\ n \text{ squarefree}}} 1 = \sum_{n \leqslant \mathrm{N}} \sum_{d^2 \mid n} \mu(d) &= \sum_{d \leqslant \sqrt{\mathrm{N}}} \mu(d) \sum_{\substack{n \leqslant \mathrm{N} \\ d^2 \mid n}} 1 \\ &= \sum_{d \leqslant \sqrt{\mathrm{N}}} \mu(d) \left(\frac{\mathrm{N}}{d^2} + \mathrm{O}(1) \right) \\ &= \mathrm{N} \sum_{d} \frac{\mu(d)}{d^2} + \mathrm{O}(\sqrt{\mathrm{N}}).\end{aligned}$$

Obviously, this is much shorter, although one needs to know the formula (1.4), which was implicitly derived in the previous proof.[2] But there is something quite important to be gained from the probabilistic viewpoint, which might be missed by reading too quickly the second proof. Indeed, in formulas like (1.3) (or many others), the precise nature of the underlying probability space Ω_{N} is quite hidden – as is customary in probability where this is often not really relevant. In our situation, this suggests naturally to study similar problems for *different* sequences of integer-valued random variables rather than taking integers uniformly between 1 and N.

This has indeed been done, and in many different ways. But even before looking at any example, we can predict that some new – interesting – phenomena will arise when doing so. Indeed, even if our first proof of Proposition 1.3.3 was written in a very general probabilistic language, it did use one special feature of Ω_{N}: it only contains integers $n \leqslant \mathrm{N}$, and even more particularly, it does not contain any element divisible by d^2 for d larger than $\sqrt{\mathrm{N}}$. (More probabilistically, the probability $\mathbf{P}_{\mathrm{N}}(d^2 \text{ divides } n)$ is then zero.)

[2] Readers who are already well versed in analytic number theory might find it useful to translate back and forth various estimates written in probabilistic style in this book.

Now consider the following extension of the problem, which is certainly one of the first that may come to mind beyond our initial setting: we fix an irreducible polynomial $\mathrm{P} \in \mathbf{Z}[\mathrm{X}]$ of degree $m \geqslant 1$, and consider new Bernoulli random variables $\mathsf{B}_{\mathrm{P},\mathrm{N}}$ which are indicators of the event that $\mathrm{P}(n)$ is squarefree on Ω_{N} (instead of n itself). Asking about the limit of these random variables means asking for the "probability" that $\mathrm{P}(n)$ is squarefree, when $1 \leqslant n \leqslant \mathrm{N}$. But although there is an elementary analogue of Theorem 1.3.1, it is easy to see that this does *not* give enough control of

$$\mathbf{P}_{\mathrm{N}}(d^2 \text{ divides } \mathrm{P}(n))$$

when d is "too large" compared with N. And this explains partly why, in fact, as of 2020 at least, *there is not even a* single *irreducible polynomial* $\mathrm{P} \in \mathbf{Z}[\mathrm{X}]$ *of degree* 4 *or higher for which it is known that* $\mathrm{P}(n)$ *is squarefree infinitely often.*

Exercise 1.3.5 (1) Let $k \geqslant 2$ be an integer. Compute the "probability," in the same sense as in Proposition 1.3.3, that an integer n is k-free, that is, that there is no integer $m \geqslant 2$ such that m^k divides n.

(2) Compute the "probability" that two integers n_1 and n_2 are coprime, in the sense of taking the corresponding Bernoulli random variables on $\Omega_{\mathrm{N}} \times \Omega_{\mathrm{N}}$ and their limit as $\mathrm{N} \to +\infty$.

Exercise 1.3.6 Let $\mathrm{P} \in \mathbf{Z}[\mathrm{X}]$ be an irreducible polynomial of degree $m \geqslant 1$. For $q \geqslant 1$, let π_q be the projection from $\mathbf{Z}$ to $\mathbf{Z}/q\mathbf{Z}$ as before.

(1) Show that for any $q \geqslant 1$, the random variables $\mathrm{X}_{\mathrm{N}}(n) = \pi_q(\mathrm{P}(n))$ converge in law to a probability measure $\mu_{\mathrm{P},q}$ on $\mathbf{Z}/q\mathbf{Z}$. Is $\mu_{\mathrm{P},q}$ uniform?

(2) Find values of T, depending on N and as large as possible, such that

$$\mathbf{P}_{\mathrm{N}}(\mathrm{P}(n) \text{ is not divisible by } p^2 \text{ for } p \leqslant \mathrm{T}) > 0.$$

How large should T be so that this implies straightforwardly that

$$\{n \geqslant 1 \mid \mathrm{P}(n) \text{ is squarefree}\}$$

is infinite?

(3) Prove that the set

$$\{n \geqslant 1 \mid \mathrm{P}(n) \text{ is } (m+1)\text{-free}\}$$

is infinite.

We conclude this section with another very important feature of Theorem 1.3.1 from the probabilistic point of view, namely, its link with independence.

If q_1 and q_2 are positive integers which are coprime, then the Chinese Remainder Theorem implies that the map

$$\begin{cases} \mathbf{Z}/q_1q_2\mathbf{Z} \longrightarrow \mathbf{Z}/q_1\mathbf{Z} \times \mathbf{Z}/q_2\mathbf{Z}, \\ x \mapsto (x \,(\mathrm{mod}\, q_1), x \,(\mathrm{mod}\, q_2)) \end{cases}$$

is a bijection (in fact, a ring isomorphism). Under this bijection, the uniform probability measure $\mu_{q_1q_2}$ on $\mathbf{Z}/q_1q_2\mathbf{Z}$ corresponds to the product measure $\mu_{q_1} \otimes \mu_{q_2}$. In particular, the random variables $x \mapsto x \,(\mathrm{mod}\, q_1)$ and $x \mapsto x \,(\mathrm{mod}\, q_2)$ on $\mathbf{Z}/q_1q_2\mathbf{Z}$ are independent.

The interpretation of this is that the random variables π_{q_1} and π_{q_2} on Ω_N are *asymptotically independent* as $N \to +\infty$, in the sense that

$$\lim_{N\to+\infty} \mathbf{P}_N(\pi_{q_1}(n) = a \text{ and } \pi_{q_2}(n) = b) = \frac{1}{q_1q_2}$$
$$= \Big(\lim_{N\to+\infty} \mathbf{P}_N(\pi_{q_1}(n) = a)\Big) \times \Big(\lim_{N\to+\infty} \mathbf{P}_N(\pi_{q_2}(n) = b)\Big)$$

for all $(a,b) \in \mathbf{Z}^2$. Intuitively, one would say that *divisibility by q_1 and q_2 are independent*, and especially that *divisibility by distinct primes are independent events*. We summarize this in the following extremely useful proposition:

Proposition 1.3.7 *For $N \geqslant 1$, let $\Omega_N = \{1, \dots, N\}$ with the uniform probability measure $\mathbf{P}_N$. Fix a finite set S of pairwise coprime integers.*

As $N \to +\infty$, the vector $(\pi_q)_{q\in S}$ seen as random vector on Ω_N with values in

$$X_S = \prod_{q\in S} \mathbf{Z}/q\mathbf{Z}$$

converges in law to a vector of independent and uniform random variables. In fact, for any function

$$f : X_S \longrightarrow \mathbf{C},$$

we have

$$\left|\mathbf{E}(f((\pi_q)_{q\in S})) - \mathbf{E}(f)\right| \leqslant \frac{2}{N}\|f\|_1. \tag{1.5}$$

Proof This is just an elaboration of the previous discussion. Let r be the product of the elements of S. Then the Chinese Remainder Theorem gives a ring-isomorphism $X_S \longrightarrow \mathbf{Z}/r\mathbf{Z}$ such that the uniform measure μ_r on the right-hand side corresponds to the product of the uniform measures on X_S.

Thus f can be identified with a function $g\colon \mathbf{Z}/r\mathbf{Z} \longrightarrow \mathbf{C}$, and its expectation to the expectation of g according to μ_r. By Theorem 1.3.1, we get

$$\left|\mathbf{E}(f((\pi_q)_{q\in\mathrm{S}})) - \mathbf{E}(f)\right| = \left|\mathbf{E}(g(\pi_r)) - \mathbf{E}(g)\right| \leqslant \frac{2\|g\|_1}{\mathrm{N}},$$

which is the desired result since f and g have also the same ℓ^1 norm. □

Remark 1.3.8 (1) Note that the random variables obtained by reduction modulo two coprime integers are not exactly independent: it is not true in general that

$$\mathbf{P}_{\mathrm{N}}(\pi_{q_1}(n) = a \text{ and } \pi_{q_2}(n) = b) = \mathbf{P}_{\mathrm{N}}(\pi_{q_1}(n) = a)\, \mathbf{P}_{\mathrm{N}}(\pi_{q_2}(n) = b).$$

This is the source of many interesting aspects of probabilistic number theory where classical ideas and concepts of probability for sequences of independent random variables are generalized or "tested" in a context where independence only holds in an asymptotic or approximate sense.

(2) There is one subtle point that appears in quantitative applications of Theorem 1.3.1 and Proposition 1.3.7 that is worth mentioning. Given an integer $q \geqslant 1$, certain functions f on $\mathbf{Z}/q\mathbf{Z}$ might have a large norm $\|f\|_1$, and yet they may have expressions as linear combinations of functions $\widetilde{f}$ on certain spaces $\mathbf{Z}/d\mathbf{Z}$, where d is a divisor of q, which have much smaller norms $\|\widetilde{f}\|_1$. Taking such possibilities into account and arguing modulo d instead of modulo q may lead to stronger estimates for the error

$$\mathbf{E}_{\mathrm{N}}(f(\pi_q(n))) - \mathbf{E}(f)$$

than those we have written down in terms of $\|f\|_1$. This is, for instance, especially clear if we take f to be a nonzero constant, in which case the difference is actually 0, but $\|f\|_1$ is of size q.

One can incorporate formally these improvements by using a different norm than $\|f\|_1$, as we now explain.

Let $q \geqslant 1$ be an integer. Let Φ_q be the set of functions $\varphi_{d,a}\colon \mathbf{Z}/q\mathbf{Z} \to \mathbf{C}$ which are characteristic functions of classes $x \equiv a \pmod d$ for some positive divisor $d \mid q$ and some $a \in \mathbf{Z}/d\mathbf{Z}$ (these are well-defined functions modulo q). In particular, the function $\varphi_{q,a}$ is just the delta function at a in $\mathbf{Z}/q\mathbf{Z}$, and $\varphi_{1,0}$ is the constant function 1.

For an arbitrary function $f\colon \mathbf{Z}/q\mathbf{Z} \to \mathbf{C}$, let

$$\|f\|_{c,1} = \inf\left\{\sum_{d\mid q}\ \sum_{a \,(\mathrm{mod}\, d)} |\lambda_{d,a}| \ \Big|\ f = \sum_{d\mid q}\ \sum_{a \,(\mathrm{mod}\, d)} \lambda_{d,a}\varphi_{d,a}\right\}.$$

This defines a norm on the space of functions on $\mathbf{Z}/q\mathbf{Z}$ (the subscript c refers to *congruences*); the norm $\|f\|_{c,1}$ measures how simply the function f may be expressed as a linear combination of indicator functions of congruence classes modulo divisors of q.[3] Note that $\|f\|_{c,1} \leqslant \|f\|_1$, because one always has the representation

$$f = \sum_{a \in \mathbf{Z}/q\mathbf{Z}} f(a)\varphi_{q,a}.$$

Now the estimates (1.1) and (1.5) can be improved to

$$\left|\mathbf{E}(f(\mathrm{X_N})) - \mathbf{E}(f)\right| \leqslant \frac{2}{\mathrm{N}}\|f\|_{c,1}, \tag{1.6}$$

$$\left|\mathbf{E}(f((\pi_q)_{q \in \mathrm{S}})) - \mathbf{E}(f)\right| \leqslant \frac{2}{\mathrm{N}}\|f\|_{c,1}, \tag{1.7}$$

respectively. Indeed, it suffices (using linearity and the triangle inequality) to check this for $f = \varphi_{d,a}$ for some divisor $d \mid q$ and some $a \in \mathbf{Z}/d\mathbf{Z}$ (with $\|\varphi_{d,a}\|_{c,1}$ replaced by 1 in the right-hand side), in which case the difference (in the first case) is

$$\frac{1}{\mathrm{N}} \sum_{\substack{n \leqslant \mathrm{N} \\ n \equiv a \,(\mathrm{mod}\, d)}} 1 - \frac{1}{q} \sum_{\substack{x \,(\mathrm{mod}\, q) \\ x \equiv a \,(\mathrm{mod}\, d)}} 1 = \frac{1}{\mathrm{N}} \sum_{\substack{n \leqslant \mathrm{N} \\ n \equiv a \,(\mathrm{mod}\, d)}} 1 - \frac{1}{d},$$

which reduces to the case of single element modulo d, for which we now apply Theorem 1.3.1.

Another corollary of these elementary statements identifies the limiting distribution of the valuations of integers. To state it, we denote by $\mathrm{S_N}$ the identity random variable on the probability space $\Omega_\mathrm{N} = \{1, \ldots, \mathrm{N}\}$ with uniform probability measure of Theorem 1.3.1.

Corollary 1.3.9 *For p prime, let v_p denote the p-adic valuation on $\mathbf{Z}$. The random vectors $(v_p(\mathrm{S_N}))_p$ converge in law, in the sense of finite distributions, to a sequence $(\mathrm{V}_p)_p$ of independent geometric random variables with*

$$\mathbf{P}(\mathrm{V}_p = k) = \left(1 - \frac{1}{p}\right)\frac{1}{p^k}$$

for $k \geqslant 0$. In other words, for any finite set of primes S and any nonnegative integers $(k_p)_{p \in \mathrm{S}}$, we have

$$\lim_{\mathrm{N} \to +\infty} \mathbf{P}_\mathrm{N}(v_p(\mathrm{S_N}) = k_p \textit{ for } p \in \mathrm{S}) = \prod_{p \in \mathrm{S}} \mathbf{P}(\mathrm{V}_p = k_p).$$

[3] In terms of functional analysis, this means that this is a quotient norm of the ℓ^1 norm on the space with basis Φ_q.

Proof For a given prime p and integer $k \geqslant 0$, the condition that $v_p(n) = k$ means that $n \pmod{p^{k+1}}$ belongs to the subset in $\mathbf{Z}/p^{k+1}\mathbf{Z}$ of residue classes of the form bp^k where $1 \leqslant b \leqslant p-1$; by Theorem 1.3.1, we therefore have

$$\lim_{N\to+\infty} \mathbf{P}_N(v_p(S_N) = k) = \frac{p-1}{p^{k+1}} = \mathbf{P}(V_p = k).$$

Proposition 1.3.7 then shows that this extends to any finite set of primes. □

Example 1.3.10 Getting quantitative estimates in this corollary is a good example of Remark 1.3.8 (2). We illustrate this point in the simplest case, which will be used in Section 2.2.

Consider two primes $p \neq q$ and the probability

$$\mathbf{P}_N(v_p(S_N) = v_q(S_N) = 1).$$

The indicator function φ of this event is naturally defined modulo p^2q^2, and its norm $\|\varphi\|_1$ is the number of integers modulo p^2q^2 that are multiples of pq, but not of p^2 or q^2. By inclusion-exclusion, this means that $\|\varphi\|_1 = (p-1)(q-1)$. On the other hand, we have $\varphi = \varphi_1 - \varphi_2 - \varphi_3 + \varphi_4$ where

- the function φ_1 is defined modulo pq as the indicator of the class 0;
- the function φ_2 is defined modulo p^2q as the indicator of the class 0;
- the function φ_3 is defined modulo pq^2 as the indicator of the class 0;
- the function φ_4 is defined modulo p^2q^2 as the indicator of the class 0.

Hence, in the notation of Remark 1.3.8 (2), we have $\|\varphi\|_{c,1} \leqslant 4$; using this remark, or by applying Theorem 1.3.1 four times, we get

$$\mathbf{P}_N(v_p(S_N) = v_q(S_N) = 1) = \frac{1}{pq}\Big(1-\frac{1}{p}\Big)\Big(1-\frac{1}{q}\Big) + O\Big(\frac{1}{N}\Big),$$

instead of having an error term of size pq/N, as suggested by a direct application of (1.1).

1.4 Another Prototype: The Distribution of the Euler Function

Although Proposition 1.3.7 is extremely simple, it is the only necessary arithmetic ingredient in the proof of a result that is another prototype of probabilistic number theory in our sense. This is a theorem proved by Schoenberg [108] in 1928, which therefore predates the Erdős–Kac Theorem by about ten years (although Schoenberg phrased the result quite differently, since this date is also before Kolmogorov's formalization of probability theory).

The Euler "totient" function is defined for integers $n \geqslant 1$ by $\varphi(n) = |(\mathbf{Z}/n\mathbf{Z})^\times|$ (the number of invertible residue classes modulo n). By the Chinese Remainder Theorem (see Example C.1.8), this function is multiplicative, in the sense that $\varphi(n_1 n_2) = \varphi(n_1)\varphi(n_2)$ for n_1 coprime to n_2. Computing $\varphi(p^k) = p^k - p^{k-1} = p^k(1 - 1/p)$ for p prime and $k \geqslant 1$, one deduces that

$$\frac{\varphi(n)}{n} = \prod_{p \mid n} \left(1 - \frac{1}{p}\right)$$

for all integers $n \geqslant 1$ (where the product is over primes p dividing n).

Now define random variables F_N on $\Omega_\mathrm{N} = \{1, \ldots, \mathrm{N}\}$ (with the uniform probability measure as before) by

$$\mathsf{F}_\mathrm{N}(n) = \frac{\varphi(n)}{n}.$$

We will prove that the sequence $(\mathsf{F}_\mathrm{N})_{\mathrm{N} \geqslant 1}$ converges in law, and identify its limiting distribution. For this purpose, let $(\mathrm{B}_p)_p$ be a sequence of independent Bernoulli random variables, indexed by primes, with

$$\mathbf{P}(\mathrm{B}_p = 1) = \frac{1}{p} \quad \text{and} \quad \mathbf{P}(\mathrm{B}_p = 0) = 1 - \frac{1}{p}$$

(such random variables will also occur prominently in the next chapter).

Proposition 1.4.1 *The random variables* F_N *converge in law to the random variable given by*

$$\mathrm{F} = \prod_p \left(1 - \frac{\mathrm{B}_p}{p}\right),$$

where the infinite product ranges over all primes and converges almost surely.

This proposition is not only a good illustration of limiting behavior of arithmetic random variables, but the proof that we give, which emphasizes probabilistic methods, is an excellent introduction to a number of techniques that will occur later in more complicated contexts. Before we begin, note how the limiting random variable is highly nongeneric, and in fact retains some arithmetic information, since it is a product over primes. In particular, although the arithmetic content does not go beyond Proposition 1.3.7, this theorem is certainly not an obvious fact.

Proof For $M \geqslant 1$, we denote by $F_{N,M}$ the random variable on Ω_N defined by

$$F_{N,M}(n) = \prod_{\substack{p|n \\ p\leqslant M}} \left(1 - \frac{1}{p}\right).$$

It is natural to think of these as approximations to F_N. On the other hand, for a fixed M, these are finite products and hence easier to handle. We will use a fairly simple "perturbation lemma" to prove the convergence in law of the sequence $(F_N)_{N\geqslant 1}$ from the understanding of the behavior of $F_{N,M}$. The lemma is precisely Proposition B.4.4, which the reader should read now.[4]

First, we fix $M \geqslant 1$. Since only primes $p \leqslant M$ occur in the definition of $F_{N,M}$, it follows from Proposition 1.3.7 that the random variables $F_{N,M}$ converge in law as $N \to +\infty$ to the random variable

$$F_M = \prod_{p\leqslant M} \left(1 - \frac{B_p}{p}\right).$$

Thus Assumption (1) in Proposition B.4.4 is satisfied. We proceed to check Assumption (2), which concerns the approximation of F_N by $F_{N,M}$ on average.

We write $E_{N,M} = F_N - F_{N,M}$. The expectation of $|E_{N,M}|$ is given by

$$\begin{aligned} \mathbf{E}_N(|E_{N,M}|) &= \frac{1}{N}\sum_{n\leqslant N}\left|\prod_{p|n}\left(1-\frac{1}{p}\right) - \prod_{\substack{p|n \\ p\leqslant M}}\left(1-\frac{1}{p}\right)\right| \\ &\leqslant \frac{1}{N}\sum_{n\leqslant N}\left|\prod_{\substack{p|n \\ p>M}}\left(1-\frac{1}{p}\right) - 1\right|. \end{aligned}$$

For a given n, expanding the product, we see that the quantity

$$\prod_{\substack{p|n \\ p>M}}\left(1-\frac{1}{p}\right) - 1$$

is bounded by the sum of $1/d$ over integers $d \geqslant 2$ which are squarefree, divide n, and have all prime factors $> M$; let D_n be the set of such integers. In particular, we always have $M < d \leqslant N$ if $d \in D_n$.

[4] Note that a similar argument reappears in a much more sophisticated context in Chapter 5 (see the proof of Theorem 5.2.2).

Thus

$$\mathrm{E_N}(|\mathrm{E_{N,M}}|) \leqslant \frac{1}{\mathrm{N}} \sum_{n \leqslant \mathrm{N}} \sum_{d \in \mathrm{D}_n} \frac{1}{d} \leqslant \sum_{\mathrm{M}<d\leqslant \mathrm{N}} \frac{1}{d} \times \frac{1}{\mathrm{N}} \sum_{\substack{n\leqslant \mathrm{N}\\ n\equiv 0\,(\mathrm{mod}\, d)}} 1$$

$$\leqslant \sum_{\mathrm{M}<d\leqslant \mathrm{N}} \frac{1}{d^2} \leqslant \frac{1}{\mathrm{M}}$$

for all $\mathrm{N} \geqslant \mathrm{M}$. Assumption (2) of Proposition B.4.4 follows immediately, and we conclude that $(\mathrm{F_N})_{\mathrm{N}\geqslant 1}$ converges in law, and that its limit is the limit in law F of the random variables $\mathrm{F_M}$ as $\mathrm{M} \to +\infty$. The last thing to check in order to finish the proof is that the random product

$$\prod_p \Bigl(1 - \frac{\mathrm{B}_p}{p}\Bigr) \tag{1.8}$$

over primes converges almost surely, and has the same law as F. The almost sure convergence follows from Kolmogorov's Three Series Theorem, applied to the logarithm of this product, which is a sum

$$\sum_p \mathrm{Y}_p, \qquad \mathrm{Y}_p = \log\Bigl(1 - \frac{\mathrm{B}_p}{p}\Bigr)$$

of independent random variables. Note that $\mathrm{Y}_p \leqslant 0$ and that it only takes the values 0 (with probability $1 - 1/p$) and $\log(1 - 1/p)$ (with probability $1/p$), so that

$$\mathbf{E}(\mathrm{Y}_p) = \frac{1}{p}\log\Bigl(1 - \frac{1}{p}\Bigr) \sim -\frac{1}{p^2},$$

$$\mathbf{V}(\mathrm{Y}_p) = \mathbf{E}(\mathrm{Y}_p^2) - \mathbf{E}(\mathrm{Y}_p)^2 = \frac{1}{p}\log\Bigl(1 - \frac{1}{p}\Bigr)^2 - \frac{1}{p^2}\log\Bigl(1 - \frac{1}{p}\Bigr)^2 \ll \frac{1}{p^3},$$

which implies by Theorem B.10.1 that the random series $\sum \mathrm{Y}_p$ converges almost surely, and hence so does its exponential, which is the product (1.8). Now, from this convergence almost surely, it is immediate that the law of the random product is also the law of F. □

In Section 2.2 of the next chapter, we will state and prove a theorem due to Erdős and Wintner that implies the existence of limiting distributions for much more general multiplicative functions.

Remark 1.4.2 The distribution function of the arithmetic function $n \mapsto \varphi(n)/n$ is the function defined for $x \in \mathbf{R}$ by

$$f(x) = \mathbf{P}(\mathrm{F} \leqslant x).$$

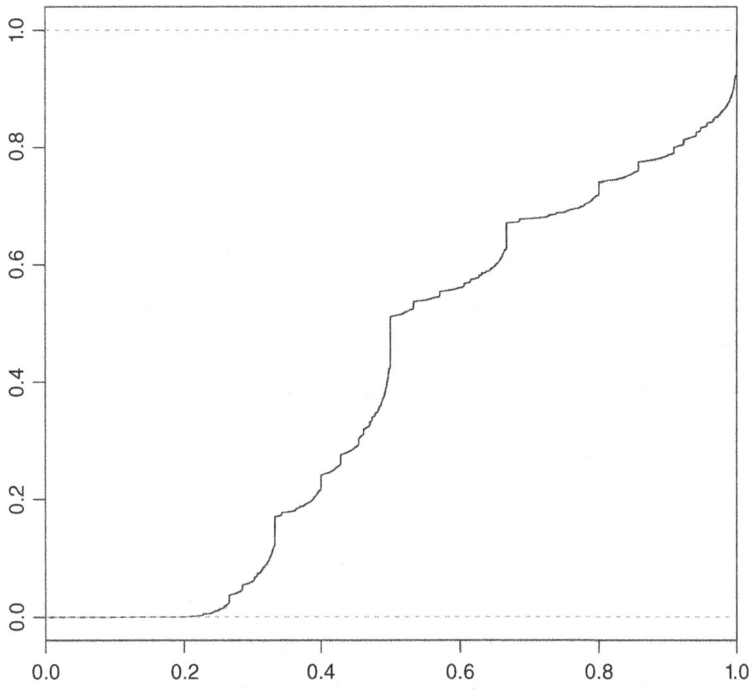

Figure 1.1 Empirical plot of the distribution function of $\varphi(n)/n$ for $n \leqslant 10^6$.

This function has been extensively studied, and is still the object of current research. It is a concrete example of a function exhibiting unusual properties in real analysis: it was proved by Schoenberg [108, 109] that f is continuous and strictly increasing, and by Erdős [34] that it is purely singular, that is, that there exists a set N of Lebesgue measure 0 in $\mathbf{R}$ such that $\mathbf{P}(\mathrm{F} \in \mathrm{N}) = 1$; this means that the function f is differentiable for all $x \notin \mathrm{N}$, with derivative equal to 0 (Exercise 1.4.4 explains the proof).

In Figure 1.1, we plot the "empirical" values of f coming from integers $n \leqslant 10^6$.

In the next two exercises, we use the notation of Proposition 1.4.1.

Exercise 1.4.3 Prove probabilistically that

$$\lim_{\mathrm{N}\to+\infty} \mathbf{E}_{\mathrm{N}}(\mathsf{F}_{\mathrm{N}}) = \frac{1}{\zeta(2)} \quad \text{and}$$

$$\lim_{\mathrm{N}\to+\infty} \mathbf{E}_{\mathrm{N}}(\mathsf{F}_{\mathrm{N}}^{-1}) = \prod_p \left(1 + \frac{1}{p(p-1)}\right) = \frac{\zeta(2)\zeta(3)}{\zeta(6)},$$

where

$$\zeta(s) = \prod_p (1 - p^{-s})^{-1}$$

is the Riemann zeta function (see Corollary C.1.5 for the product expression). In other words, we have

$$\lim_{N\to+\infty} \frac{1}{N} \sum_{n\leqslant N} \frac{\varphi(n)}{n} = \frac{1}{\zeta(2)} \quad \text{and} \quad \lim_{N\to+\infty} \frac{1}{N} \sum_{n\leqslant N} \frac{n}{\varphi(n)} = \frac{\zeta(2)\zeta(3)}{\zeta(6)}.$$

Recover these formulas using Möbius inversion (as in the "direct" proof of Proposition 1.3.3).

Exercise 1.4.4 (1) Prove that the support of the law of F is [0, 1]. [**Hint**: Use Proposition B.10.8.]

By the Jessen–Wintner Purity Theorem (see, e.g., [20, Th. 3.26]), this fact implies that the function f is purely singular (in the sense of Remark 1.4.2), provided there exists a set N of Lebesgue measure 0 such that $\mathbf{P}(F \in N) > 0$. In turn, by elementary properties of absolutely continuous probability measures, this follows if there exists $\alpha > 0$ and, for any $\varepsilon > 0$, a Borel set $I_\varepsilon \subset [0, 1]$ such that

(1) we have $\mathbf{P}(F \in I_\varepsilon) \geqslant \alpha$ for all ε small enough; and
(2) the Lebesgue measure of I_ε tends to 0 as $\varepsilon \to 0$.

The next questions will establish the existence of such sets. We define $G = \log(F)$, and for $M \geqslant 2$, we let G_M denote the partial sum

$$G_M = \sum_{p\leqslant M} \log\left(1 - \frac{B_p}{p}\right).$$

(2) Prove that for any $\delta > 0$, we have

$$\mathbf{P}(|G - G_M| > \delta) \ll \frac{1}{\delta M}$$

for any $M > 0$.

(3) For any finite set T of primes $p \leqslant M$, with characteristic function χ_T, prove that

$$\mathbf{P}(B_p = \chi_T(p) \text{ for } p \leqslant M) \gg \frac{1}{\log M} \times \prod_{p\in T} \frac{1}{p}.$$

Hint: Use the Mertens Formula (Proposition C.3.1).

(4) Let $\mathcal{T}_M$ be a set of subsets T of the set of primes $p \leqslant M$, and let X_M be the event

$$\{ \text{ there exists } T \in \mathcal{T}_M \text{ such that } B_p = \chi_T(p) \text{ for } p \leqslant M \}.$$

Show that

$$\mathbf{P}(X_M) \gg \frac{1}{\log M} \sum_{T \in \mathcal{T}_M} \prod_{p \in T} \frac{1}{p}.$$

(5) Let $\delta > 0$ be some auxiliary parameter and

$$I_M = \bigcup_{T \in \mathcal{T}_M} \Big[\sum_{p \in T} \log(1 - 1/p) - \delta, \sum_{p \in T} \log(1 - 1/p) + \delta, \Big].$$

Show that the Lebesgue measure of I_M is $\leqslant 2\delta |\mathcal{T}_M|$ and that

$$\mathbf{P}(G \in I_M) \gg \frac{1}{\log M} \sum_{T \in \mathcal{T}_M} \prod_{p \in T} \frac{1}{p} - \frac{1}{\delta M}.$$

(6) Conclude by finding a choice of $\delta > 0$ and $\mathcal{T}_M$ such that the Lebesgue measure of I_M tends to 0 as $M \to +\infty$ whereas $\mathbf{P}(G \in I_M) \gg 1$ for M large enough.

1.5 Generalizations

Theorem 1.3.1 and Proposition 1.3.7 are obviously very simple statements. However, Proposition 1.4.1 has already shown that they should not be disregarded as trivial (and our careful presentation should – maybe – not be considered as overly pedantic). A further and even stronger sign in this direction is the fact that if one considers other natural sequences of probability measures on the integers, instead of the uniform measures on $\{1, \ldots, N\}$, one quickly encounters very delicate questions, and indeed fundamental open problems.

We have already mentioned the generalization related to polynomial values $P(n)$ for some fixed polynomial $P \in \mathbf{Z}[X]$. Here are two other natural sequences of measures that have been studied.

1.5.1 Primes

Maybe the most important variant consists in replacing the space Ω_N of positive $n \leqslant N$ by the subset Π_N of prime numbers $p \leqslant N$ (with the uniform probability measure on these finite sets). According to the Prime Number Theorem (Theorem C.3.3), there are about $N/(\log N)$ primes in Π_N. In this case, the qualitative analogue of Theorem 1.3.1 is given by the theorem of Dirichlet, Hadamard and de la Vallée Poussin on primes in arithmetic progression (Theorem C.3.7), which implies that, for any fixed $q \geqslant 1$, the random variables π_q on Π_N converge in law to the probability measure on $\mathbf{Z}/q\mathbf{Z}$ which is the uniform measure on the subset $(\mathbf{Z}/q\mathbf{Z})^\times$ of invertible residue classes (this change of the measure compared with the case of integers is simply due to the obvious fact that at most one prime may be divisible by the integer q).

It is *expected* that a bound similar to (1.1) should be true. More precisely, there *should* exist a constant $C \geqslant 0$ such that

$$\left|\mathbf{E}_{\Pi_N}(f(\pi_q)) - \mathbf{E}(f)\right| \leqslant \frac{C(\log qN)^2}{\sqrt{N}} \|f\|_1, \tag{1.9}$$

but that statement, once it is translated to more standard notation, is very close to the Generalized Riemann Hypothesis for Dirichlet L-functions (which is Conjecture C.5.8).[5] Even a similar bound with $\sqrt{N}$ replaced by N^θ for any fixed $\theta > 0$ is not known, and would be a sensational breakthrough. Note that here the function f is defined on $(\mathbf{Z}/q\mathbf{Z})^\times$ and we have

$$\mathbf{E}(f) = \frac{1}{\varphi(q)} \sum_{a \in (\mathbf{Z}/q\mathbf{Z})^\times} f(a),$$

with $\varphi(q) = |(\mathbf{Z}/q\mathbf{Z})^\times|$ denoting the Euler function (see Example C.1.8).

However, weaker versions of (1.9), amounting roughly to a version valid on average over $q \leqslant \sqrt{N}$, are known: the Bombieri–Vinogradov Theorem states that, for any constant $A > 0$, there exists $B > 0$ such that we have

$$\sum_{q \leqslant \sqrt{N}/(\log N)^B} \max_{a \in (\mathbf{Z}/q\mathbf{Z})^\times} \left|\mathbf{P}_{\Pi_N}(\pi_q = a) - \frac{1}{\varphi(q)}\right| \ll \frac{1}{(\log N)^A}, \tag{1.10}$$

where the implied constant depends only on A (see, e.g., [59, Ch. 17]). In many applications, this is essentially as useful as (1.9).

Exercise 1.5.1 Compute the "probability" that $p - 1$ be squarefree, for p prime. (This can be done using the Bombieri–Vinogradov Theorem, for instance.)

[Further references: Friedlander and Iwaniec [43]; Iwaniec and Kowalski [59].]

1.5.2 Random walks

A more recent (and extremely interesting) type of problem arises from taking measures on $\mathbf{Z}$ derived from *random walks* on certain discrete groups.

[5] It implies it for nontrivial Dirichlet characters.

For simplicity, we only consider a special case. Let $m \geqslant 2$ be an integer, and let $G = SL_m(\mathbf{Z})$ be the group of $m \times m$ matrices with integral coefficients and determinant 1. This is a complicated infinite (countable) group, but it is known to have finite generating sets. We fix one such set S, and assume that $1 \in S$ and $S = S^{-1}$ for convenience. (A well-known example is the set S consisting of 1 and the elementary matrices $1 + E_{i,j}$ for $1 \leqslant i \neq j \leqslant m$, where $E_{i,j}$ is the matrix where only the (i, j)th coefficient is nonzero, and equal to 1, and their inverses $1 - E_{i,j}$; the fact that these generate $SL_n(\mathbf{Z})$ can be seen from the row-and-column operation reduction algorithm for such matrices.)

The generating set S defines then a random walk $(\gamma_n)_{n\geqslant 0}$ on G: let $(\xi_n)_{n\geqslant 1}$ be a sequence of independent S-valued random variables (defined on some probability space Ω) such that $\mathbf{P}(\xi_n = s) = 1/|S|$ for all n and all $s \in S$. Then we let

$$\gamma_0 = 1, \qquad \gamma_{n+1} = \gamma_n \xi_{n+1}.$$

Fix some (nonconstant) polynomial function F of the coefficients of an element $g \in G$, so $F \in \mathbf{Z}[(g_{i,j})]$ (for instance, $F(g) = g_{1,1}$, or $F(g) = \mathrm{Tr}(g)$ for $g = (g_{i,j})$ in G). We can then study the analogue of Theorem 1.3.1 when applied to the random variables $\pi_q(F(\gamma_n))$ as $n \to +\infty$, or in other words, the distribution of $F(g)$ modulo q, as g varies in G according to the distribution of the random walk.

Let $G_q = SL_m(\mathbf{Z}/q\mathbf{Z})$ be the finite special linear group. It is an elementary exercise, using finite Markov chains and the surjectivity of the projection map $G \longrightarrow G_q$, to check that the sequence of random variables $(\pi_q(F(\gamma_n)))_{n\geqslant 0}$ converges in law as $n \to +\infty$. Indeed, its limit is a random variable F_q on $\mathbf{Z}/q\mathbf{Z}$ defined by

$$\mathbf{P}(F_q = x) = \frac{1}{|G_q|}|\{g \in G_q \mid F(g) = x\}|,$$

for all $x \in \mathbf{Z}/q\mathbf{Z}$, where we view F as also defining a function $F\colon G_q \longrightarrow \mathbf{Z}/q\mathbf{Z}$. (In other words, F_q is distributed like the direct image under F of the uniform measure on G_q.)

In fact, elementary Markov chain theory (or direct computations) shows that there exists a constant $c_q > 1$ such that for any function $f\colon G_q \longrightarrow \mathbf{C}$, we have

$$\left|\mathbf{E}(f(\pi_q(\gamma_n)) - \mathbf{E}(f)\right| \leqslant \frac{\|f\|_1}{c_q^n}, \tag{1.11}$$

in analogy with (1.1), with

$$\|f\|_1 = \sum_{g\in G_q} |f(g)|.$$

This is a very good result for a fixed q (note that the number of elements reached by the random walk after n steps also grows exponentially with n). For applications, our previous discussion already shows that it will be important to exploit (1.11) for q varying with n, and uniformly over a wide range of q. This requires an understanding of the variation of the constant c_q with q. It is a rather deep fact (Property (τ) of Lubotzky for $\mathrm{SL}_2(\mathbf{Z})$, and Property (T) of Kazhdan for $\mathrm{SL}_m(\mathbf{Z})$ if $m \geqslant 3$) that there exists $c > 1$, depending only on m, such that $c_q \geqslant c$ for all $q \geqslant 1$. Thus we do get a uniform bound

$$\left|\mathbf{E}(f(\pi_q(\gamma_n)) - \mathbf{E}(f)\right| \leqslant \frac{\|f\|_1}{c^n}$$

valid for all $n \geqslant 1$ and all $q \geqslant 1$. This is related to the theory (and applications) of *expander graphs*.

[Further references: Breuillard and Oh [21], Kowalski [65], [67].]

1.6 Outline of the Book

Here is now a quick outline of the main results that we will prove in the text. For detailed statements, we refer to the introductory sections of the corresponding chapters.

Chapter 2 presents first the Erdős–Wintner Theorem on the limiting distribution of additive functions, before discussing the Erdős–Kac Theorem. These are good examples to begin with, because they are the most natural starting point for probabilistic number theory, and remain quite lively topics of contemporary research. This will lead to natural appearances of the Gaussian distribution as well as Poisson distributions.

Chapters 3 and 4 are concerned with the distribution of values of the Riemann zeta function. We discuss results outside of the critical line (due to Bohr–Jessen, Bagchi and Voronin) in the first of these chapters, and consider deeper results on the critical line (due to Selberg, but following a recent presentation of Radziwiłł and Soundararajan) in the second. The limit theorems one obtains can have rather unorthodox limiting distributions (random Euler products, sometimes viewed as random functions, and – conjecturally – also eigenvalues of random unitary matrices of large size).

Chapter 5 takes up a fascinating topic in the distribution of prime numbers: the *Chebychev bias*, which attempts to compare the number of primes $\leqslant x$ in various residue classes modulo a fixed integer $q \geqslant 1$, and to see if some classes are "more equal" than others. Our treatment follows the basic paper of Rubinstein and Sarnak.

In Chapter 6, we consider the distribution, in the complex plane, of polygonal paths joining partial sums of Kloosterman sums, following work of the author and W. Sawin [79, 12]. Here we will use convergence in law in Banach spaces and some elementary probability in Banach spaces, and the limit object that arises will be a very special random Fourier series.

In all of these chapters, we usually only discuss in detail one specific example of fairly general results and theories: just the additive function $\omega(n)$ instead of more general additive functions, just the Riemann zeta function instead of more general L-functions, and specific families of exponential sums. However, we will briefly mention some of the natural generalizations of the results presented.

Similarly, since our objective in this book is explicitly to write an *introduction* to the topic of probabilistic number theory, we did not attempt to cover the most refined results or the cutting-edge of research, or to discuss all possible topics. For the same reason, we do not discuss in depth the applications of our main results, although we usually mention at least some of them. Besides the discussion in Chapter 7 of other areas of interaction between probability theory and number theory, the reader is invited to read the short survey by Perret-Gentil [92].

At the end of the book are appendices that discuss the results of complex analysis, probability theory and number theory that we use in the main chapters of the book. In general, these are presented with some examples and detailed references, but without complete proofs, at least when they can be considered to be standard parts of their respective fields. We do not expect every reader to already be familiar with all of these facts, and in order to make it possible to read the text relatively linearly, each chapter begins with a list of the main results from these appendices that it will require, with the corresponding reference (when no reference is given, this means that the result in question will be presented within the chapter itself). We also note that the number-theoretic results in Appendix C are stated in the "classical" style of analytic number theory, without attempting to fit them to a probabilistic interpretation.

2

Classical Probabilistic Number Theory

Probability tools	Arithmetic tools
Definition of convergence in law (Section B.3)	Integers in arithmetic progressions (Section 1.3)
Convergence in law using auxiliary parameters (Prop. B.4.4)	Mertens and Chebychev estimate (Prop. C.3.1)
Central Limit Theorem (Th. B.7.2)	Additive and multiplicative functions (Section C.1, C.2)
Gaussian random variables (Section B.7)	
The method of moments (Th. B.5.5)	
Poisson random variables (Section B.9)	

2.1 Introduction

This chapter contains some of the earliest theorems of probabilistic number theory. We will prove the Erdős–Kac Theorem, but first we consider an even more classical topic: the distribution of multiplicative and additive arithmetic functions. The essential statements predate the Erdős–Kac Theorem, and can be taken to be the beginning of true probabilistic number theory. As we will see, the limiting distributions that are obtained are far from generic.

2.2 Distribution of Arithmetic Functions

The classical problem of the distribution of the values of arithmetic functions concerns the limiting behavior of (arithmetic) random variables of the form $g(\mathrm{S_N})$, where g is an additive or multiplicative function, and $\mathrm{S_N}$ is the identity random variable on the probability space $\Omega_{\mathrm{N}} = \{1, \ldots, \mathrm{N}\}$ with uniform probability measure. We saw an example in Proposition 1.4.1, but we will now prove a much more general statement.

In fact, in the additive case (see Section C.2 for the definition of additive functions), there is a remarkable *characterization* of those additive functions g for which the sequence $(g(\mathrm{S_N}))_{\mathrm{N}}$ converges in law as $\mathrm{N} \to +\infty$. Arithmetically, it may be surprising that it depends on no more than Theorem 1.3.1 (or Corollary 1.3.9), and the simplest upper bound of the right order of magnitude for the numbers of primes less than a given quantity (Chebychev's estimate); this was not even needed for Proposition 1.4.1.

Theorem 2.2.1 *Let g be a complex-valued additive function such that the series*

$$\sum_{|g(p)|\leqslant 1} \frac{g(p)}{p}, \qquad \sum_{|g(p)|\leqslant 1} \frac{|g(p)|^2}{p}, \qquad \sum_{|g(p)|>1} \frac{1}{p}$$

converge. Then the sequence of random variables $(g(\mathrm{S_N}))_{\mathrm{N}}$ converges in law to the series over primes

$$\sum_p g(p^{\mathrm{V}_p}), \tag{2.1}$$

where $(\mathrm{V}_p)_p$ is a sequence of independent geometric random variables with

$$\mathbf{P}(\mathrm{V}_p = k) = \left(1 - \frac{1}{p}\right)\frac{1}{p^k}$$

for $k \geqslant 0$.

Recall that, in terms of p-adic valuations of integers, we can write

$$g(n) = \sum_p g\left(p^{v_p(n)}\right)$$

for any integer $n \geqslant 1$. Since the sequence of p-adic valuations converges in law to the sequence (V_p) (Corollary 1.3.9), the formula (2.1) for the limiting distribution appears as a completely natural expression.

Proof We write $g = g^\flat + g^\sharp$ where both summands are additive functions, and

$$g^\flat(p^k) = \begin{cases} g(p) & \text{if } k = 1 \text{ and } |g(p)| \leqslant 1, \\ 0 & \text{otherwise.} \end{cases}$$

Thus $g^\sharp(p) = 0$ for a prime p unless $|g(p)| > 1$. We denote by (B_p) the Bernoulli random variable indicator function of the event $\{V_p = 1\}$; we have

$$\mathbf{P}(B_p = 1) = \frac{1}{p}\Big(1 - \frac{1}{p}\Big).$$

We will prove that the vectors $(g^\flat(S_N), g^\sharp(S_N))$ converge in law to

$$\Big(\sum_p g^\flat(p^{V_p}), \sum_p g^\sharp(p^{V_p})\Big),$$

and the desired conclusion then follows by composing with the continuous addition map $\mathbf{C}^2 \to \mathbf{C}$ (i.e., applying Proposition B.3.2).

We will apply Proposition B.4.4 to the random vectors $G_N = (g^\flat(S_N), g^\sharp(S_N))$ (with values in $\mathbf{C}^2$), with the approximations $G_N = G_{N,M} + E_{N,M}$, where

$$G_{N,M} = \Big(\sum_{p \leqslant M} g^\flat(p^{v_p(S_N)}), \sum_{p \leqslant M} g^\sharp(p^{v_p(S_N)})\Big).$$

Let $M \geqslant 1$ be fixed. The random vectors $G_{N,M}$ are finite sums, and are expressed as obviously continuous functions of the valuations v_p of the elements of Ω_N, for $p \leqslant M$. Since the vector of these valuations converges in law to $(V_p)_{p \leqslant M}$ by Corollary 1.3.9, applying composition with a continuous map (Proposition B.3.2 again), it follows that $(G_{N,M})_N$ converges in law as $N \to +\infty$ to the vector

$$\Big(\sum_{p \leqslant M} g^\flat(p^{V_p}), \sum_{p \leqslant M} g^\sharp(p^{V_p})\Big).$$

It is therefore enough to verify that Assumption (2) of Proposition B.4.4 holds, and we may do this separately for each of the two coordinates of the vector (by taking the norm on $\mathbf{C}^2$ in the proposition to be the maximal of the modulus of the two coordinates).

We begin with the second coordinate involving $g^\sharp$. For any $\delta > 0$, and $2 \leqslant M < N$, we have

$$\begin{aligned}\mathbf{P}_N\Big(\Big|\sum_{M<p\leqslant N} g^\sharp(p^{v_p(S_N)})\Big| > \delta\Big) &\leqslant \sum_{M<p\leqslant N} \mathbf{P}_N(v_p(S_N) \geqslant 2) \\ &\quad + \sum_{\substack{M<p\leqslant N \\ |g(p)|>1}} \mathbf{P}_N(v_p(S_N) = 1) \\ &\leqslant \sum_{p>M} \frac{1}{p^2} + \sum_{\substack{p>M \\ |g(p)|>1}} \frac{1}{p} \qquad (2.2)\end{aligned}$$

(simply because, if the sum is nonzero, at least one term must be nonzero, and the probability of a union of countably many sets is bounded by the sums of the probabilities of the individual sets).

Since the right-hand side converges to 0 as $\mathrm{M} \to +\infty$ (by assumption), this verifies that the variant discussed in Remark B.4.5 of the assumption of Proposition B.4.4 holds (note that the series

$$\sum_{p \leqslant \mathrm{M}} g^{\sharp}(p^{\mathrm{V}_p})$$

converges in law by a straightforward application of Kolmogorov's Three Series Theorem, which is stated in Remark B.10.2 – indeed, since $|g^{\sharp}| \geqslant 1$, it suffices to observe that

$$\sum_{p \leqslant \mathrm{M}} \mathbf{P}(|g^{\sharp}(p_p^{\mathrm{V}})| \geqslant 2) < +\infty,$$

which follows by arguing as in (2.2)).

We next handle $g^{\flat}$. We denote by $\mathsf{B}_{\mathrm{N},p}$ the Bernoulli random variable indicator of the event $\{v_p(\mathsf{S}_\mathrm{N}) = 1\}$, and define

$$\varpi_\mathrm{N}(p) = \mathbf{P}_\mathrm{N}(\mathsf{B}_{\mathrm{N},p} = 1) = \mathbf{P}_\mathrm{N}(v_p(\mathsf{S}_\mathrm{N}) = 1).$$

We also write $\varpi(p) = \mathbf{P}(\mathrm{B}_p = 1)$. Note that

$$\varpi_\mathrm{N}(p) \leqslant \frac{1}{p} \quad \text{and} \quad \varpi_\mathrm{N}(p) = \frac{1}{p}\left(1 - \frac{1}{p}\right) + \mathrm{O}\left(\frac{1}{\mathrm{N}}\right) = \varpi(p) + \mathrm{O}\left(\frac{1}{\mathrm{N}}\right).$$

The first coordinate of $\mathsf{E}_{\mathrm{N},\mathrm{M}}$ is

$$\mathsf{H}_{\mathrm{N},\mathrm{M}} = \sum_{p > \mathrm{M}} g^{\flat}(p^{\mathrm{V}_p}) = \sum_{p > \mathrm{M}} g^{\flat}(p)\mathsf{B}_{\mathrm{N},p}$$

(which is a finite sum, so convergence issues do not arise). We will prove that

$$\lim_{\mathrm{M} \to +\infty} \limsup_{\mathrm{N} \to +\infty} \mathbf{E}_\mathrm{N}(|\mathsf{H}_{\mathrm{N},\mathrm{M}}|^2) = 0,$$

which will also us to conclude.

By expanding the square, we have

$$\begin{aligned} \mathbf{E}_\mathrm{N}(|\mathsf{H}_{\mathrm{N},\mathrm{M}}|^2) &= \mathbf{E}_\mathrm{N}\left(\Big|\sum_{p > \mathrm{M}} g^{\flat}(p)\mathsf{B}_{\mathrm{N},p}\Big|^2\right) \\ &= \sum_{p_1, p_2 > \mathrm{M}} \mathbf{E}_\mathrm{N}\left(\overline{g^{\flat}(p_1)}g^{\flat}(p_2)\mathsf{B}_{\mathrm{N},p_1}\mathsf{B}_{\mathrm{N},p_2}\right). \qquad (2.3) \end{aligned}$$

The contribution of the diagonal terms $p_1 = p_2$ to (2.3) is

$$\sum_{p>M} |g^\flat(p)|^2 \varpi_N(p) \leqslant \sum_{p>M} \frac{|g^\flat(p)|^2}{p}.$$

We have

$$\mathbf{E}_N(\mathsf{B}_{N,p_1}\mathsf{B}_{N,p_2}) = \mathbf{P}_N(v_{p_1}(\mathsf{S}_N) = v_{p_2}(\mathsf{S}_N) = 1) = \varpi(p_1)\varpi(p_2) + O\Big(\frac{1}{N}\Big)$$

(by Example 1.3.10), so that the nondiagonal terms become

$$\sum_{\substack{p_1,p_2>M\\ p_1\neq p_2}} \overline{g^\flat(p_1)}g^\flat(p_2)\varpi(p_1)\varpi(p_2) + O\Big(\frac{1}{N}\sum_{\substack{p_1,p_2>M\\ p_1p_2\leqslant N}} |g^\flat(p_1)||g^\flat(p_2)|\Big). \tag{2.4}$$

The first term S_1 in this sum is

$$S_1 = \Big|\sum_{p>M} g^\flat(p)\varpi(p)\Big|^2 - \sum_{p>M} |g^\flat(p)|^2\varpi(p)^2 \leqslant \Big|\sum_{p>M} g^\flat(p)\varpi(p)\Big|^2$$

$$= \Big|\sum_{p>M} \frac{g^\flat(p)}{p}\Big(1-\frac{1}{p}\Big)\Big|^2,$$

where the right-hand side of the last equality is convergent because of the assumptions of the theorem, so that the left-hand side is also finite.

Next, since $|g^\flat(p)| \leqslant 1$ for all primes, the second term S_2 in (2.4) satisfies

$$S_2 \ll \frac{1}{N}\sum_{\substack{p_1,p_2>M\\ p_1p_2\leqslant N}} 1 \ll \frac{\log\log N}{\log N}$$

for all $M \geqslant 1$ by Chebychev's estimate of Proposition C.3.1 (extended to products of two primes as in Exercise C.3.2 (2)). Finally, from the convergence assumptions, this means that

$$\limsup_{N\to+\infty} \mathbf{E}_N(|\mathsf{H}_{N,M}|^2) \ll \Big|\sum_{p>M} \frac{g^\flat(p)}{p}\Big|^2 + \sum_{p>M} \frac{|g^\flat(p)|^2}{p} \to 0$$

as $M \to +\infty$, and this concludes the proof. □

Remark 2.2.2 The result above is due to Erdős [33]; the fact that the converse assertion also holds (namely, that if the sequence $(g(\mathrm{S_N}))_\mathrm{N}$ converges in law, then the three series

$$\sum_{|g(p)|\leqslant 1} \frac{g(p)}{p}, \qquad \sum_{|g(p)|\leqslant 1} \frac{|g(p)|^2}{p}, \qquad \sum_{|g(p)|>1} \frac{1}{p}$$

are convergent) is known as the Erdős–Wintner Theorem [36]. The reader may be interested in thinking about proving this; see, for example, [115, pp. 327–328] for the details.

Although it is of course customary and often efficient to pass from additive functions to multiplicative functions by taking the logarithm, this is not always possible. For instance, the (multiplicative) Möbius function $\mu(n)$ *does* have the property that the sequence $(\mu(\mathrm{S_N}))_\mathrm{N}$ converges in law to a random variable taking values 0, 1 and -1 with probabilities which are equal, respectively, to

$$1 - \frac{6}{\pi^2}, \qquad \frac{3}{\pi^2}, \qquad \frac{3}{\pi^2}.$$

The limiting probability that $\mu(n) = 0$ comes from the elementary Proposition 1.3.3, but the fact that, among the values 1 and -1, the asymptotic probability is equal, is quite a bit deeper: it turns out to be "elementarily" equivalent to the Prime Number Theorem in the form

$$\pi(x) \sim \frac{x}{\log x}$$

as $x \to +\infty$ (see, e.g., [59, §2.1] for the proof). However, there is no additive function $\log \mu(n)$, so we cannot even begin to speak of its potential limiting distribution!

2.3 The Erdős–Kac Theorem

We begin by recalling the statement (see Theorem 1.1.1), in its probabilistic phrasing:

Theorem 2.3.1 (Erdős–Kac Theorem) *For* $\mathrm{N} \geqslant 1$, *let* $\Omega_\mathrm{N} = \{1, \ldots, \mathrm{N}\}$ *with the uniform probability measure* $\mathbf{P}_\mathrm{N}$. *Let* $\mathrm{X_N}$ *be the random variable*

$$n \mapsto \frac{\omega(n) - \log\log \mathrm{N}}{\sqrt{\log\log \mathrm{N}}}$$

on Ω_N *for* $\mathrm{N} \geqslant 3$. *Then* $(\mathrm{X_N})_{\mathrm{N}\geqslant 3}$ *converges in law to a standard Gaussian random variable, that is, to a Gaussian random variable with expectation* 0 *and variance* 1.

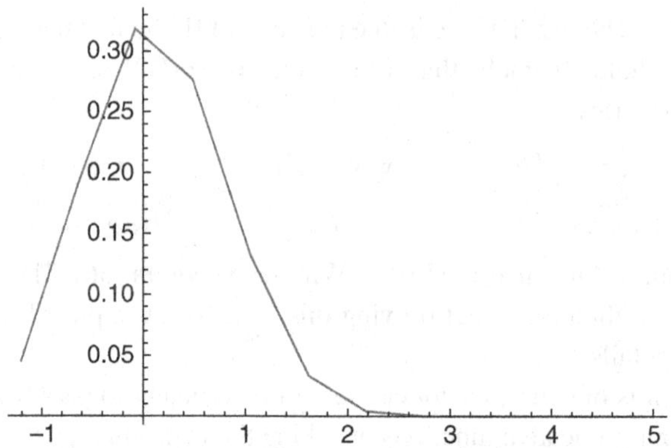

Figure 2.1 The normalized number of prime divisors for $n \leqslant 10^{10}$.

Figure 2.1 shows a plot of the empirical density of X_N for $N = 10^{10}$: one can see something that could be the shape of the Gaussian density appearing, but the fit is very far from perfect (we will comment later why this could be expected).

The original proof of Theorem 2.3.1 is due to Erdős and Kac in 1939 [35]. We will explain a proof following the work of Granville and Soundararajan [51] and of Billingsley [9, p. 394]. As usual, the presentation emphasizes the probabilistic nature of the argument.

As before, we begin by explaining why the statement can be considered to be unsurprising. This is an elaboration of the type of heuristic argument that we used to justify the limit in Theorem 2.2.1.

The arithmetic function ω is additive. Write

$$\omega(n) = \sum_p \mathsf{B}_p(n)$$

for $n \in \Omega_N$, where B_p is as usual the Bernoulli random variable on Ω_N that is the characteristic function of the event $p \mid n$. Using Proposition 1.3.7, the natural probabilistic guess for a limit (if there was one) would be the series

$$\sum_p \mathrm{B}_p,$$

where (B_p) are independent Bernoulli random variables, as in Proposition 1.4.1. But this series diverges almost surely: indeed, the series

$$\sum_p \mathbf{E}(\mathrm{B}_p) = \sum_p \frac{1}{p}$$

diverges by the basic Mertens estimate from prime number theory, namely,

$$\sum_{p \leqslant N} \frac{1}{p} = \log \log N + O(1)$$

for $N \geqslant 3$ (see Proposition C.3.1 in Appendix C), so that the divergence follows from Kolmogorov's Theorem, B.10.1 (or indeed an application of the Borel–Cantelli Lemma; see Exercise B.10.4).

One can however refine the formula for ω by observing that $n \in \Omega_N$ has no prime divisor larger than N, so that we also have

$$\omega(n) = \sum_{p \leqslant N} \mathsf{B}_p(n) \tag{2.5}$$

for $n \in \Omega_N$. Correspondingly, we may expect that the probabilistic distribution of ω on Ω_N will be similar to that of the sum

$$\sum_{p \leqslant N} \mathrm{B}_p. \tag{2.6}$$

But the latter is a sum of *independent* (though not identically distributed) random variables, and its asymptotic behavior is therefore well understood. In fact, a simple case of the Central Limit Theorem (see Theorem B.7.2) implies that the renormalized random variables

$$\frac{\displaystyle\sum_{p \leqslant N} \mathrm{B}_p - \sum_{p \leqslant N} p^{-1}}{\sqrt{\displaystyle\sum_{p \leqslant N} p^{-1}(1 - p^{-1})}}$$

converge in law to a standard Gaussian random variable. It is then to be expected that the arithmetic sums (2.5) are sufficiently close to (2.6) so that a similar renormalization of ω on Ω_N will lead to the same limit, and this is exactly the statement of Theorem 2.3.1 (by the Mertens Formula again).

We now begin the rigorous proof. We will prove convergence in law using the method of moments, as explained in Section B.3 of Appendix B, specifically in Theorem B.5.5 and Remark B.5.9. This is definitely not the only way to confirm the heuristic above, but it may be the simplest.

More precisely, we will proceed as follows:

(1) We show, using Theorem 1.3.1, that for any fixed integer $k \geqslant 0$, we have

$$\mathbf{E}_N(X_N^k) = \mathbf{E}(\mathrm{X}_N^k) + o(1),$$

where (X_N) is the same renormalized random variable described above, namely,

$$X_N = \frac{Z_N - \mathbf{E}(Z_N)}{\sqrt{\mathbf{V}(Z_N)}}$$

with

$$Z_N = \sum_{p \leqslant N} B_p. \tag{2.7}$$

(2) As we already mentioned, the Central Limit Theorem applies to the sequence (X_N), and shows that it converges in law to a standard Gaussian random variable $\mathcal{N}$.

(3) It follows that

$$\lim_{N \to +\infty} \mathbf{E}_N(X_N^k) = \mathbf{E}(\mathcal{N}^k),$$

and hence, by the method of moments (Theorem B.5.5), we conclude that X_N converges in law to $\mathcal{N}$. (Interestingly, we do not need to know the value of the moments $\mathbf{E}(\mathcal{N}^k)$ for this argument to apply.)

This sketch indicates that the Erdős–Kac Theorem is really a result of very general nature that should be valid for many random integers, and not merely for a uniformly chosen integer in Ω_N. Note that only Step 1 has real arithmetic content. As we will see, that arithmetic content is concentrated on two results: Theorem 1.3.1, which makes the link with probability theory, and the Mertens estimate, which is only required in the form of the divergence of the series

$$\sum_p \frac{1}{p}$$

(at least if one is ready to use its partial sums

$$\sum_{p \leqslant N} \frac{1}{p}$$

for renormalization, instead of the asymptotic value $\log \log N$).

We now implement this strategy. As will be seen, some tweaks will be required. (The reader is invited to check that omitting those tweaks leads, at the very least, to a much more complicated-looking problem!)

Step 1 (Truncation). This is a classical technique that applies here, and is used to shorten and simplify the sum in (2.7), in order to control the error terms in the next step. We consider the random variables B_p on Ω_{N} as above, that is, $\mathsf{B}_p(n) = 1$ if p divides n and $\mathsf{B}_p(n) = 0$ otherwise. Let

$$\sigma_{\mathrm{N}} = \sum_{p \leqslant \mathrm{N}} \frac{1}{p}.$$

We only need recall at this point that $\sigma_{\mathrm{N}} \to +\infty$ as $\mathrm{N} \to +\infty$. We then define

$$\mathrm{Q} = \mathrm{N}^{1/(\log\log \mathrm{N})^{1/3}} \tag{2.8}$$

and

$$\tilde{\omega}(n) = \sum_{\substack{p \mid n \\ p \leqslant \mathrm{Q}}} 1 = \sum_{p \leqslant \mathrm{Q}} \mathsf{B}_p(n) \quad \text{and} \quad \tilde{\omega}_0(n) = \sum_{p \leqslant \mathrm{Q}} \Big(\mathsf{B}_p(n) - \frac{1}{p}\Big)$$

viewed as random variables on Ω_{N}. The point of this truncation is the following: first, for $n \in \Omega_{\mathrm{N}}$, we have

$$\tilde{\omega}(n) \leqslant \omega(n) \leqslant \tilde{\omega}(n) + (\log\log \mathrm{N})^{1/3},$$

simply because if $\alpha > 0$ and if $p_1, \ldots, p_m$ are primes $\geqslant \mathrm{N}^{\alpha}$ dividing $n \leqslant \mathrm{N}$, then we get

$$\mathrm{N}^{m\alpha} \leqslant p_1 \cdots p_m \leqslant \mathrm{N},$$

and hence $m \leqslant \alpha^{-1}$. Second, for any $\mathrm{N} \geqslant 1$ and any $n \in \Omega_{\mathrm{N}}$, we get by definition of σ_{N} the identity

$$\begin{aligned} \tilde{\omega}_0(n) &= \tilde{\omega}(n) - \sum_{p \leqslant \mathrm{Q}} \frac{1}{p} \\ &= \omega(n) - \sigma_{\mathrm{N}} + \mathrm{O}((\log\log \mathrm{N})^{1/3}) \end{aligned} \tag{2.9}$$

because the Mertens formula

$$\sum_{p \leqslant x} \frac{1}{p} = \log\log x + \mathrm{O}(1)$$

and the definition of σ_{N} show that

$$\sum_{p \leqslant \mathrm{Q}} \frac{1}{p} = \sum_{p \leqslant \mathrm{N}} \frac{1}{p} + \mathrm{O}(\log\log\log \mathrm{N}) = \sigma_{\mathrm{N}} + \mathrm{O}(\log\log\log \mathrm{N}).$$

Now define

$$\tilde{\mathsf{X}}_{\mathrm{N}}(n) = \frac{\tilde{\omega}_0(n)}{\sqrt{\sigma_{\mathrm{N}}}}$$

as random variables on Ω_N. We will prove that $\tilde{X}_N$ converges in law to $\mathcal{N}$. The elementary Lemma B.5.3 of Appendix B (applied using (2.9)) then shows that the random variables

$$n \mapsto \frac{\omega(n) - \sigma_N}{\sqrt{\sigma_N}}$$

converge in law to $\mathcal{N}$. Finally, applying the same lemma one more time using the Mertens formula we obtain the Erdős–Kac Theorem.

It remains to prove the convergence of $\tilde{X}_N$. We fix a nonnegative integer k, and our target is to prove the limit

$$\mathbf{E}_N(\tilde{X}_N^k) \to \mathbf{E}(\mathcal{N}^k) \tag{2.10}$$

as $N \to +\infty$. Once this is proved for all k, then the method of moments shows that (X_N) converges in law to the standard normal random variable $\mathcal{N}$.

Remark 2.3.2 We might also have chosen to perform a truncation at $p \leqslant N^\alpha$ for some fixed $\alpha \in]0,1[$. However, in that case, we would need to adjust the value of α depending on k in order to obtain (2.10), and then passing from the truncated variables to the original ones would require some minor additional argument. Note that the function $(\log\log N)^{1/3}$ which is used to define the truncation could be replaced by any function going to infinity slower than $(\log\log N)^{1/2}$.

Step 2 (Moment computation). We now begin the proof of (2.10). We use the definition of $\tilde{\omega}_0(n)$ and expand the kth power in $\mathbf{E}_N(\tilde{X}_N^k)$ to derive

$$\mathbf{E}_N(\tilde{X}_N^k) = \frac{1}{\sigma_N^{k/2}} \sum_{p_1 \leqslant Q} \cdots \sum_{p_k \leqslant Q} \mathbf{E}_N\Bigl(\Bigl(\mathsf{B}_{p_1} - \frac{1}{p_1}\Bigr) \cdots \Bigl(\mathsf{B}_{p_k} - \frac{1}{p_k}\Bigr)\Bigr)$$

(where we omit for simplicity the subscripts N for the arithmetic random variables B_{p_i}). The crucial point is that the random variable

$$\Bigl(\mathsf{B}_{p_1} - \frac{1}{p_1}\Bigr) \cdots \Bigl(\mathsf{B}_{p_k} - \frac{1}{p_k}\Bigr) \tag{2.11}$$

can be expressed as $f(\pi_q)$ for some modulus $q \geqslant 1$ and some function $f : \mathbf{Z}/q\mathbf{Z} \longrightarrow \mathbf{C}$, so that the basic result of Theorem 1.3.1 may be applied to each summand.

To be precise, the value at $n \in \Omega_N$ of the random variable (2.11) only depends on the residue class x of n in $\mathbf{Z}/q\mathbf{Z}$, where q is the least common multiple of $p_1, \ldots, p_k$. In fact, this value is equal to $f(x)$, where

$$f(x) = \Bigl(\delta_{p_1}(x) - \frac{1}{p_1}\Bigr) \cdots \Bigl(\delta_{p_k}(x) - \frac{1}{p_k}\Bigr)$$

with δ_{p_i} denoting the characteristic function of the residues classes modulo q which are 0 modulo p_i. It is clear that $|f(x)| \leqslant 1$, as product of terms which are all $\leqslant 1$, and hence we have the bound

$$\|f\|_1 \leqslant q$$

(this is extremely imprecise, but suffices for now). From this we get

$$\left|\mathbf{E}_{\mathrm{N}}\left(\left(\mathrm{B}_{p_1} - \frac{1}{p_1}\right) \cdots \left(\mathrm{B}_{p_k} - \frac{1}{p_k}\right)\right) - \mathbf{E}(f)\right| \leqslant \frac{2q}{\mathrm{N}} \leqslant \frac{2\mathrm{Q}^k}{\mathrm{N}}$$

by Theorem 1.3.1.

But by the definition of f, we also see that

$$\mathbf{E}(f) = \mathbf{E}\left(\left(\mathrm{B}_{p_1} - \frac{1}{p_1}\right) \cdots \left(\mathrm{B}_{p_k} - \frac{1}{p_k}\right)\right),$$

where the random variables (B_p) form a sequence of *independent* Bernoulli random variables with $\mathbf{P}(\mathrm{B}_p = 1) = 1/p$ (the (B_p) for p dividing q are realized concretely as the characteristic functions δ_p on $\mathbf{Z}/q\mathbf{Z}$ with uniform probability measure).

Therefore we derive

$$\begin{aligned}
\mathbf{E}_{\mathrm{N}}(\tilde{\mathrm{X}}_{\mathrm{N}}^k) &= \frac{1}{\sigma_{\mathrm{N}}^{k/2}} \sum_{p_1 \leqslant \mathrm{Q}} \cdots \sum_{p_k \leqslant \mathrm{Q}} \left\{\mathbf{E}\left(\left(\mathrm{B}_{p_1} - \frac{1}{p_1}\right) \cdots \left(\mathrm{B}_{p_k} - \frac{1}{p_k}\right)\right)\right. \\
&\qquad \left. + \mathrm{O}(\mathrm{Q}^k \mathrm{N}^{-1})\right\} \\
&= \left(\frac{\tau_{\mathrm{N}}}{\sigma_{\mathrm{N}}}\right)^{k/2} \mathbf{E}(\mathrm{X}_{\mathrm{N}}^k) + \mathrm{O}(\mathrm{Q}^{2k}\mathrm{N}^{-1}) \\
&= \left(\frac{\tau_{\mathrm{N}}}{\sigma_{\mathrm{N}}}\right)^{k/2} \mathbf{E}(\mathrm{X}_{\mathrm{N}}^k) + o(1)
\end{aligned}$$

by our choice (2.8) of Q, where

$$\mathrm{X}_{\mathrm{N}} = \frac{1}{\sqrt{\tau_{\mathrm{N}}}} \sum_{p \leqslant \mathrm{Q}} \left(\mathrm{B}_p - \frac{1}{p}\right).$$

and

$$\tau_{\mathrm{N}} = \sum_{p \leqslant \mathrm{Q}} \frac{1}{p}\left(1 - \frac{1}{p}\right) = \sum_{p \leqslant \mathrm{Q}} \mathbf{V}(\mathrm{B}_p)$$

Step 3 (Conclusion). We now note that the version of the Central Limit Theorem which is recalled in Theorem B.7.2 applies to the random variables (B_p), and implies precisely that X_{N} converges in law to $\mathcal{N}$. But moreover, the sequence (X_N) satisfies the uniform integrability assumption in the converse of the method of moments (see Example B.5.7, applied to the variables $B_p - 1/p$, which are independent and bounded by 1), and hence we have in particular

$$\mathbf{E}(X_N^k) \longrightarrow \mathbf{E}(\mathcal{N}^k).$$

Since $\tau_N \sim \sigma_N$ by the Mertens formula, we deduce that $\mathbf{E}_N(\tilde{X}_N^k)$ converges also to $\mathbf{E}(\mathcal{N}^k)$, which was our desired goal (2.10).

Exercise 2.3.3 One can avoid appealing to the converse of the method of moments by directly using the combinatorics involved in proofs of the Central Limit Theorem based on moments, which directly imply the convergence of moments for (X_N). Find such a proof in this special case. (See, for instance, [9, p. 391]; note that one must then know what are the moments of Gaussian random variables,; these are recalled in Proposition B.7.3.)

Exercise 2.3.4 Consider the probability spaces $\Omega_N^\flat$ consisting of integers $1 \leqslant n \leqslant N$ that are squarefree, with the uniform probability measure. Prove a version of the Erdős–Kac Theorem for the number of prime factors of an element of $\Omega_N^\flat$.

Exercise 2.3.5 For an integer $N \geqslant 1$, let $m(N)$ denote the set of integers that occur in the multiplication table for integers $1 \leqslant n \leqslant N$:

$$m(N) = \{k = ab \mid 1 \leqslant a \leqslant N, \quad 1 \leqslant b \leqslant N\} \subset \Omega_{N^2}.$$

Prove that $\mathbf{P}_{N^2}(m(N)) \to 0$, that is, that

$$\lim_{N \to +\infty} \frac{|m(N)|}{N^2} = 0.$$

This result is the basic statement concerning the "multiplication table" problem of Erdős; the precise asymptotic behavior of $|m(N)|$ has been determined by K. Ford [41] (improving results of Tenenbaum): we have

$$\frac{|m(N)|}{N^2} \asymp (\log N)^{-\alpha} (\log \log N)^{-3/2},$$

where

$$\alpha = 1 - \frac{1 + \log \log 2}{\log 2}.$$

See also the work of Koukoulopoulos [64] for generalizations.

Exercise 2.3.6 Let $\Omega(n)$ be the number of prime divisors of an integer $n \geqslant 1$, counted *with* multiplicity (so $\Omega(12) = 3$).[1] Prove that

$$\mathbf{P}_{\mathrm{N}}\Big(\Omega(n) - \omega(n) \geqslant (\log\log \mathrm{N})^{1/4}\Big) \leqslant (\log\log \mathrm{N})^{-1/4},$$

and deduce that the random variables

$$n \mapsto \frac{\Omega(n) - \log\log \mathrm{N}}{\sqrt{\log\log \mathrm{N}}}$$

also converge in law to $\mathcal{N}$.

Exercise 2.3.7 Try to prove the Erdős–Kac Theorem using the same "approximation" approach used in the proof of the Erdős–Wintner Theorem; what seems to go wrong (suggesting – if not proving – that one really should use different tools)?

2.4 Convergence without Renormalization

One important point that is made clear by the proof of the Erdős–Kac Theorem is that, although one might think that a statement about the behavior of the number of prime factors of integers tells us something about the distribution of primes (which are those integers n with $\omega(n) = 1$), the Erdős–Kac Theorem *provides no such information.* This can be seen mechanically from the proof, where the truncation step means in particular that primes are simply discarded unless they are smaller than the truncation level Q, or intuitively from the fact that the statement itself implies that "most" integers of size about N have $\log\log \mathrm{N}$ prime factors. For instance, as $\mathrm{N} \to +\infty$, we have

$$\mathbf{P}_{\mathrm{N}}\Big(|\omega(n) - \log\log \mathrm{N}| > a\sqrt{\log\log \mathrm{N}}\Big) \longrightarrow \mathbf{P}(|\mathcal{N}| > a)$$
$$\leqslant \sqrt{\frac{2}{\pi}} \int_a^{+\infty} e^{-x^2/2} dx \leqslant e^{-a^2/4}.$$

The problem lies in the normalization used to obtain a definite theorem of convergence in law: this "crushes" to some extent the more subtle aspects of the distribution of values of $\omega(n)$, especially with respect to extreme values. One can however still study this function probabilistically, but one must use less generic methods, to go beyond the "universal" behavior given by the Central Limit Theorem. There are at least two possible approaches in this direction, and we now briefly survey some of the results.

[1] We only use this function in this section and hope that confusion with Ω_{N} will be avoided.

Both methods have in common a switch in probabilistic focus: instead of looking for a Gaussian approximation of a normalized version of $\omega(n)$, one looks for a *Poisson approximation* of the *un-normalized* function.

Recall (see also Section B.9 in Appendix B) that a Poisson distribution with real parameter $\lambda \geqslant 0$ satisfies

$$\mathbf{P}(\lambda = k) = e^{-\lambda}\frac{\lambda^k}{k!}$$

for any integer $k \geqslant 0$. It turns out that an inductive computation using the Prime Number Theorem leads to the asymptotic formula

$$\begin{aligned}\frac{1}{\mathrm{N}}|\{n \leqslant \mathrm{N} \mid \omega(n) = k\}| &\sim \frac{1}{(k-1)!}\frac{(\log\log \mathrm{N})^{k-1}}{\log \mathrm{N}}\\ &= e^{-\log\log \mathrm{N}}\frac{(\log\log \mathrm{N})^{k-1}}{(k-1)!}\end{aligned}$$

for any fixed integer $k \geqslant 1$. This suggests that a better probabilistic approximation to the arithmetic function $\omega(n)$ on Ω_{N} is a Poisson distribution with parameter $\log\log \mathrm{N}$. The Erdős–Kac Theorem would then be, in essence, a consequence of the simple fact that a sequence (X_n) of Poisson random variables with parameters $\lambda_n \to +\infty$ has the property that

$$\frac{\mathrm{X}_n - \lambda_n}{\sqrt{\lambda_n}} \to \mathcal{N}, \tag{2.12}$$

as explained in Proposition B.9.1. Figure 2.2 shows the density of the values of $\omega(n)$ for $n \leqslant 10^{10}$ and the corresponding Poisson density. (The values of the probabilities for consecutive integers are joined by line segments for readability.)

Remark 2.4.1 The fact that the approximation error in such a statement is typically of size comparable to $\lambda_n^{-1/2}$ explains why one can expect that the convergence to a Gaussian in the Erdős–Kac Theorem should be extremely slow, since in that case the normalizing factor is of size $\log\log \mathrm{N}$, and goes to infinity very slowly.

To give a rigorous meaning to these ideas of Poisson approximation of $\omega(n)$, one must first give a precise definition, which can not be a straightforward convergence property, because the parameter of the Poisson approximation is not fixed.

Harper [53] (to the author's knowledge) was the first to implement explicitly such an idea. He derived an explicit upper bound for the *total variation*

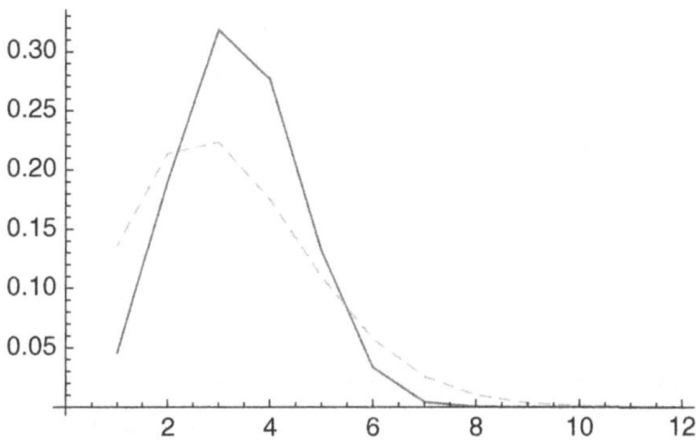

Figure 2.2 The number of prime divisors for $n \leqslant 10^{10}$ (solid line) compared with a Poisson distribution.

distance between a truncated version of $\omega(n)$ on Ω_{N} and a suitable Poisson random variable, namely, between

$$\sum_{\substack{p|n \\ p\leqslant \mathrm{Q}}} 1, \quad \text{where } \mathrm{Q} = \mathrm{N}^{1/(3\log\log \mathrm{N})^2}$$

and a Poisson random variable Po_{N} with parameter

$$\lambda_{\mathrm{N}} = \sum_{p\leqslant \mathrm{Q}} \frac{1}{\mathrm{N}} \left\lfloor \frac{\mathrm{N}}{p} \right\rfloor$$

(so that the Mertens formula implies that $\lambda_{\mathrm{N}} \sim \log\log \mathrm{N}$).

Precisely, Harper proves that for *any* subset A of the nonnegative integers, we have

$$\left| \mathbf{P}_{\mathrm{N}}\Big(\sum_{\substack{p|n \\ p\leqslant \mathrm{Q}}} 1 \in \mathrm{A} \Big) - \mathbf{P}(\mathrm{Po}_{\mathrm{N}} \in \mathrm{A}) \right| \ll \frac{1}{\log\log \mathrm{N}},$$

and moreover that the decay rate $(\log\log \mathrm{N})^{-1}$ is best possible. This requires some additional arithmetic information than the proof of Theorem 2.3.1 (essentially some form of sieve), but the arithmetic ingredients remain to a large extent elementary. On the other hand, new ingredients from probability theory are involved, especially cases of Stein's Method for Poisson approximation.

A second approach starts from a proof of the Erdős–Kac Theorem due to Rényi and Turán [100], which is the implementation of the Lévy Criterion for convergence in law. Precisely, they prove that

$$\mathbf{E}_{\mathrm{N}}(e^{it\omega(n)}) = (\log \mathrm{N})^{e^{it}-1}(\Phi(t) + o(1)) \tag{2.13}$$

for any $t \in \mathbf{R}$ as $N \to +\infty$ (in fact, uniformly for $t \in \mathbf{R}$ – note that the function here is 2π-periodic), with a factor $\Phi(t)$ given by

$$\Phi(t) = \frac{1}{\Gamma(e^{it})} \prod_p \left(1 - \frac{1}{p}\right)^{e^{it}} \left(1 + \frac{e^{it}}{p-1}\right), \tag{2.14}$$

where the product over all primes is absolutely convergent. Recognizing that the term $(\log N)^{e^{it}-1}$ is the characteristic function of a Poisson random variable Po_N with parameter $\log\log N$, one can then obtain the Erdős–Kac Theorem by the same computation that leads to (2.12), combined with the continuity of Φ that shows that

$$\Phi\left(\frac{t}{\sqrt{\log\log N}}\right) \longrightarrow \Phi(0) = 1$$

as $N \to +\infty$.

The computation that leads to (2.13) is now interpreted as an instance of the Selberg–Delange method (see [115, II.5, Th. 3] for the general statement, and [115, II.6, Th. 1] for the special case of interest here).

It should be noted that the proof of (2.13) is quite a bit deeper than the proof of Theorem 2.3.1, and this is at it should be, because this formula contains precise information about the extreme values of $\omega(n)$, which we saw are not relevant to the Erdős–Kac Theorem. Indeed, taking $t = \pi$ and observing that $\Phi(\pi) = 0$ (because of the pole of the Gamma function), we obtain

$$\frac{1}{N} \sum_{n \leqslant N} (-1)^{\omega(n)} = \mathbf{E}(e^{-i\pi\omega(n)}) = o\left(\frac{1}{(\log N)^2}\right).$$

It is well known (as for the partial sums of the Möbius function, mentioned in Remark 2.2.2) that this implies elementarily the Prime Number Theorem

$$\sum_{p \leqslant N} 1 \sim \frac{N}{\log N}$$

(see again [59, §2.1]).

The link between the formula (2.13) and Poisson distribution was noticed in joint work with Nikeghbali [77]. Among other things, we remarked that it implies easily a bound for the Kolmogorov–Smirnov distance between $n \mapsto \omega(n)$ on Ω_N and a Poisson random variable Po_N. Additional work with A. Barbour [5] leads to bounds in total variation distance, and to even better (but non-Poisson) approximations. Another suggestive remark is that if we consider the independent random variables that appear in the proof of the Erdős–Kac theorem, namely,

$$X_N = \sum_{p \leqslant N} \left(B_p - \frac{1}{p} \right),$$

where (B_p) is a sequence of independent Bernoulli random variables with $\mathbf{P}(B_p = 1) = 1/p$, then we have (by a direct computation) the following analogue of (2.13):

$$\mathbf{E}(e^{itX_N}) = (\log N)^{e^{it}-1} \left(\prod_p \left(1 - \frac{1}{p}\right)^{e^{it}} \left(1 + \frac{e^{it}}{p-1}\right) + o(1) \right).$$

It is natural to ask then if there is a similar meaning to the factor $1/\Gamma(e^{it})$ that also appears in (2.14). And there is: for $N \geqslant 1$, define ℓ_N as the random variable on the symmetric group $\mathfrak{S}_N$ that maps a permutation σ to the number of cycles in its canonical cyclic representation (where we count fixed points as cycles of length 1, so, for instance, we have $\ell_N(1) = N$). Then, giving $\mathfrak{S}_N$ the uniform probability measure, we have

$$\mathbf{E}(e^{it\ell_N}) = N^{e^{it}-1} \left(\frac{1}{\Gamma(e^{it})} + o(1) \right), \tag{2.15}$$

corresponding to a Poisson distribution with parameter $\log N$ this time. This is not an isolated property: see the survey paper of Granville [48] for many significant analogies between (multiplicative) properties of integers and random permutations.[2]

Remark 2.4.2 Observe that (2.13) would be true *if* we had a decomposition

$$\omega(n) = \mathrm{Po}_N(n) + Y_N(n)$$

as random variables on Ω_N, where Y_N is independent of Po_N and converges in law to a random variable with characteristic function Φ. However, this is not in fact the case, because Φ is not a characteristic function of a probability measure! (It is unbounded on $\mathbf{R}$.)

Exercise 2.4.3 The goal of this exercise is to give a proof of the formula (2.15). We assume basic familiarity with the notion of tensor product of vector spaces and symmetric powers of vector spaces, and elementary representation theory of finite groups.

For $N \geqslant 1$, we define ℓ_N as a random variable on $\mathfrak{S}_N$ as above.

[2] Some readers might also enjoy the comic-book version [49].

(1) Show that the formula (2.15) follows from the exact expression

$$\mathbf{E}(e^{it\ell_N}) = \prod_{j=1}^{N}\Bigl(1-\frac{1}{j}+\frac{e^{it}}{j}\Bigr)$$

valid for all $N \geqslant 1$ and all $t \in \mathbf{R}$. [**Hint**: Use the formula

$$\frac{1}{\Gamma(z+1)} = \prod_{k\geqslant 1}\Bigl(1+\frac{z}{k}\Bigr)\Bigl(1+\frac{1}{k}\Bigr)^{-z},$$

which is valid for all $z \in \mathbf{C}$ (this is due to Euler).]

(2) Show that (1) is also equivalent with the formula

$$\mathbf{E}(m^{\ell_N}) = \prod_{j=1}^{N}\Bigl(1-\frac{1}{j}+\frac{m}{j}\Bigr) \tag{2.16}$$

for all $N \geqslant 1$ and all *integers* $m \geqslant 0$.

(3) Let $m \geqslant 0$ be a fixed integer. Let V be an m-dimensional complex vector space. For any $N \geqslant 1$, there is a homomorphism

$$\varrho_N\colon \mathfrak{S}_N \to \mathrm{GL}(V\otimes\cdots\otimes V) = \mathrm{GL}(V^{\otimes N})$$

(with N tensor factors) such that $\sigma \in \mathfrak{S}_N$ is sent to the unique automorphism of the tensor power $V^{\otimes N}$ which satisfies

$$x_1\otimes\cdots\otimes x_N \mapsto x_{\sigma(1)}\otimes\cdots\otimes x_{\sigma(N)}$$

for all $(x_1,\ldots,x_N) \in V^{\otimes N}$. (This is a representation of $\mathfrak{S}_N$ on the vector space $V^{\otimes N}$; note that this space has dimension m^N.)

(4) Show that for any $\sigma \in \mathfrak{S}_N$, the trace of the automorphism $\varrho_N(\sigma)$ of $V^{\otimes N}$ is equal to $m^{\ell_N(\sigma)}$.

(5) Deduce that the formula (2.16) holds. [**Hint**: Use the fact that for any representation $\varrho\colon G \to \mathrm{GL}(E)$ of a finite group on a finite-dimensional **C**-vector space, the average of the trace of $\varrho(g)$ over $g \in G$ is equal to the dimension of the space of vectors $x \in E$ that are invariant, that is, that satisfy $\varrho(g)(x) = x$ for all $g \in G$ (see, e.g., [70, Prop. 4.3.1] for this); then identify this space to compute its dimension.]

(6) Deduce also from (2.16) that there exists a sequence $(B_j)_{j\geqslant 1}$ of independent Bernoulli random variables such that we have an equality in law

$$\ell_N = B_1+\cdots+B_N$$

for all $N \geqslant 1$, and $\mathbf{P}(B_j = 1) = 1/j$ for all $j \geqslant 1$. (This decomposition is often obtained by what is called the "Chinese Restaurant Process" in the probabilistic literature; see, for instance, [2, Example 2.4].)

2.5 Final Remarks

Classically, the Erdős–Wintner and the Erdős–Kac Theorem (and related topics) are presented in a different manner, which is well illustrated in the book of Tenenbaum [115, III.1, III.2]. This emphasizes the notion of *density* of sets of integers, namely, quantities like

$$\limsup_{N\to+\infty} \frac{1}{N}|\{1 \leqslant n \leqslant N \mid n \in A\}|$$

for a given set A, or the associated liminf, or the limit when it exists. Convergence in law is then often encapsulated in the existence of these limits for sets of the form

$$A = \{n \geqslant 1 \mid f(n) \leqslant x\},$$

the limit $F(x)$ (which is only assumed to exist for continuity points of F) being a "distribution function," that is, $F(x) = \mathbf{P}(X \leqslant x)$ for some real-valued random variable X.

Our emphasis on a more systematic probabilistic presentation has the advantage of leading more naturally to the use of purely probabilistic techniques and insights. This will be especially relevant when we consider random variables with values in more complicated sets than **R** (as we will do in the next chapters), in which case the analogue of distribution functions becomes awkward or simply doesn't exist. Our point of view is also more natural when we come to consider arithmetic random variables Y_N on Ω_N that genuinely depend on N, in the sense that there doesn't exist an arithmetic function f such that Y_N is the restriction of f to Ω_N for all $N \geqslant 1$.

Among the many generalizations of the Erdős–Kac Theorem (and related results for more general arithmetic functions), we wish to mention Billingsley's work [8, Th. 4.1, Example 1, p. 764] that obtains a functional version where the convergence in law is toward *Brownian motion* (we refer to Billingsley's very accessible text [7] for a first presentation of Brownian motion, and to the book of Revuz and Yor [104] for a complete modern treatment): for $0 \leqslant t \leqslant 1$, define a random variable $\widetilde{X}_N$ on Ω_N with values in the Banach space $C([0,1])$ of continuous functions on $[0,1]$ by putting $\widetilde{X}_N(n)(0) = 0$ and

$$\widetilde{X}_N(n)\left(\frac{\log\log k}{\log\log N}\right) = \frac{1}{(\log\log N)^{1/2}}\left(\sum_{\substack{p\mid n\\ p\leqslant k}} 1 - \log\log k\right)$$

for $2 \leqslant k \leqslant N$, and by linear interpolation between such points. Then Billingsley proves that $\widetilde{X}_N$ converges in law to Brownian motion as $N \to +\infty$.

Another very interesting limit theorem of Billingsley (see [6] and also [10, Th. I.4.5]) deals with the distribution of *all* the prime divisors of an integer $n \in \Omega_{\mathrm{N}}$, and establishes convergence in law of a suitable normalization of these. Precisely, let X be the compact topological space

$$\mathrm{X} = \prod_{k \geqslant 1} [0, 1].$$

For all integers $n \geqslant 1$, denote by

$$p_1 \geqslant p_2 \geqslant \cdots \geqslant p_{\Omega(n)}$$

the prime divisors of n, counted with multiplicity and in nonincreasing order. Moreover, define $p_k = 1$ if $k > \Omega(n)$. Define then an X-valued random variable $\mathsf{D}_{\mathrm{N}} = (\mathsf{D}_{\mathrm{N},k})_{k \geqslant 1}$, where

$$\mathsf{D}_{\mathrm{N},k}(n) = \frac{\log p_k}{\log n}$$

for $n \in \Omega_{\mathrm{N}}$ (in other words, we have $p_k = n^{\mathsf{D}_{\mathrm{N},k}(n)}$). Then Billingsley proved that the random variables D_{N} converge, as $\mathrm{N} \to +\infty$, to a measure on X, which is called the Poisson–Dirichlet distribution (with parameter 1). This measure is quite an interesting one, and occurs also (among other places) in a similar limit theorem for random variables encoding the length of the cycles occurring in a random permutation, again ordered to be nonincreasing (another example of the connections between prime factorizations and permutations which were mentioned in the previous section 2.4).

A shorter proof of this limit theorem was given by Donnelly and Grimmett [27]. It is based on the remark that the Poisson–Dirichlet measure is the image under a certain continuous map of the natural measure on X under which the components of elements of X form a sequence of independent uniformly distributed random variables on $[0, 1]$; arithmetically, it turns out to depend only on the estimate

$$\sum_{p \leqslant \mathrm{N}} \frac{\log p}{p} = \log \mathrm{N} + \mathrm{O}(1),$$

which is at the same level of depth as the Mertens formula (see C.3.1 (3)).

[Further references: Tenenbaum [115].**]**

3

The Distribution of Values of the Riemann Zeta Function, I

Probability tools	Arithmetic tools
Definition of convergence in law (Section B.3)	Dirichlet series (Section A.4)
Kolmogorov's Theorem for random series (Th. B.10.1)	Riemann zeta function (Section C.4)
Weyl Criterion and Kronecker's Theorem (Section B.6, Th. B.6.5)	Fundamental theorem of arithmetic
Menshov–Rademacher Theorem (Th. B.10.5)	Mean square of $\zeta(s)$ outside the critical line (Prop. C.4.1)
Lipschitz test functions (Prop. B.4.1)	Euler product (Lemma C.1.4)
Support of a random series (Prop. B.10.8)	Strong Mertens estimate and Prime Number Theorem (Cor. C.3.4)

3.1 Introduction

The Riemann zeta function is defined first for complex numbers s such that $\operatorname{Re}(s) > 1$, by means of the series

$$\zeta(s) = \sum_{n \geqslant 1} \frac{1}{n^s}.$$

It plays an important role in prime number theory, arising because of the famous Euler product formula, which expresses $\zeta(s)$ as a product over primes, in this region: we have

$$\zeta(s) = \prod_p (1 - p^{-s})^{-1} \tag{3.1}$$

if $\mathrm{Re}(s) > 1$ (see Corollary C.1.5). By standard properties of series of holomorphic functions (note that $s \mapsto n^s = e^{s \log n}$ is entire for any $n \geqslant 1$), the Riemann zeta function is holomorphic for $\mathrm{Re}(s) > 1$. It is of crucial importance however that it admits an analytic continuation to $\mathbf{C} - \{1\}$, with furthermore a simple pole at $s = 1$ with residue 1.

This analytic continuation can be performed simultaneously with the proof of the *functional equation*: the function defined by

$$\xi(s) = \pi^{-s/2}\Gamma(s/2)\zeta(s)$$

satisfies

$$\xi(1-s) = \xi(s)$$

and has simple poles with residue 1 at $s = 0$ and $s = 1$. Since the inverse of the Gamma function is an entire function (Proposition A.3.2), the analytic continuation of the Riemann zeta function follows immediately.

However, for many purposes (including the results of this chapter), it is enough to know that $\zeta(s)$ has analytic continuation for $\mathrm{Re}(s) > 0$, and this can be checked quickly using the following computation, based on summation by parts (Lemma A.1.1): using the notation $\langle x \rangle$ for the fractional part of a real number x, namely, the unique real number in $[0, 1[$ such that $x - \langle x \rangle \in \mathbf{Z}$ for $\mathrm{Re}(s) > 1$,[1] we have

$$\begin{aligned}\sum_{n \geqslant 1} \frac{1}{n^s} &= s \int_1^{+\infty} \Bigl(\sum_{1 \leqslant n \leqslant t} 1 \Bigr) t^{-s-1} dt \\ &= s \int_1^{+\infty} (t - \langle t \rangle) t^{-s-1} dt \\ &= s \int_1^{+\infty} t^{-s} dt - s \int_1^{+\infty} \langle t \rangle t^{-s-1} dt = \frac{s}{s-1} - s \int_1^{+\infty} \langle t \rangle t^{-s-1} dt.\end{aligned}$$

The rational function $s \mapsto s/(s-1)$ has a simple pole at $s = 1$ with residue 1. Also, since $0 \leqslant \langle t \rangle \leqslant 1$, the integral defining the function

$$s \mapsto s \int_1^{+\infty} \langle t \rangle t^{-s-1} dt$$

is absolutely convergent, and therefore this function is holomorphic, for $\mathrm{Re}(s) > 0$. The expression above then shows that the Riemann zeta function is meromorphic, with a simple pole at $s = 1$ with residue 1, for $\mathrm{Re}(s) > 0$.

[1] A more standard notation would be $\{x\}$, but this clashes with the notation used for set constructions.

Since $\zeta(s)$ is quite well behaved for $\mathrm{Re}(s) > 1$, and since the Gamma function is a very well-known function, the functional equation $\zeta(1-s) = \zeta(s)$ shows that one can understand the behavior of $\zeta(s)$ for s outside of the *critical strip*

$$\mathrm{S} = \{s \in \mathbf{C} \mid 0 \leqslant \mathrm{Re}(s) \leqslant 1\}.$$

The Riemann Hypothesis is a fundamental (still conjectural) statement about the Riemann zeta function in the critical strip: it states that if $s \in \mathrm{S}$ satisfies $\zeta(s) = 0$, then the real part of s must be $1/2$. Because holomorphic functions (with relatively slow growth, a property true for ζ, although this requires some argument to prove) are essentially characterized by their zeros (just like polynomials are!), the proof of this conjecture would enormously expand our understanding of the properties of the Riemann zeta function. Although it remains open, this should motivate our interest in the distribution of values of the zeta function. Another motivation is that it contains crucial information about primes, which will be very visible in Chapter 5.

We first focus our attention to a vertical line $\mathrm{Re}(s) = \tau$, where τ is a fixed real number such that $\tau \geqslant 1/2$ (the case $\tau \leqslant 1$ will be the most interesting, but some statements do not require this assumption). We consider real numbers $\mathrm{T} \geqslant 2$ and define the probability space $\Omega_{\mathrm{T}} = [-\mathrm{T}, \mathrm{T}]$ with the uniform probability measure $dt/(2\mathrm{T})$. We then view

$$t \mapsto \zeta(\tau + it)$$

as a random variable $\mathsf{Z}_{\tau,\mathrm{T}}$ on $\Omega_{\mathrm{T}} = [-\mathrm{T}, \mathrm{T}]$. These are arithmetically defined random variables. Do they have some specific, interesting, asymptotic behavior?

The answer to this question turns out to depend on τ, as the following first result of Bohr and Jessen reveals:

Theorem 3.1.1 (Bohr–Jessen) *Let $\tau > 1/2$ be a fixed real number. Define $\mathsf{Z}_{\tau,\mathrm{T}}$ as the random variable $t \mapsto \zeta(\tau + it)$ on Ω_{T}. There exists a probability measure μ_τ on $\mathbf{C}$ such that $\mathsf{Z}_{\tau,\mathrm{T}}$ converges in law to μ_τ as $\mathrm{T} \to +\infty$. Moreover, the support of μ_τ is compact if $\tau > 1$, and is equal to $\mathbf{C}$ if $1/2 < \tau \leqslant 1$.*

We will describe precisely the measure μ_τ in Section 3.2: it is a highly nongeneric probability distribution, whose definition (and hence properties) retains a significant amount of arithmetic, in contrast with the Erdős–Kac Theorem, where the limit is a very generic distribution.

Theorem 3.1.1 is in fact a direct consequence of a result due to Voronin [119] and Bagchi [4], which extends it in a very surprising direction. Instead of fixing $\tau \in]1/2, 1[$ and looking at the distribution of the single values $\zeta(\tau + it)$ as t varies, we consider for such τ some radius r such that the disc

$$\mathrm{D} = \{s \in \mathbf{C} \mid |s - \tau| \leqslant r\}$$

is contained in the interior of the critical strip, and we look for $t \in \mathbf{R}$ at the *functions*

$$\zeta_{\mathrm{D},t} : \begin{cases} \mathrm{D} & \longrightarrow & \mathbf{C}, \\ s & \mapsto & \zeta(s + it), \end{cases}$$

which are "vertical translates" of the Riemann zeta function restricted to D. For each $\mathrm{T} \geqslant 0$, we view $t \mapsto \zeta_{\mathrm{D},t}$ as a random variable (say, $\mathsf{Z}_{\mathrm{D,T}}$) on $([-\mathrm{T}, \mathrm{T}], dt/(2\mathrm{T}))$ with values in the space H(D) of functions which are holomorphic in the interior of D and continuous on its boundary. Bagchi's remarkable result is a convergence in law in this space, that is, a functional limit theorem: there exists a probability measure ν on H(D) such that the random variables $\mathsf{Z}_{\mathrm{D,T}}$ converge in law to ν as $\mathrm{T} \to +\infty$. Computing the support of ν (which is a nontrivial task) leads to a proof of Voronin's Universality Theorem: for any function $f \in \mathrm{H(D)}$ which does not vanish on D, and for any $\varepsilon > 0$, there exists $t \in \mathbf{R}$ such that

$$\|\zeta(\cdot + it) - f\|_\infty < \varepsilon,$$

where the norm is the supremum norm on D. In other words, up to arbitrarily small error, all holomorphic functions f (that do not vanish) can be seen by looking at some vertical translate of the Riemann zeta function!

We illustrate this fact in Figure 3.1, which presents density plots of $|\zeta(s + it)|$ for various values of $t \in \mathbf{R}$, as functions of s in the square $[3/4 - 1/8, 3/4 + 1/8] \times [-1/8, 1/8]$. Voronin's Theorem implies that, for suitable t, such a picture will be indistinguishable from that associated to *any* holomorphic function on this square that never vanishes there.

We will prove the Bohr–Jessen–Bagchi theorems in the next section, and use in particular the computation of the support of Bagchi's limiting distribution for translates of the Riemann zeta function to prove Voronin's Universality Theorem in Section 3.3.

3.2 The Theorems of Bohr–Jessen and of Bagchi

We begin by stating a precise version of Bagchi's Theorem. In the remainder of this chapter, we denote by Ω_{T} the probability space $([-\mathrm{T}, \mathrm{T}], dt/(2\mathrm{T}))$ for $\mathrm{T} \geqslant 1$. We will often write $\mathbf{E}_{\mathrm{T}}(\cdot)$ and $\mathbf{P}_{\mathrm{T}}(\cdot)$ for the corresponding expectation and probability.

Figure 3.1 The modulus of $\zeta(s+it)$ for s in the square $[3/4-1/8, 3/4+1/8] \times [-1/8, 1/8]$, for $t = 0$, 21000, 58000, and 75000.

Theorem 3.2.1 (Bagchi [4]) *Let τ be such that $1/2 < \tau$. If $1/2 < \tau < 1$, let $r > 0$ be such that*

$$\mathrm{D} = \{s \in \mathbf{C} \mid |s - \tau| \leqslant r\} \subset \{s \in \mathbf{C} \mid 1/2 < \mathrm{Re}(s) < 1\},$$

and if $\tau \geqslant 1$, let D *be any compact subset of $\{s \in \mathbf{C} \mid \mathrm{Re}(s) \geqslant 1\}$ such that $\tau \in \mathrm{D}$.*

Consider the H(D)*-valued random variables $\mathsf{Z}_{\mathrm{D,T}}$ defined by*

$$t \mapsto (s \mapsto \zeta(s+it))$$

on Ω_{T}. Let $(\mathrm{X}_p)_p$ be a sequence of independent random variables, indexed by the primes, which are identically distributed, with distribution uniform on the unit circle $\mathbf{S}^1 \subset \mathbf{C}^\times$.

Then we have convergence in law $\mathsf{Z}_{\mathrm{D,T}} \longrightarrow \mathrm{Z}_{\mathrm{D}}$ as $\mathrm{T} \to +\infty$, where Z_{D} is the random Euler product defined by

$$\mathrm{Z}_{\mathrm{D}}(s) = \prod_p (1 - p^{-s}\mathrm{X}_p)^{-1}.$$

In this theorem, the space H(D) is viewed as a Banach space (hence a metric space, so that convergence in law makes sense) with the norm

$$\|f\|_\infty = \sup_{z \in \mathrm{D}} |f(z)|.$$

We can already see that Theorem 3.2.1 is (much) stronger than the convergence in law component of Theorem 3.1.1, which we now prove assuming this result:

Corollary 3.2.2 *Fix τ such that $1/2 < \tau$. As $\mathrm{T} \to +\infty$, the random variables $\mathsf{Z}_{\tau,\mathrm{T}}$ of Theorem 3.1.1 converge in law to the random variable $\mathrm{Z_D}(\tau)$, where* D *is either a disc*

$$\mathrm{D} = \{s \in \mathbf{C} \mid |s - \tau| \leqslant r\}$$

contained in the interior of the critical strip, if $\tau < 1$, or any compact subset of $\{s \in \mathbf{C} \mid \mathrm{Re}(s) \geqslant 1\}$ such that $\tau \in \mathrm{D}$.

Proof Fix D as in the statement. Tautologically, we have

$$\mathsf{Z}_{\tau,\mathrm{T}} = \zeta_{\mathrm{D,T}}(\tau)$$

or $\mathsf{Z}_{\tau,\mathrm{T}} = e_\tau \circ \zeta_{\mathrm{D,T}}$, where

$$e_\tau \begin{cases} \mathrm{H(D)} & \longrightarrow & \mathbf{C}, \\ f & \mapsto & f(\tau) \end{cases}$$

is the evaluation map. This map is continuous on H(D), so it follows by composition (Proposition B.3.2 in Appendix B) that the convergence in law $\mathrm{Z_{D,T}} \longrightarrow \mathrm{Z_D}$ of Bagchi's Theorem implies the convergence in law of $\mathsf{Z}_{\tau,\mathrm{T}}$ to the random variable $e_\tau \circ \mathrm{Z_D}$, which is simply $\mathrm{Z_D}(\tau)$. □

In order to prove the final part of Theorem 3.1.1, and to derive Voronin's Universality Theorem, we need to understand the support of the limit $\mathrm{Z_D}$ in Bagchi's Theorem. We will prove in Section 3.3:

Theorem 3.2.3 (Bagchi, Voronin) *Let τ be such that $1/2 < \tau < 1$, and r such that*

$$\mathrm{D} = \{s \in \mathbf{C} \mid |s - \tau| \leqslant r\} \subset \{s \in \mathbf{C} \mid 1/2 < \mathrm{Re}(s) < 1\}.$$

The support of $\mathrm{Z_D}$ contains

$$\mathrm{H(D)}^\times = \{f \in \mathrm{H(D)} \mid f(z) \neq 0 \textit{ for all } z \in \mathrm{D}\}$$

and is equal to $\mathrm{H(D)}^\times \cup \{0\}$.

In particular, for any function $f \in \mathrm{H(D)}^\times$, and for any $\varepsilon > 0$, there exists $t \in \mathbf{R}$ such that

$$\sup_{s \in \mathrm{D}} |\zeta(s + it) - f(s)| < \varepsilon. \tag{3.2}$$

It is then obvious that if $1/2 < \tau < 1$, the support of the Bohr–Jessen random variable $\mathrm{Z_D}(\tau)$ is equal to $\mathbf{C}$.

Exercise 3.2.4 Prove that the support of the Bohr–Jessen random variable $Z_D(1)$ is also equal to $\mathbf{C}$.

We now begin the proof of Theorem 3.2.1 by giving some intuition for the result and in particular for the shape of the limiting distribution. Indeed, this very elementary argument will suffice to prove Bagchi's Theorem in the case $\tau > 1$. This turns out to be similar to the intuition behind the Erdős–Kac Theorem. We begin with the Euler product

$$\zeta(s+it) = \prod_p (1-p^{-s-it})^{-1},$$

which is valid for $\mathrm{Re}(s) > 1$. We can express this also (formally, we "compute the logarithm"; see Proposition A.2.2 (2)) in the form

$$\zeta(s+it) = \exp\Big(-\sum_p \log(1-p^{-s-it})\Big). \tag{3.3}$$

This displays the Riemann zeta function on Ω_T as the exponential of a sum involving the sequence (indexed by primes) of random variables $(\mathsf{X}_{p,T})_p$ such that

$$\mathsf{X}_{p,T}(t) = p^{-it},$$

each taking value in the unit circle $\mathbf{S}^1$. To understand how the zeta function will behave statistically on Ω_T, the first step is to understand the limiting behavior of this sequence.

This has a very simple answer:

Proposition 3.2.5 *For* $T \geqslant 0$, *let* $\mathsf{X}_T = (\mathsf{X}_{p,T})_p$ *be the sequence of random variables on* Ω_T *given by*

$$t \mapsto (p^{-it})_p.$$

Then X_T *converges in law as* $T \to +\infty$ *to a sequence* $X = (X_p)_p$ *of independent random variables, each of which is uniformly distributed on* $\mathbf{S}^1$.

Bagchi's Theorem is therefore to be understood as saying that we can "pass to the limit" in the formula (3.3) to obtain a convergence in law of $\zeta(s+it)$, for $s \in D$, to

$$\exp\Big(-\sum_p \log(1-p^{-s}X_p)\Big),$$

viewed as a random function.

This sketch is of course incomplete in general, the foremost objection being that we are especially interested in the zeta function outside of the region of absolute convergence, so the meaning of (3.3) is unclear. But we will see that nevertheless enough connections remain to carry the argument through.

We isolate the crucial part of the proof of Proposition 3.2.5 as a lemma, since we will also use it in the proof of Selberg's Theorem in the next chapter (see Section 4.2).

Lemma 3.2.6 *Let $r > 0$ be a real number. We have*

$$|\mathbf{E}_{\mathrm{T}}(r^{-it})| \leqslant \min\left(1, \frac{1}{\mathrm{T}|\log r|}\right). \tag{3.4}$$

In particular, if $r = n_1/n_2$ for some positive integers $n_1 \neq n_2$, then we have

$$\mathbf{E}_{\mathrm{T}}(r^{-it}) \ll \min\left(1, \frac{\sqrt{n_1 n_2}}{\mathrm{T}}\right), \tag{3.5}$$

where the implied constant is absolute.

Proof of Lemma 3.2.6 Since $|r^{-it}| = 1$, we see that the expectation is always $\leqslant 1$. If $r \neq 1$, then we get

$$\mathbf{E}_{\mathrm{T}}(r^{-it}) = \frac{1}{2\mathrm{T}}\left[\frac{i}{\log r} r^{-it}\right]_{-\mathrm{T}}^{\mathrm{T}} = \frac{i(r^{i\mathrm{T}} - r^{-i\mathrm{T}})}{2\mathrm{T}(\log r)},$$

which has modulus at most $|\log r|^{-1}\mathrm{T}^{-1}$, hence the first bound holds.

Assume now that $r = n_1/n_2$ with $n_1 \neq n_2$ positive integers. Assume that $n_2 > n_1 \geqslant 1$. Then $n_2 \geqslant n_1 + 1$, and hence

$$\left|\log \frac{n_1}{n_2}\right| \geqslant \left|\log\left(1 + \frac{1}{n_1}\right)\right| \gg \frac{1}{n_1} \geqslant \frac{1}{\sqrt{n_1 n_2}}.$$

If $n_2 < n_1$, we exchange the role of n_1 and n_2, and since both sides of the bound (3.5) are symmetric in terms of n_1 and n_2, the result follows. □

Proof of Proposition 3.2.5 It is convenient here to view the sequences $(\mathrm{X}_{p,\mathrm{T}})_p$ and $(\mathrm{X}_p)_p$ as two random variables on Ω_{T}, taking value in the infinite product

$$\widehat{\mathbf{S}}^1 = \prod_p \mathbf{S}^1$$

of copies of the unit circle indexed by primes. Note that $\widehat{\mathbf{S}}^1$ is a compact abelian group (with componentwise product).

In this interpretation, the limit (or more precisely the law of (X_p)) is simply the probability Haar measure on the group $\widehat{\mathbf{S}}^1$ (see Section B.6). This allows us to prove convergence in law using the well-known Weyl Criterion: the statement of the proposition is equivalent with the property that

$$\lim_{\mathrm{T}\to+\infty} \mathbf{E}_{\mathrm{T}}(\chi(\mathsf{X}_{p,\mathrm{T}})) = 0 \tag{3.6}$$

for any nontrivial continuous unitary character $\chi : \widehat{\mathbf{S}}^1 \longrightarrow \mathbf{S}^1$. An elementary property of compact groups shows that for any such character there exists a finite nonempty subset S of primes, and for each $p \in \mathrm{S}$ some integer $m_p \in \mathbf{Z} - \{0\}$, such that

$$\chi(z) = \prod_{p\in\mathrm{S}} z_p^{m_p}$$

for any $z = (z_p)_p \in \widehat{\mathbf{S}}^1$ (see Example B.6.2(2)). We then have by definition

$$\mathbf{E}_{\mathrm{T}}(\chi(\mathsf{X}_{p,\mathrm{T}})) = \frac{1}{2\mathrm{T}} \int_{-\mathrm{T}}^{\mathrm{T}} \prod_{p\in\mathrm{S}} p^{-itm_p} dt = \frac{1}{2\mathrm{T}} \int_{-\mathrm{T}}^{\mathrm{T}} r^{-it} dt,$$

where $r > 0$ is the rational number given by

$$r = \prod_{p\in\mathrm{S}} p^{m_p}.$$

Since we have $r \neq 1$ (because S is not empty and $m_p \neq 0$), we obtain $\mathbf{E}_{\mathrm{T}}(\chi(\mathsf{X}_{p,\mathrm{T}})) \to 0$ as $\mathrm{T} \to +\infty$ from (3.4). □

As a corollary, Bagchi's Theorem follows formally for $\tau > 1$ and D contained in the set of complex numbers with real part > 1. This is once more a very simple fact which is often not specifically discussed, but which gives an indication and a motivation for the more difficult study in the critical strip.

Special case of Theorem 3.2.1 for $\tau > 1$ Assume that $\tau > 1$ and that D is a compact subset containing τ contained in $\{s \in \mathbf{C} \mid \mathrm{Re}(s) > 1\}$. We view $\mathsf{X}_{\mathrm{T}} = (\mathsf{X}_{p,\mathrm{T}})$ as random variables with values in the topological space $\widehat{\mathbf{S}}^1$, as before. This is also (as a countable product of metric spaces) a metric space. We claim that the map

$$\varphi \begin{cases} \widehat{\mathbf{S}}^1 & \longrightarrow & \mathrm{H}(\mathrm{D}), \\ (x_p) & \mapsto & \left(s \mapsto -\sum_p \log(1 - x_p p^{-s})\right) \end{cases}$$

is continuous (see Definition A.2.1 again for the definition of the logarithm here). If this is so, then the composition principle (see Proposition B.3.2) and Proposition 3.2.5 imply that $\varphi(\mathsf{X}_{\mathrm{T}})$ converges in law to the H(D)-valued random variable $\varphi(\mathrm{X})$, where $\mathrm{X} = (\mathrm{X}_p)$ with the X_p uniform and independent on $\mathbf{S}^1$. But this is exactly the statement of Bagchi's Theorem for D.

Now we check the claim. Fix $\varepsilon > 0$. Let $\mathrm{T} > 0$ be some parameter to be chosen later in terms of ε. For any $x = (x_p)$ and $y = (y_p)$ in $\widehat{\mathbf{S}}^1$, we have

$$\|\varphi(x) - \varphi(y)\|_\infty \leqslant \sum_{p \leqslant \mathrm{T}} \|\log(1 - x_p p^{-s}) - \log(1 - y_p p^{-s})\|_\infty$$
$$+ \sum_{p > \mathrm{T}} \|\log(1 - x_p p^{-s})\|_\infty + \sum_{p > \mathrm{T}} \|\log(1 - y_p p^{-s})\|_\infty.$$

Because D is compact in the half-plane $\mathrm{Re}(s) > 1$, the minimum of the real part of $s \in \mathrm{D}$ is some real number $\sigma_0 > 1$. Since $|x_p| = |y_p| = 1$ for all primes, and since

$$|\log(1 - z)| \leqslant 2|z|$$

for $|z| \leqslant 1/2$ (Proposition A.2.2 (3)), it follows that

$$\sum_{p > \mathrm{T}} \|\log(1 - x_p p^{-s})\|_\infty + \sum_{p > \mathrm{T}} \|\log(1 - y_p p^{-s})\|_\infty \leqslant 4 \sum_{p > \mathrm{T}} p^{-\sigma_0} \ll \mathrm{T}^{1-\sigma_0}.$$

We fix T so that $\mathrm{T}^{1-\sigma_0} < \varepsilon/2$. Now the map

$$(x_p)_{p \leqslant \mathrm{T}} \mapsto \sum_{p \leqslant \mathrm{T}} \|\log(1 - x_p p^{-s}) - \log(1 - y_p p^{-s})\|_\infty$$

is obviously continuous, and therefore uniformly continuous since the domain is a compact set. This function has value 0 when $x_p = y_p$ for $p \leqslant \mathrm{T}$, so there exists $\delta > 0$ such that

$$\sum_{p \leqslant \mathrm{T}} |\log(1 - x_p p^{-s}) - \log(1 - y_p p^{-s})| < \frac{\varepsilon}{2}$$

if $|x_p - y_p| \leqslant \delta$ for $p \leqslant \mathrm{T}$. Therefore, provided that

$$\max_{p \leqslant \mathrm{T}} |x_p - y_p| \leqslant \delta,$$

we have

$$\|\varphi(x) - \varphi(y)\|_\infty \leqslant \varepsilon.$$

This proves the (uniform) continuity of φ. □

We now begin the proof of Bagchi's Theorem in the critical strip. The argument follows partly his original proof [4], which is quite different from the Bohr–Jessen approach, with some simplifications. Here are the main steps of the proof:

- we prove convergence almost surely of the random Euler product, and of its formal Dirichlet series expansion; this also shows that they define random *holomorphic* functions;
- we prove that both the Riemann zeta function and the limiting Dirichlet series are, in suitable mean sense, limits of smoothed partial sums of their respective Dirichlet series;
- we then use an elementary argument to conclude using Proposition 3.2.5.

We fix from now on a sequence $(\mathrm{X}_p)_p$ of independent random variables all uniformly distributed on $\mathbf{S}^1$. We often view the sequence (X_p) as an $\widehat{\mathbf{S}}^1$-valued random variable, as in the proof of Proposition 3.2.5. Furthermore, for any positive integer $n \geqslant 1$, we define

$$\mathrm{X}_n = \prod_{p \mid n} \mathrm{X}_p^{v_p(n)}, \tag{3.7}$$

where $v_p(n)$ is the p-adic valuation of n. Thus (X_n) is a sequence of $\mathbf{S}^1$-valued random variables.

Exercise 3.2.7 Prove that the sequence $(\mathrm{X}_n)_{n\geqslant 1}$ is neither independent nor symmetric.

Exercise 3.2.8 The following exercise provides the starting point of recent probabilistic approaches to the problem of estimating the so-called *pseudo-moments* of the Riemann zeta function (see the thesis of M. Gerspach [46]), although it is often proved using different approaches, such as the ergodic theorem for flows.

For any real numbers $q \geqslant 0$ and $x \geqslant 1$ and any sequence of complex numbers (a_n), prove that the limit

$$\lim_{\mathrm{T}\to+\infty} \frac{1}{\mathrm{T}} \int_{\mathrm{T}}^{2\mathrm{T}} \Big| \sum_{n \leqslant x} a_n n^{-it} \Big|^q dt$$

exists and that it is equal to

$$\mathbf{E}\Big(\Big| \sum_{n \leqslant x} a_n \mathrm{X}_n \Big|^q \Big).$$

We first show that the limiting random functions are indeed well defined as H(D)-valued random variables.

Proposition 3.2.9 *Let $\tau \in]1/2, 1[$, and let $U_\tau = \{s \in \mathbf{C} \mid \mathrm{Re}(s) > \tau\}$.*

(1) *The random Euler product defined by*

$$Z(s) = \prod_p (1 - X_p p^{-s})^{-1}$$

converges almost surely for any $s \in U_\tau$. For any compact subset $K \subset U_\tau$, the random function

$$Z_K : \begin{cases} K & \longrightarrow \mathbf{C}, \\ s & \mapsto Z(s) \end{cases}$$

is an $H(K)$-valued random variable.

(2) *The random Dirichlet series defined by*

$$\tilde{Z} = \sum_{n \geqslant 1} X_n n^{-s}$$

converges almost surely for any $s \in U_\tau$. For any compact subset $K \subset U_\tau$, the random function $\tilde{Z}_K : s \mapsto \tilde{Z}(s)$ on K is an $H(K)$-valued random variable.

(3) *We have $\tilde{Z}_K = Z_K$ almost surely.*

Proof (1) For $N \geqslant 1$ and $s \in K$, we have by definition

$$\sum_{p \leqslant N} \log(1 - X_p p^{-s})^{-1} = \sum_{p \leqslant N} \frac{X_p}{p^s} + \sum_{k \geqslant 2} \sum_{p \leqslant N} \frac{X_{p^k}}{p^{ks}}.$$

Since $\mathrm{Re}(s) > 1/2$ for $s \in K$, the series

$$\sum_{k \geqslant 2} \sum_p \frac{X_{p^k}}{p^{ks}}$$

converges absolutely for $s \in U_\tau$. By Lemma A.4.1, its sum is therefore an $H(K)$-valued random variable for any compact subset K of U_τ.

Fix now $\tau_1 < \tau$ such that $\tau_1 > \frac{1}{2}$. We can apply Kolmogorov's Theorem, B.10.1, to the independent random variables $(X_p p^{-\tau_1})$, since

$$\sum_p \mathbf{V}(p^{-\tau_1} X_p) = \sum_p \frac{1}{p^{2\tau_1}} < +\infty.$$

Thus the series

$$\sum_p \frac{X_p}{p^{\tau_1}}$$

converges almost surely. By Lemma A.4.1 again, it follows that

$$P(s) = \sum_p \frac{X_p}{p^s}$$

converges almost surely for all $s \in U_\tau$, and is holomorphic on U_τ. By restriction, its sum is an H(K)-valued random variable for any K compact in U_τ.

These facts show that the sequence of partial sums

$$\sum_{p \leqslant N} \log(1 - X_p p^{-s})^{-1}$$

converges almost surely as $N \to +\infty$ to a random holomorphic function on K. Taking the exponential, we obtain the almost sure convergence of the random Euler product to a random holomorphic function Z_K on K.

(2) The argument is similar, except that the sequence $(X_n)_{n \geqslant 1}$ is not independent. However, it is orthonormal: if $n \neq m$, we have

$$\mathbf{E}(X_n \overline{X}_m) = 0 \quad \text{and} \quad \mathbf{E}(|X_n|^2) = 1$$

(indeed X_n and X_m may be viewed as characters of $\widehat{\mathbf{S}}^1$, and they are distinct if $n \neq m$, so that this is the orthogonality property of characters of compact groups). We can then apply the Menshov–Rademacher Theorem, B.10.5, to (X_n) and $a_n = n^{-\tau_1}$: since

$$\sum_{n \geqslant 1} |a_n|^2 (\log n)^2 = \sum_{n \geqslant 1} \frac{(\log n)^2}{n^{2\tau_1}} < +\infty,$$

the series $\sum X_n n^{-\tau_1}$ converges almost surely, and Lemma A.4.1 shows that $\tilde{Z}$ converges almost surely on U_τ, and defines a holomorphic function there. Restricting to K leads to $\tilde{Z}_K$ as H(K)-valued random variable.

Finally, to prove that $Z_K = \tilde{Z}_K$ almost surely, we may replace K by the compact subset

$$K_1 = \{s \in \mathbf{C} \mid \tau_1 \leqslant \sigma \leqslant A, \quad |t| \leqslant B\},$$

with $A \geqslant 2$ and B chosen large enough to ensure that $K \subset K_1$. The previous argument shows that the random Euler product and Dirichlet series converge almost surely on K_1. But K_1 contains the open set

$$V = \{s \in \mathbf{C} \mid 1 < \mathrm{Re}(s) < 2, \quad |t| < B\},$$

where the Euler product and Dirichlet series converge absolutely, so that Lemma C.1.4 proves that the random holomorphic functions Z_{K_1} and $\tilde{Z}_{K_1}$ are equal when restricted to V. By analytic continuation (and continuity), they are equal also on K_1, hence *a posteriori* on K. ☐

We will prove Bagchi's Theorem using the random Dirichlet series, which is easier to handle than the Euler product. However, we will still denote it $\mathrm{Z}(s)$, which is justified by the last part of the proposition.

Some additional properties of this random Dirichlet series are now needed. Most importantly, we need to find a finite approximation that also applies to the Riemann zeta function. This will be done using *smooth partial sums*.

First we need to check that $\mathrm{Z}(s)$ is of polynomial growth on average on vertical strips.

Lemma 3.2.10 *Let* $\mathrm{Z}(s)$ *be the random Dirichlet series* $\sum \mathrm{X}_n n^{-s}$ *defined and holomorphic almost surely for* $\mathrm{Re}(s) > 1/2$*. For any* $\sigma_1 > 1/2$*, we have*

$$\mathbf{E}(|\mathrm{Z}(s)|) \ll 1 + |s|$$

uniformly for all s *such that* $\mathrm{Re}(s) \geqslant \sigma_1$*.*

Proof The series

$$\sum_{n \geqslant 1} \frac{\mathrm{X}_n}{n^{\sigma_1}}$$

converges almost surely. Therefore the partial sums

$$\mathrm{S}_u = \sum_{n \leqslant u} \frac{\mathrm{X}_n}{n^{\sigma_1}}$$

are bounded almost surely.

By summation by parts (Lemma A.1.1), it follows that for any s with real part $\sigma > \sigma_1$, we have

$$\mathrm{Z}(s) = (s - \sigma_1) \int_1^{+\infty} \frac{\mathrm{S}_u}{u^{s-\sigma_1+1}} du,$$

where the integral converges almost surely. Hence

$$|\mathrm{Z}(s)| \leqslant (1 + |s|) \int_1^{+\infty} \frac{|\mathrm{S}_u|}{u^{\sigma-\sigma_1+1}} du.$$

Fubini's Theorem (for nonnegative functions) and the Cauchy–Schwarz inequality then imply

$$\begin{aligned}
\mathbf{E}(|\mathrm{Z}(s)|) &\leqslant (1 + |s|) \int_1^{+\infty} \mathbf{E}(|\mathrm{S}_u|) \frac{du}{u^{\sigma-\sigma_1+1}} \\
&\leqslant (1 + |s|) \int_1^{+\infty} \mathbf{E}(|\mathrm{S}_u|^2)^{1/2} \frac{du}{u^{\sigma-\sigma_1+1}} \\
&= (1 + |s|) \int_1^{+\infty} \left(\sum_{n \leqslant u} \frac{1}{n^{2\sigma_1}} \right)^{1/2} \frac{du}{u^{\sigma-\sigma_1+1}}
\end{aligned}$$

using the orthonormality of the variables X_n. The integrand is $\ll u^{-\frac{1}{2}-\sigma}$, hence the integral converges uniformly for $\sigma \geqslant \sigma_1$. □

We can then deduce a good result on average approximation by partial sums. We refer to Section A.3 for the definition and properties of the Mellin transform.

Proposition 3.2.11 *Let* $\varphi\colon [0, +\infty[\longrightarrow [0,1]$ *be a smooth function with compact support such that* $\varphi(0) = 1$. *Let* $\widehat{\varphi}$ *denote its Mellin transform. For* $N \geqslant 1$, *define the* $H(D)$-*valued random variable*

$$Z_{D,N} = \sum_{n\geqslant 1} X_n \varphi\left(\frac{n}{N}\right) n^{-s}.$$

There exists $\delta > 0$ *such that*

$$\mathbf{E}(\|Z_D - Z_{D,N}\|_\infty) \ll N^{-\delta}$$

for $N \geqslant 1$.

We recall that the norm $\|\cdot\|_\infty$ refers to the sup norm on the compact set D.

Proof The first step is to apply the smoothing process of Proposition A.4.3 in Appendix A. The random Dirichlet series

$$Z(s) = \sum_{n\geqslant 1} X_n n^{-s}$$

converges almost surely for $\mathrm{Re}(s) > 1/2$. For $\sigma > 1/2$ and any $\delta > 0$ such that

$$-\delta + \sigma \geqslant 1/2,$$

we have therefore almost surely the representation

$$Z_D(s) - Z_{D,N}(s) = -\frac{1}{2i\pi}\int_{(-\delta)} Z(s+w)\widehat{\varphi}(w)N^w dw \tag{3.8}$$

for $s \in D$. (Figure 3.2 may help understand the location of the regions involved in the proof.)

Note that here and below, it is important that the "almost surely" property holds for *all* s; this is simply because we work with random variables taking values in $H(D)$, and not with particular evaluations of these random functions at a specific $s \in D$.

We need to control the supremum norm on D, since this is the norm on the space $H(D)$. For this purpose, we use Cauchy's integral formula.

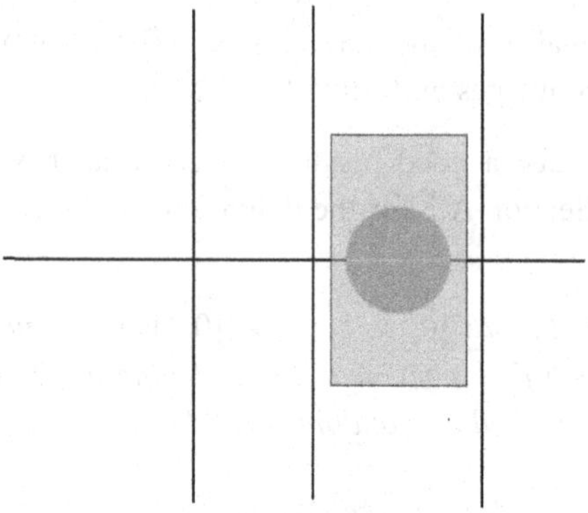

Figure 3.2 Regions and contours in the proof of Proposition 3.2.11.

Let S be a compact segment in $]1/2, 1[$ such that the fixed rectangle $\mathrm{R} = \mathrm{S} \times [-1/2, 1/2] \subset \mathbf{C}$ contains D in its interior. Then, almost surely, for any v in D, Cauchy's Theorem gives

$$\mathrm{Z_D}(v) - \mathrm{Z_{D,N}}(v) = \frac{1}{2i\pi} \int_{\partial \mathrm{R}} (\mathrm{Z_D}(s) - \mathrm{Z_{D,N}}(s)) \frac{ds}{s - v},$$

where the boundary of R is oriented counterclockwise. The definition of R ensures that $|s - v|^{-1} \gg 1$ for $v \in \mathrm{D}$ and $s \in \partial\mathrm{R}$, so that the random variable $\|\mathrm{Z_D} - \mathrm{Z_{D,N}}\|_\infty$ satisfies

$$\|\mathrm{Z_D} - \mathrm{Z_{D,N}}\|_\infty \ll \int_{\partial \mathrm{R}} |\mathrm{Z_D}(s) - \mathrm{Z_{D,N}}(s)|\, |ds|.$$

Using (3.8) and writing $w = -\delta + iu$ with $u \in \mathbf{R}$, we obtain

$$\|\mathrm{Z_D} - \mathrm{Z_{D,N}}\|_\infty \ll \mathrm{N}^{-\delta} \int_{\partial \mathrm{R}} \int_{\mathbf{R}} |\mathrm{Z}(-\delta + \sigma + i(t + u))|\, |\widehat{\varphi}(-\delta + iu)||ds|du.$$

Therefore, taking the expectation, and using Fubini's Theorem (for nonnegative functions), we get

$$\begin{aligned}
&\mathbf{E}(\|\mathrm{Z_D} - \mathrm{Z_{D,N}}\|_\infty) \\
&\quad \ll \mathrm{N}^{-\delta} \int_{\partial \mathrm{R}} \int_{\mathbf{R}} \mathbf{E}\left(|\mathrm{Z}(-\delta + \sigma + i(t + u))|\right) |\widehat{\varphi}(-\delta + iu)||ds|du \\
&\quad \ll \mathrm{N}^{-\delta} \sup_{s=\sigma+it \in \mathrm{R}} \int_{\mathbf{R}} \mathbf{E}\left(|\mathrm{Z}(-\delta + \sigma + i(t + u))|\right) |\widehat{\varphi}(-\delta + iu)|du.
\end{aligned}$$

We therefore need to bound

$$\int_{\mathbf{R}} \mathbf{E}\left(|\mathrm{Z}(-\delta + \sigma + i(t + u))|\right) |\widehat{\varphi}(-\delta + iu)|du$$

for some fixed $\sigma + it$ in the compact rectangle R. We take

$$\delta = \frac{1}{2}(\min \mathrm{S} - 1/2),$$

which is > 0 since S is compact in $]1/2, 1[$, so that

$$-\delta + \sigma > 1/2 \quad \text{and} \quad 0 < \delta < 1.$$

Since $\widehat{\varphi}$ decays faster than any polynomial at infinity in vertical strips, and

$$\mathbf{E}(|\mathrm{Z}(s)|) \ll 1 + |s|$$

uniformly for $s \in \mathrm{R}$ by Lemma 3.2.10, we have

$$\int_{\mathbf{R}} \mathbf{E}\left(|\mathrm{Z}(-\delta + \sigma + i(t+u))|\right) \, |\widehat{\varphi}(-\delta + iu)| du \ll 1$$

uniformly for $s = \sigma + it \in \mathrm{R}$, and the result follows. □

The last preliminary result is a similar approximation result for the translates of the Riemann zeta function by smooth partial sums of its Dirichlet series.

Proposition 3.2.12 *Let* $\varphi\colon [0, +\infty[\longrightarrow [0, 1]$ *be a smooth function with compact support such that* $\varphi(0) = 1$. *Let* $\widehat{\varphi}$ *denote its Mellin transform. For* $\mathrm{N} \geqslant 1$, *define*[2]

$$\zeta_{\mathrm{N}}(s) = \sum_{n \geqslant 1} \varphi\left(\frac{n}{\mathrm{N}}\right) n^{-s},$$

and define $\mathsf{Z}_{\mathrm{N,T}}$ *to be the* H(D)*-valued random variable*

$$t \mapsto (s \mapsto \zeta_{\mathrm{N}}(s + it)).$$

There exists $\delta > 0$ *such that*

$$\mathbf{E}_{\mathrm{T}}(\|\mathsf{Z}_{\mathrm{D,T}} - \mathsf{Z}_{\mathrm{N,T}}\|_\infty) \ll \mathrm{N}^{-\delta} + \mathrm{N}\mathrm{T}^{-1}$$

for $\mathrm{N} \geqslant 1$ *and* $\mathrm{T} \geqslant 1$.

Note that ζ_{N} is an entire function, since φ has compact support, so that the range of the sum is in fact finite. The meaning of the statement is that the smoothed partial sums ζ_{N} give very uniform and strong approximations to the vertical translates of the Riemann zeta function.

Proof We will write Z_{T} for $\mathsf{Z}_{\mathrm{D,T}}$ for simplicity. We begin by applying the smoothing process of Proposition A.4.3 in Appendix A in the case $a_n = 1$.

[2] There should be no confusion with $\mathsf{Z}_{\mathrm{D,T}}$.

For $\sigma > 1/2$ and any $\delta > 0$ such that $-\delta+\sigma \geqslant 1/2$, we have (as in the previous proof) the representation

$$\zeta(s) - \zeta_N(s) = -\frac{1}{2i\pi}\int_{(-\delta)} \zeta(s+w)\widehat{\varphi}(w)N^w dw - N^{1-s}\widehat{\varphi}(1-s), \quad (3.9)$$

where the second term on the right-hand side comes from the fact that the Riemann zeta function has a pole at $s = 1$ with residue 1.

As before, let S be a compact segment in $]1/2, 1[$ such that the fixed rectangle $R = S \times [-1/2, 1/2] \subset \mathbf{C}$ contains D in its interior. Then for any v with $\mathrm{Re}(v) > 1/2$ and $t \in \mathbf{R}$, Cauchy's Theorem gives

$$\zeta(v+it) - \zeta_N(v+it) = \frac{1}{2i\pi}\int_{\partial R} (\zeta(s+it) - \zeta_N(s+it))\frac{ds}{s-v},$$

where the boundary of R is oriented counterclockwise; using $|s-v|^{-1} \gg 1$ for $v \in D$ and $s \in \partial R$, we deduce that the random variable $\|Z_T - Z_{N,T}\|_\infty$, which takes the value

$$\sup_{s\in D} |\zeta(s+it) - \zeta_N(s+it)|$$

at $t \in \Omega_T$, satisfies

$$\|Z_T - Z_{N,T}\|_\infty \ll \int_{\partial R} |\zeta(s+it) - \zeta_N(s+it)||ds|$$

for $t \in \Omega_T$. Taking the expectation with respect to t (i.e., integrating over $t \in [-T, T]$) and applying Fubini's Theorem for nonnegative functions leads to

$$\begin{aligned}\mathbf{E}_T\left(\|Z_T - Z_{N,T}\|_\infty\right) &\ll \int_{\partial R} \mathbf{E}_T\left(|\zeta(s+it) - \zeta_N(s+it)|\right)|ds| \\ &\ll \sup_{s\in\partial R} \mathbf{E}_T\left(|\zeta(s+it) - \zeta_N(s+it)|\right). \quad (3.10)\end{aligned}$$

We take again $\delta = \frac{1}{2}(\min S - 1/2) > 0$, so that $0 < \delta < 1$. For any fixed $s = \sigma + it \in \partial R$, we have

$$-\delta + \sigma \geqslant \frac{1}{2} + \delta > \frac{1}{2}.$$

Applying (3.9) and using again Fubini's Theorem, we obtain

$$\begin{aligned}&\mathbf{E}_T\left(|\zeta(s+it) - \zeta_N(s+it)|\right) \\ &\ll N^{-\delta}\int_{\mathbf{R}} |\widehat{\varphi}(-\delta+iu)|\,\mathbf{E}_T\left(|\zeta(\sigma-\delta+i(t+u))|\right)du \\ &\quad + N^{1-\sigma}\,\mathbf{E}_T(|\widehat{\varphi}(1-s-it)|).\end{aligned}$$

The rapid decay of $\widehat{\varphi}$ on vertical strips shows that the second term (arising from the pole) is $\ll \mathrm{NT}^{-1}$. In the first term, since $\sigma - \delta \geqslant \min(\mathrm{S}) - \delta \geqslant \frac{1}{2} + \delta$, we have

$$\mathbf{E}_{\mathrm{T}}\left(|\zeta(\sigma - \delta + i(t+u))|\right) = \frac{1}{2\mathrm{T}} \int_{-\mathrm{T}}^{\mathrm{T}} |\zeta(\sigma - \delta + i(t+u))| dt$$
$$\ll 1 + \frac{|u|}{\mathrm{T}} \ll 1 + |u| \tag{3.11}$$

by Proposition C.4.1 in Appendix C. Hence

$$\mathbf{E}_{\mathrm{T}}\left(|\zeta(s+it) - \zeta_{\mathrm{N}}(s+it)|\right) \ll \mathrm{N}^{-\delta} \int_{\mathbf{R}} |\widehat{\varphi}(-\delta + iu)|(1+|u|) du + \mathrm{NT}^{-1}. \tag{3.12}$$

Now the fast decay of $\widehat{\varphi}(s)$ on the vertical line $\mathrm{Re}(s) = -\delta$ shows that the last integral is bounded, and we conclude from (3.10) that

$$\mathbf{E}_{\mathrm{T}}\left(\|\mathsf{Z}_{\mathrm{T}} - \mathsf{Z}_{\mathrm{N,T}}\|_{\infty}\right) \ll \mathrm{N}^{-\delta} + \mathrm{NT}^{-1},$$

as claimed. □

Finally we can prove Theorem 3.2.1:

Proof of Bagchi's Theorem By Proposition B.4.1, it is enough to prove that for any bounded and Lipschitz function $f : \mathrm{H(D)} \longrightarrow \mathbf{C}$, we have

$$\mathbf{E}_{\mathrm{T}}(f(\mathsf{Z}_{\mathrm{D,T}})) \longrightarrow \mathbf{E}(f(\mathsf{Z}_{\mathrm{D}}))$$

as $\mathrm{T} \to +\infty$. We may use the Dirichlet series expansion of Z_{D} according to Proposition 3.2.9, (2).

Since D is fixed, we omit it from the notation for simplicity, denoting $\mathsf{Z}_{\mathrm{T}} = \mathsf{Z}_{\mathrm{D,T}}$ and $\mathrm{Z} = \mathsf{Z}_{\mathrm{D}}$. Fix some integer $\mathrm{N} \geqslant 1$ to be chosen later. We denote

$$\mathsf{Z}_{\mathrm{T,N}} = \sum_{n \geqslant 1} n^{-s-it} \varphi\left(\frac{n}{\mathrm{N}}\right)$$

(viewed as a random variable defined for $t \in [-\mathrm{T}, \mathrm{T}]$) and

$$\mathsf{Z}_{\mathrm{N}} = \sum_{n \geqslant 1} \mathrm{X}_n n^{-s} \varphi\left(\frac{n}{\mathrm{N}}\right)$$

the smoothed partial sums of the Dirichlet series as in Propositions 3.2.12 and 3.2.11.

We then write

$$
\begin{aligned}
|\mathbf{E}_T(f(Z_T)) - \mathbf{E}(f(Z))| \leqslant {} & |\mathbf{E}_T(f(Z_T) - f(Z_{T,N}))| \\
& + |\mathbf{E}_T(f(Z_{T,N})) - \mathbf{E}(f(Z_N))| \\
& + |\mathbf{E}(f(Z_N) - f(Z))|.
\end{aligned}
$$

Since f is a Lipschitz function on H(D), there exists a constant $C \geqslant 0$ such that

$$|f(x) - f(y)| \leqslant C\|x - y\|_\infty$$

for all $x, y \in H(D)$. Hence we have

$$
\begin{aligned}
|\mathbf{E}_T(f(Z_T)) - \mathbf{E}(f(Z))| \leqslant {} & C\,\mathbf{E}_T(\|Z_T - Z_{T,N}\|_\infty) \\
& + |\mathbf{E}_T(f(Z_{T,N})) - \mathbf{E}(f(Z_N))| + C\,\mathbf{E}(\|Z_N - Z\|_\infty).
\end{aligned}
$$

Fix $\varepsilon > 0$. Propositions 3.2.12 and 3.2.11 together show that there exists some $N \geqslant 1$ and some constant $C_1 \geqslant 0$ such that

$$\mathbf{E}_T(\|Z_T - Z_{T,N}\|_\infty) < \varepsilon + \frac{C_1 N}{T}$$

for all $T \geqslant 1$ and

$$\mathbf{E}(\|Z_N - Z\|_\infty) < \varepsilon.$$

We *fix* such a value of N. By Proposition 3.2.5 and composition, the random variables $Z_{T,N}$ (which are Dirichlet polynomials) converge in law to Z_N as $T \to +\infty$. Since $N/T \to 0$ also for $T \to +\infty$, we deduce that for all T large enough, we have

$$|\mathbf{E}_T(f(Z_T)) - \mathbf{E}(f(Z))| < 4\varepsilon.$$

This finishes the proof. □

Exercise 3.2.13 Prove that if $\sigma > 1/2$ is fixed, then we have almost surely

$$\lim_{T\to+\infty} \frac{1}{2T}\int_{-T}^{T} |Z(\sigma + it)|^2 dt = \zeta(2\sigma).$$

[**Hint**: Use the Birkhoff–Khintchine pointwise ergodic theorem for flows; see, e.g., [30, §8.6.1].]

Before we continue toward the computation of the support of Bagchi's measure, and hence the proof of Voronin's Theorem, we can use the current available information to obtain bounds on the probability that the Riemann zeta function is "large" on the subset D. More precisely, it is natural to discuss the probability that the logarithm of the modulus of translates of the zeta function is large, since this will also detect how close it might approach zero.

Proposition 3.2.14 *Let σ_0 be the infimum of the real part of s for $s \in \mathrm{D}$. There exists a positive constant $c > 0$, depending on* D, *such that for any* $\mathrm{A} > 0$, *we have*

$$\limsup_{\mathrm{T}\to+\infty} \mathbf{P}_{\mathrm{T}}(\| \log |\mathrm{Z}_{\mathrm{D},\mathrm{T}}| \, \|_\infty > \mathrm{A}) \leqslant c \exp\left(-c^{-1}\mathrm{A}^{1/(1-\sigma_0)}(\log \mathrm{A})^{1/(2(1-\sigma_0))}\right).$$

Proof Convergence in law implies that

$$\limsup_{\mathrm{T}\to+\infty} \mathbf{P}_{\mathrm{T}}(\| \log |\mathrm{Z}_{\mathrm{D},\mathrm{T}}| \, \|_\infty > \mathrm{A}) \leqslant \mathbf{P}_{\mathrm{T}}(\| \log |\mathrm{Z}_{\mathrm{D}}| \, \|_\infty > \mathrm{A})$$

and

$$\log |\mathrm{Z}_{\mathrm{D}}| = \sum_p \mathrm{Re}\left(\frac{\mathrm{X}_p}{p^s}\right) + \mathrm{O}(1),$$

where the implied constant depends on D. In addition, we have

$$\mathbf{P}_{\mathrm{T}}\left(\left\|\sum_p \mathrm{Re}\left(\frac{\mathrm{X}_p}{p^s}\right)\right\|_\infty > \mathrm{A}\right) \leqslant \mathbf{P}_{\mathrm{T}}\left(\left\|\sum_p \frac{\mathrm{X}_p}{p^s}\right\|_\infty > \mathrm{A}\right).$$

Since $\sigma_0 > \frac{1}{2}$ and the random variables (X_p) are independent and bounded by 1, we can therefore estimate the right-hand side of this last inequality using the variant of Proposition B.11.13 discussed in Remark B.11.14 (2) for the Banach space H(D), and hence conclude the proof. □

Remark 3.2.15 It is also possible to obtain lower bounds for these probabilities, by evaluating at a fixed element of D (see Theorem 6.3.1 for a similar argument, although the shape of the lower bound is different).

3.3 The Support of Bagchi's Measure

Our goal in this section is to explain the proof of Theorem 3.2.3, which is due to Bagchi [4, Ch. 5]. Since it involves results of complex analysis that are quite far from the main interest of this book, we will only treat in detail the part of the proof that involves arithmetic, giving references for the other results that are used.

The support is easiest to compute using the random Euler interpretation of the random Dirichet series, because it is essentially a sum of independent random variables. To be precise, define

$$\mathrm{P}(s) = \sum_{p} \frac{\mathrm{X}_p}{p^s} \quad \text{and} \quad \tilde{\mathrm{P}}(s) = \sum_{p} \sum_{k \geqslant 1} \frac{\mathrm{X}_p^k}{p^{ks}}$$

(see the proof of Proposition 3.2.9). The series converge almost surely for $\mathrm{Re}(s) > 1/2$. We claim that the support of the distribution of $\tilde{\mathrm{P}}$, when viewed as an H(D)-valued random variable, is equal to H(D). Let us first assume this.

Since $\mathrm{Z} = \exp(\tilde{\mathrm{P}})$, we deduce by composition (see Lemma B.2.1) that the support of Z is the closure of the set of functions of the form e^g, where $g \in \mathrm{H(D)}$. But this last set is precisely $\mathrm{H(D)}^\times$, and Lemma A.5.5 in Appendix A shows that its closure in H(D) is $\mathrm{H(D)}^\times \cup \{0\}$.

Finally, to prove the approximation property (3.2), which is the original version of Voronin's Universality Theorem, we simply apply Lemma B.3.3 to the family of random variables $\mathrm{Z_T}$, which gives the much stronger statement that for any $\varepsilon > 0$, we have

$$\liminf_{\mathrm{T}\to+\infty} \lambda\left(\{t \in [-\mathrm{T},\mathrm{T}] \mid \sup_{s\in \mathrm{D}} |\zeta(s+it) - f(s)| < \varepsilon\}\right) > 0,$$

where λ denotes Lebesgue measure.

From Proposition B.10.8 in Appendix B, the following proposition will imply that the support of the random Dirichlet series P is H(D). The statement is slightly more general to help with the last step afterward.

Proposition 3.3.1 *Let τ be such that $1/2 < \tau < 1$. Let $r > 0$ be such that*

$$\mathrm{D} = \{s \in \mathbf{C} \mid |s - \tau| \leqslant r\} \subset \{s \in \mathbf{C} \mid 1/2 < \mathrm{Re}(s) < 1\}.$$

Let N be an arbitrary positive real number. The set of all series

$$\sum_{p>\mathrm{N}} \frac{x_p}{p^s} \quad \text{with} \quad (x_p) \in \widehat{\mathbf{S}}^1,$$

which converge in H(D), is dense in H(D).

We will deduce the proposition from the density criterion of Theorem A.5.1 in Appendix A, applied to the space H(D) and the sequence (f_p) with $f_p(s) = p^{-s}$ for p prime. Since $\|f_p\|_\infty = p^{-\sigma_1}$, where $\sigma_1 = \tau - r > 1/2$, the condition

$$\sum_{p} \|f_p\|_\infty^2 < +\infty$$

holds. Furthermore, Proposition 3.2.9 certainly shows that there exist *some* $(x_p) \in \widehat{\mathbf{S}}^1$ such that the series $\sum_p x_p f_p$ converges in H(D). Hence the conclusion of Theorem A.5.1 is what we seek, and we only need to check the following lemma to establish the last hypothesis required to apply it:

Lemma 3.3.2 *Let $\mu \in C(\bar{D})'$ be a continuous linear functional. Let*

$$g(z) = \mu(s \mapsto e^{sz})$$

be its Laplace transform. If

$$\sum_p |g(\log p)| < +\infty, \tag{3.13}$$

then we have $g = 0$.

Indeed, the point is that $\mu(f_p) = \mu(s \mapsto p^{-s}) = g(\log p)$, so that the assumption (3.13) concerning g is precisely (A.3).

This is a statement that has some arithmetic content, as we will see, and indeed the proof involves the Prime Number Theorem.

Proof Let

$$\varrho = \limsup_{r \to +\infty} \frac{\log |g(r)|}{r},$$

which is finite by Lemma A.5.2 (1). By Lemma A.5.2 (3), it suffices to prove that $\varrho \leqslant 1/2$ to conclude that $g = 0$. To do this, we will use Theorem A.5.3, that provides access to the value of ϱ by "sampling" g along certain sequences of real numbers tending to infinity.

The idea is that (3.13) implies that $|g(\log p)|$ cannot be often of size at least $1/p = e^{-\log p}$, since the series $\sum p^{-1}$ diverges. Since the sequence $\log p$ increase slowly, this makes it possible to find real numbers $r_k \to +\infty$ growing linearly and such that $|g(r_k)| \leqslant e^{-r_k}$, and from this and Theorem A.5.3 we will get a contradiction.

To be precise, we first note that for $y \in \mathbf{R}$, we have

$$|g(iy)| \leqslant \|\mu\| \, \|s \mapsto e^{iys}\|_\infty \leqslant \|\mu\| e^{r|y|}$$

(since the maximum of the absolute value of the imaginary part of $s \in \bar{D}$ is r), and therefore

$$\limsup_{\substack{y \in \mathbf{R} \\ |y| \to +\infty}} \frac{\log |g(iy)|}{|y|} \leqslant r.$$

We put $\alpha = r \leqslant 1/4$. Then the first condition of Theorem A.5.3 holds for the function g. We also take $\beta = 1$ so that $\alpha\beta < \pi$.

For any $k \geqslant 0$, let I_k be the set of primes p such that $e^k \leqslant p < e^{k+1}$. By the Mertens Formula (C.4), or the Prime Number Theorem, we have

$$\sum_{p \in I_k} \frac{1}{p} \sim \frac{1}{k}$$

as $k \to +\infty$. Let further A be the set of those $k \geqslant 0$ for which the inequality

$$|g(\log p)| \geqslant \frac{1}{p}$$

holds for *all* primes $p \in \mathrm{I}_k$, and let B be its complement among the nonnegative integers. We then note that

$$\sum_{k \in \mathrm{A}} \frac{1}{k} \ll \sum_{k \in \mathrm{A}} \sum_{p \in \mathrm{I}_k} \frac{1}{p} \ll \sum_{k \in \mathrm{A}} \sum_{p \in \mathrm{I}_k} |g(\log p)| < +\infty.$$

This shows that B is infinite. For $k \in \mathrm{B}$, let p_k be a prime in I_k such that $|g(\log p_k)| < p_k^{-1}$. Let $r_k = \log p_k$. We then have

$$\limsup_{k \to +\infty} \frac{\log |g(r_k)|}{r_k} \leqslant -1.$$

Since $p_k \in \mathrm{I}_k$, we have

$$r_k = \log p_k \sim k.$$

Furthermore, if we order B in increasing order, the fact that

$$\sum_{k \notin \mathrm{B}} \frac{1}{k} < +\infty$$

implies that the kth element n_k of B satisfies $n_k \sim k$.

Now we consider the sequence formed from the r_{2k}, arranged in increasing order. We have $r_{2k}/k \to 2$ from the above. Moreover, since $r_k \in \mathrm{I}_k$, we have

$$r_{2k+2} - r_{2k} \geqslant 1,$$

by construction, hence $|r_{2k} - r_{2l}| \gg |k - l|$. Since $|g(r_{2k})| \leqslant e^{-r_{2k}}$ for all $k \in \mathrm{B}$, we can apply Theorem A.5.3 to this increasing sequence and we get

$$\varrho = \limsup_{k \to +\infty} \frac{\log |g(r_{2k})|}{r_{2k}} \leqslant -1 < 1/2,$$

as desired. □

There remains a last lemma to prove, that allows us to go from the support of the series $\mathrm{P}(s)$ of independent random variables to that of the full series $\tilde{\mathrm{P}}(s)$.

Lemma 3.3.3 *The support of* $\tilde{\mathrm{P}}(s)$ *is* $\mathrm{H}(\mathrm{D})$.

Proof We can write

$$\tilde{\mathrm{P}} = -\sum_{p} \log(1 - \mathrm{X}_p p^{-s}),$$

where the random variables $(\log(1-X_p p^{-s}))_p$ are independent, and the series converges almost surely in H(D). Therefore it is enough by Proposition B.10.8 to prove that the set of convergent series

$$-\sum_p \log(1 - x_p p^{-s}) \quad \text{and} \quad (x_p) \in \widehat{\mathbf{S}}^1$$

is dense in H(D).

Fix $f \in \mathrm{H(D)}$ and $\varepsilon > 0$ be fixed. For $\mathrm{N} \geqslant 1$ and any $(x_p) \in \widehat{\mathbf{S}}^1$, let

$$h_{\mathrm{N}}(s) = \sum_{p>\mathrm{N}} \sum_{k\geqslant 2} \frac{x_p^k}{kp^{ks}}.$$

This series converges absolutely for any s such that $\mathrm{Re}(s) > 1/2$ and $(x_p) \in \widehat{\mathbf{S}}^1$, and we have

$$\|h_{\mathrm{N}}\|_\infty \leqslant \sum_{p>\mathrm{N}} \sum_{k\geqslant 2} \frac{1}{kp^{k/2}} \to 0$$

as $\mathrm{N} \to +\infty$, uniformly with respect to $(x_p) \in \widehat{\mathbf{S}}^1$. Fix N such that $\|h_{\mathrm{N}}\|_\infty < \frac{\varepsilon}{2}$ for any $(x_p) \in \widehat{\mathbf{S}}^1$.

Now let $x_p = 1$ for $p \leqslant \mathrm{N}$ and define $f_0 \in \mathrm{H(D)}$ by

$$f_0(s) = f(s) + \sum_{p\leqslant \mathrm{N}} \log(1 - x_p p^{-s}).$$

For any choice of $(x_p)_{p>\mathrm{N}}$ such that the series

$$\sum_p \frac{x_p}{p^s}$$

defines an element of H(D), we can then write

$$f(s) + \sum_p \log(1 - x_p p^{-s}) = g_{\mathrm{N}}(s) + f_0(s) + h_{\mathrm{N}}(s),$$

for $s \in \mathrm{D}$, where

$$g_{\mathrm{N}}(s) = \sum_{p>\mathrm{N}} \frac{x_p}{p^s}.$$

By Proposition 3.3.1, there exists $(x_p)_{p>\mathrm{N}}$ such that the series g_{N} converges in H(D) and $\|g_{\mathrm{N}} + f_0\|_\infty < \frac{\varepsilon}{2}$. We then have

$$\Big\| f + \sum_p \log(1 - x_p p^{-s}) \Big\|_\infty < \varepsilon.$$

□

Exercise 3.3.4 This exercise uses Voronin's Theorem to deduce that the Riemann zeta function is not the solution to any algebraic differential equation.

(1) For $(a_0, \dots, a_m) \in \mathbf{C}^{m+1}$ such that $a_0 \neq 0$, prove that there exists $(b_0, \dots, b_m) \in \mathbf{C}^{m+1}$ such that we have

$$\exp\left(\sum_{k=0}^{m} b_k s^k\right) = \sum_{k=0}^{m} \frac{a_k}{k!} s^k + \mathrm{O}(s^{m+1})$$

for $s \in \mathbf{C}$.

Now fix a real number σ with $\frac{1}{2} < \sigma < 1$, and let g be a holomorphic function on $\mathbf{C}$ which does not vanish.

(2) For any $\varepsilon > 0$, prove that there exists a real number t and $r > 0$ such that

$$\sup_{|s| \leqslant r} |\zeta(s + \sigma + it) - g(s)| < \varepsilon \frac{r^k}{k!}.$$

(3) Let $n \geqslant 1$ be an integer. Prove that there exists $t \in \mathbf{R}$ such that for all integers k with $0 \leqslant k \leqslant n-1$, we have

$$|\zeta^{(k)}(\sigma + it) - g^{(k)}(0)| < \varepsilon.$$

(4) Let $n \geqslant 1$ be an integer. Prove that the image in $\mathbf{C}^n$ of the map

$$\begin{cases} \mathbf{R} \longrightarrow \mathbf{C}^n, \\ t \mapsto (\zeta(\sigma + it), \dots, \zeta^{(n-1)}(\sigma + it)) \end{cases}$$

is dense in $\mathbf{C}^n$.

(5) Using (4), prove that if $n \geqslant 1$ and $\mathrm{N} \geqslant 1$ are integers, and $\mathrm{F}_0, \dots, \mathrm{F}_\mathrm{N}$ are continuous functions $\mathbf{C}^n \to \mathbf{C}$, not all identically zero, then the function

$$\sum_{k=0}^{\mathrm{N}} s^k \mathrm{F}_k(\zeta(s), \zeta'(s), \dots, \zeta^{(n-1)}(s))$$

is not identically zero. In particular, the Riemann zeta function satisfies no algebraic differential equation.

3.4 Generalizations

If we look back at the proof of Bagchi's Theorem, and at the proof of Voronin's Theorem, we can see precisely which arithmetic ingredients appeared. They are the following:

- the crucial link between the arithmetic objects and the probabilistic model is provided by Proposition 3.2.5, which depends on the unique factorization of integers into primes; this is an illustration of the asymptotic independence of prime numbers, similarly to Proposition 1.3.7;
- the proof of Bagchi's Theorem then relies on the mean-value property (3.11) of the Riemann zeta function; this estimate has arithmetic meaning;
- the Prime Number Theorem, which appears in the proof of Voronin's Theorem, in order to control the distribution of primes in (roughly) dyadic intervals.

Note that some arithmetic features remain in the Random Dirichlet Series that arises as the limit in Bagchi's Theorem, in contrast with the Erdős–Kac Theorem, where the limit is the universal Gaussian distribution. This means, in particular, that going beyond Bagchi's Theorem to applications (as in Voronin's Theorem) still naturally involves arithmetic problems, many of which are very interesting in their interaction with probability theory (see below for a few references).

From this analysis, it shouldn't be very surprising that Bagchi's Theorem can be generalized to many other situations. The most interesting concerns perhaps the limiting behavior, in H(D), of families of L-functions of the type

$$\mathrm{L}(f,s) = \sum_{n \geqslant 1} \lambda_f(n) n^{-s},$$

where f runs over some sequence of arithmetic objects with associated L-functions, ordered in a sequence of probability spaces (which need not be continuous like Ω_{T}). We refer to [59, Ch. 5] for a survey and discussion of L-functions, and to [69] for a discussion of families of L-functions. There are some rather elementary special cases, such as the vertical translates $\mathrm{L}(\chi, s+it)$ of a fixed Dirichlet L-function $\mathrm{L}(\chi, s)$, since almost all properties of the Riemann zeta function extend quite easily to this case. Another interesting case is the finite set Ω_q of nontrivial Dirichlet characters modulo a prime number q, with the uniform probability measure. Then one can look at the distribution of the restrictions to D of the Dirichlet L-functions $\mathrm{L}(s,\chi)$ for $\chi \in \Omega_q$, and indeed one can check that Bagchi's Theorem extends to this situation.

A second example, which is treated in [72] is, still for a prime $q \geqslant 2$, the set Ω_q of holomorphic cuspidal modular forms of weight 2 and level q, either with the uniform probability measure, or with that provided by the Petersson formula ([69, 31, ex. 8]). An analogue of Bagchi's Theorem holds, but the limiting random Dirichlet series is not the same as in Theorem 3.2.1: with the Petersson average, it is

$$\prod_{p}(1 - \mathrm{X}_p p^{-s} + p^{-2s})^{-1}, \tag{3.14}$$

where (X_p) is a sequence of independent random variables, which are all distributed according to the Sato–Tate measure (the same that appears in Example B.6.1 (3)). This different limit is simply due to the form that "local spectral equidistribution" (in the sense of [69]) takes for this family (see [69, 38]). Indeed, the local spectral equidistribution property plays the role of Proposition 3.2.5. The analogue of (3.11) follows from a stronger mean-square formula, using the Cauchy–Schwarz inequality: there exists a constant $\mathrm{A} > 0$ such that, for any $\sigma_0 > 1/2$ and all $s \in \mathbf{C}$ with $\mathrm{Re}(s) \geqslant \sigma_0$, we have

$$\sum_{f\in\Omega_q} \omega_f |\mathrm{L}(f,s)|^2 \ll (1 + |s|)^{\mathrm{A}} \tag{3.15}$$

for $q \geqslant 2$, where ω_f is the Petersson-averaging weight (see [76, Prop. 5], which proves an even more difficult result where $\mathrm{Re}(s)$ can be as small as $\frac{1}{2} + c(\log q)^{-1}$).

However, extending Bagchi's Theorem to many other families of L-functions (e.g., vertical translates of an L-function of higher rank) requires restrictions, in the current state of knowledge. The reason is that the analogue of the mean-value estimates (3.11) or (3.15) is usually only known when $\mathrm{Re}(s) \geqslant \sigma_0 > 1/2$, for some σ_0 such that $\sigma_0 < 1$. Then the only domains D for which one can prove a version of Bagchi's Theorem are those contained in $\mathrm{Re}(s) > \sigma_0$.

[Further references: Titchmarsh [117], especially Chapter 11, discusses the older work of Bohr and Jessen, which has some interesting geometric aspects that are not apparent in modern treatments. Bagchi's Thesis [4] contains some generalizations as well as more information concerning the limit theorem and Voronin's Theorem.]

4

The Distribution of Values of the Riemann Zeta Function, II

Probability tools	Arithmetic tools
Definition of convergence in law (Section B.3)	Riemann zeta function (Section C.4)
Kronecker's Theorem (Th. B.6.5)	Möbius function (Def. C.1.3)
Central Limit Theorem (Th. B.7.2)	Mean-square of $\zeta(s)$ on the critical line (Prop. C.4.1)
Gaussian random variable (Section B.7)	Multiplicative functions (Section C.1)
Lipschitz test functions (Prop. B.4.1)	Euler products (Lemma C.1.4)
Method of moments (Th. B.5.5)	

4.1 Introduction

In this chapter, as indicated previously, we will continue working with the values of the Riemann zeta function, but on the critical line $s = \frac{1}{2} + it$, where the issues are much deeper.

Indeed, the analogue of Theorem 3.1.1 fails for $\tau = 1/2$, which shows that the Riemann zeta function is significantly more complicated on the critical line. However, there is a limit theorem after normalization, due to Selberg, for the logarithm of the Riemann zeta function. To state it, we specify carefully the meaning of $\log\zeta(\frac{1}{2} + it)$. We define a random variable L_{T} on Ω_{T} by putting $\mathsf{L}(t) = 0$ if $\zeta(1/2 + it) = 0$, and otherwise

$$\mathsf{L}_{\mathrm{T}}(t) = \log\zeta\left(\tfrac{1}{2} + it\right),$$

where the logarithm of zeta is the unique branch that is holomorphic on a narrow strip

$$\{s = \sigma + iy \in \mathbf{C} \mid \sigma > \tfrac{1}{2} - \delta, \quad |y - t| \leqslant \delta\}$$

for some $\delta > 0$ and satisfies $\log \zeta(\sigma + it) \to 0$ as $\sigma \to +\infty$.

Theorem 4.1.1 (Selberg) *With notation as above, the random variables*

$$\frac{\mathsf{L}_{\mathrm{T}}}{\sqrt{\frac{1}{2}\log\log \mathrm{T}}}$$

on Ω_{T} *converge in law as* $\mathrm{T} \to +\infty$ *to a standard complex Gaussian random variable.*

We will in fact only prove "half" of this theorem: we consider only the real part of $\log \zeta(\frac{1}{2} + it)$, or in other words, we consider $\log |\zeta(\frac{1}{2} + it)|$. So we (re)define the arithmetic random variables L_{T} on Ω_{T} by $\mathsf{L}_{\mathrm{T}}(t) = 0$ if $\zeta(\frac{1}{2} + it) = 0$, and otherwise $\mathsf{L}_{\mathrm{T}}(t) = \log |\zeta(\frac{1}{2} + it)|$. Note that dealing with the modulus means in particular that we need not worry about the choice of the branch of the logarithm of complex numbers. We will prove:

Theorem 4.1.2 (Selberg) *The random variables*

$$\frac{\mathsf{L}_{\mathrm{T}}}{\sqrt{\frac{1}{2}\log\log \mathrm{T}}}$$

converge in law as $\mathrm{T} \to +\infty$ *to a standard real Gaussian random variable.*

4.2 Strategy of the Proof of Selberg's Theorem

We present the recent proof of Theorem 4.1.2 due to Radziwiłł and Soundararajan [95]. In comparison with Bagchi's Theorem, the strategy has the common feature of the use of suitable approximations to ζ, and the probabilistic limiting behavior will ultimately derive from the independence and distribution of the vector $t \mapsto (p^{-it})_p$ (as in Proposition 3.2.5). However, one has to be much more careful than in the previous section.

Precisely, the approximation used by Radziwiłł and Soundararajan involves three steps:

- **Step 1**: An approximation of L_{T} by the random variable $\widetilde{\mathsf{L}}_{\mathrm{T}}$ given by $t \mapsto \log |\zeta(\sigma_0 + it)|$ for σ_0 sufficiently close to $1/2$ (where σ_0 depends on T).

- **Step 2**: For the random variable Z_{T} given by $t \mapsto \zeta(\sigma_0 + it)$, so that $\log|\mathsf{Z}_{\mathrm{T}}| = \widetilde{\mathsf{L}}_{\mathrm{T}}$, an approximation of the *inverse* $1/\mathsf{Z}_{\mathrm{T}}$ by a short Dirichlet polynomial D_{T} of the type

$$\mathsf{D}_{\mathrm{T}}(s) = \sum_{n \geqslant 1} a_{\mathrm{T}}(n)\mu(n)n^{-s},$$

where $a_{\mathrm{T}}(n)$ is zero for n large enough (again, depending on T); here $\mu(n)$ denotes the Möbius function (see Definition C.1.3), and we recall once more that it satisfies

$$\sum_{n \geqslant 1} \mu(n)n^{-s} = \frac{1}{\zeta(s)}$$

if $\mathrm{Re}(s) > 1$ (see Corollary C.1.5). At this point, we get an approximation of L_{T} by $-\log|\mathsf{D}_{\mathrm{T}}|$.
- **Step 3**: An approximation of $|\mathsf{D}_{\mathrm{T}}|$ by what is essentially a short Euler product, namely, by $\exp(-\mathrm{Re}(\mathsf{P}_{\mathrm{T}}))$, where

$$\mathsf{P}_{\mathrm{T}}(t) = \sum_{p^k \leqslant \mathrm{X}} \frac{1}{k} \frac{1}{p^{k(\sigma_0 + it)}} \tag{4.1}$$

for suitable X (again depending on T). In this definition, and in all formulas involving such sums below, the condition $p^k \leqslant \mathrm{X}$ is implicitly restricted to integers $k \geqslant 1$. At this point, L_{T} is approximated by $\mathrm{Re}(\mathsf{P}_{\mathrm{T}})$.

Finally, the last probabilistic step is to prove that the random variables

$$\frac{\mathrm{Re}(\mathsf{P}_{\mathrm{T}})}{\sqrt{\frac{1}{2}\log\log \mathrm{T}}}$$

converge in law to a standard Gaussian random variable as $\mathrm{T} \to +\infty$.

None of these steps (except the last) is easy, in comparison with the results discussed up to now, and the specific approximations that are used (namely, the choices of the coefficients $a_{\mathrm{T}}(n)$ as well as of the length parameter X) are quite subtle and by no means obvious (they can be seen to be related to sieve methods). Even the *nature* of the approximation will not be the same in the three steps!

In order to simplify the reading of the proof, we first specify the relevant parameters. We assume from now on that $\mathrm{T} \geqslant e^{e^2}$. We denote by

$$\varrho_{\mathrm{T}} = \sqrt{\frac{1}{2}\log\log \mathrm{T}} \geqslant 1$$

the normalizing factor in the theorem. We then define

$$W = (\log\log\log T)^4 \asymp (\log \varrho_T)^4, \quad \sigma_0 = \frac{1}{2} + \frac{W}{\log T} = \frac{1}{2} + O\left(\frac{(\log \varrho_T)^4}{\log T}\right), \tag{4.2}$$

$$X = T^{1/(\log\log\log T)^2} = T^{1/\sqrt{W}}. \tag{4.3}$$

Note that we omit the dependency on T in most of these notation. We will also require a further parameter

$$Y = T^{1/(\log\log T)^2} = T^{4/\varrho^4} \leqslant X. \tag{4.4}$$

We begin by stating the precise approximation statements. All parameters are now fixed as above for the remainder of this chapter. After stating the precise form of each steps of the proof, we will show how they combine to imply Theorem 4.1.2, and finally we will establish these intermediate results.

Proposition 4.2.1 (Moving outside of the critical line) *We have*

$$\mathbf{E}_T\Big(|\mathsf{L}_T - \tilde{\mathsf{L}}_T|\Big) = o(\varrho_T)$$

as $T \to +\infty$.

We now define properly the Dirichlet polynomials that appear in the second step of the approximation. It is here that the arithmetic subtlety lies, since the definition is quite delicate. We define first

$$m_1 = 100\log\log T \asymp \varrho_T \quad \text{and} \quad m_2 = 100\log\log\log T \asymp \log \varrho_T. \tag{4.5}$$

We denote by $b_T(n)$ the characteristic function of the set of squarefree integers $n \geqslant 1$ such that all prime factors of n are $\leqslant$ Y, and n has at most m_1 prime factors. We denote by $c_T(n)$ the characteristic function of the set of squarefree integers $n \geqslant 1$ such that all prime factors p of n satisfy $Y < p \leqslant X$, and n has at most m_2 prime factors. We associate to these the Dirichlet polynomials

$$B(s) = \sum_{n\geqslant 1} \mu(n) b_T(n) n^{-s} \quad \text{and} \quad C(s) = \sum_{n\geqslant 1} \mu(n) c_T(n) n^{-s}$$

for $s \in \mathbf{C}$. Finally, define $D(s) = B(s)C(s)$. The coefficient of n^{-s} in the expansion of $D(s)$ is the Dirichlet convolution

$$\sum_{de=n} b_T(d)c_T(e)\mu(d)\mu(e) = \sum_{\substack{de=n\\(d,e)=1}} b_T(d)c_T(e)\mu(d)\mu(e)$$
$$= \mu(n) \sum_{\substack{de=n\\(d,e)=1}} b_T(d)c_T(e) = \mu(n)a_T(n),$$

say, by Proposition A.4.4, where we used the fact that d and e are coprime if $b_T(d)c_T(e)$ is nonzero since the set of primes dividing an integer in the support of b_T is disjoint from the set of primes dividing an integer in the support of c_T. It follows then from this formula that $a_T(n)$ is the characteristic function of the set of squarefree integers $n \geqslant 1$ such that

(1) all prime factors of n are $\leqslant X$;

(2) there are at most m_1 prime factors p of n such that $p \leqslant Y$;

(3) there are at most m_2 prime factors p of n such that $Y < p \leqslant X$.

It is immediate, but very important, that $a_T(n) = 0$ unless n is quite small, namely,

$$n \leqslant Y^{100 \log\log T} X^{100 \log\log\log T} = T^c,$$

where

$$c = \frac{100}{\log\log T} + \frac{100}{\log\log\log T} \to 0 \quad \text{as } T \to +\infty.$$

Finally, we define the arithmetic random variable

$$\mathsf{D}_T = \mathrm{D}(\sigma_0 + it). \tag{4.6}$$

Remark 4.2.2 Although the definition of $\mathrm{D}(s)$ may seem complicated, we will see its different components coming together in the proofs of this proposition and the next.

If we consider the support of $a_T(n)$, we note that (by the Erdős–Kac Theorem, restricted to squarefree integers as in Exercise 2.3.4) the typical number of prime factors of an integer $n \leqslant Y^{m_1}$ is about $\log\log Y^{m_1} \sim \log\log T$. Therefore the integers satisfying $b_T(n) = 1$ are quite typical, and only extreme outliers (in terms of the number of prime factors) are excluded. On the other hand, the integers satisfying $c_T(n) = 1$ have much fewer prime factors than is typical, and are therefore quite rare (they are, in a weak sense, "almost prime"). This indicates that a_T is a subtle *arithmetic* truncation of the characteristic function of integers $n \leqslant T^c$, and hence that

$$\sum_{n \geqslant 1} a_T(n)\mu(n)n^{-s}$$

is an arithmetic truncation of the Dirichlet series that formally gives the inverse of $\zeta(s)$. This should be contrasted with the more traditional *analytic* truncations of $\zeta(s)$ that were used in Lemma 3.2.10 and Proposition 3.2.11.

For comparison, it is useful to note that Selberg himself used in many applications certain truncations that are roughly of the shape

$$\sum_{n\leqslant X} \frac{\mu(n)}{n^s}\left(1-\frac{\log n}{\log X}\right).$$

Proposition 4.2.3 (Dirichlet polynomial approximation) *The difference* $Z_T D_T$ *converges to* 1 *in* L^2*, that is, we have*

$$\lim_{T\to+\infty} \mathbf{E}_T\big(|1 - Z_T D_T|^2\big) = 0.$$

Proposition 4.2.4 (Euler product approximation) *The random variables* $D_T \exp(-P_T)$ *converge to* 1 *in probability, that is, for any* $\varepsilon > 0$*, we have*

$$\lim_{T\to+\infty} \mathbf{P}_T\big(|D_T \exp(P_T) - 1| > \varepsilon\big) = 0.$$

In particular, $\mathbf{P}_T(D_T = 0)$ *tends to* 0 *as* $T \to +\infty$.

Despite our probabilistic presentation, the three previous statement are really theorems of number theory, and would usually be stated without probabilistic notation. For instance, Proposition 4.2.1 means that

$$\frac{1}{T}\int_{-T}^{T} |\log|\zeta(1/2+it)| - \log|\zeta(\sigma_0+it)||dt = o(\sqrt{\log\log T}).$$

The last result finally introduces the probabilistic behavior,

Proposition 4.2.5 (Gaussian Euler products) *The random variables* $\varrho_T^{-1} P_T$ *converge in law as* $T \to +\infty$ *to a standard complex Gaussian random variable. In particular, the random variables*

$$\frac{\mathrm{Re}(P_T)}{\sqrt{\frac{1}{2}\log\log T}}$$

converge in law to a standard Gaussian random variable.

We will now explain how to combine these ingredients for the final step of the proof.

Proof of Theorem 4.1.2 Until Proposition 4.2.5 is used, this is essentially a variant of the fact that convergence in probability implies convergence in law, and that convergence in L^1 or L^2 implies convergence in probability.

For the details, fix some standard Gaussian random variable $\mathcal{N}$. Let f be a bounded Lipschitz function $\mathbf{R} \longrightarrow \mathbf{R}$, and let $C \geqslant 0$ be a real number such that

$$|f(x) - f(y)| \leqslant C|x-y|, \qquad |f(x)| \leqslant C, \qquad \text{for } x,\ y \in \mathbf{R}.$$

We consider the difference

$$\left|\mathbf{E}_T\left(f\left(\frac{\mathsf{L}_T}{\varrho_T}\right)\right) - \mathbf{E}(f(\mathcal{N}))\right|$$

and must show that this tends to 0 as $T \to +\infty$.

We estimate this quantity using the "chain" of approximations introduced above: we have

$$\begin{aligned}
&\left|\mathbf{E}_T\left(f\left(\frac{\mathsf{L}_T}{\varrho_T}\right)\right) - \mathbf{E}(f(\mathcal{N}))\right| \\
&\quad \leqslant \mathbf{E}_T\left(\left|f\left(\frac{\mathsf{L}_T}{\varrho_T}\right) - f\left(\frac{\widetilde{\mathsf{L}}_T}{\varrho_T}\right)\right|\right) + \mathbf{E}_T\left(\left|f\left(\frac{\widetilde{\mathsf{L}}_T}{\varrho_T}\right) - f\left(\frac{\log|\mathsf{D}_T|^{-1}}{\varrho_T}\right)\right|\right) \\
&\qquad + \mathbf{E}_T\left(\left|f\left(\frac{\log|\mathsf{D}_T|^{-1}}{\varrho_T}\right) - f\left(\frac{\operatorname{Re}\mathsf{P}_T}{\varrho_T}\right)\right|\right) \\
&\qquad + \left|\mathbf{E}_T\left(f\left(\frac{\operatorname{Re}\mathsf{P}_T}{\varrho_T}\right)\right) - \mathbf{E}(f(\mathcal{N}))\right|, \qquad (4.7)
\end{aligned}$$

and we discuss each of the four terms on the right-hand side using the four previous propositions (here and below, we define $|\mathsf{D}_T|^{-1}$ to be 0 if $\mathsf{D}_T = 0$).

The first term is handled straightforwardly using Proposition 4.2.1: we have

$$\mathbf{E}_T\left(\left|f\left(\frac{\mathsf{L}_T}{\varrho_T}\right) - f\left(\frac{\widetilde{\mathsf{L}}_T}{\varrho_T}\right)\right|\right) \leqslant \frac{C}{\varrho_T}\mathbf{E}_T(|\mathsf{L}_T - \widetilde{\mathsf{L}}_T|) \longrightarrow 0$$

as $T \to +\infty$.

For the second term, let $A_T \subset \Omega_T$ be the event

$$\{\mathsf{D}_T = 0\} \cup \{|\widetilde{\mathsf{L}}_T - \log|\mathsf{D}_T|^{-1}| > 1/2\}$$

and A'_T its complement. Since $\log|\mathsf{Z}_T| = \widetilde{\mathsf{L}}_T$, we then have

$$\mathbf{E}_T\left(\left|f\left(\frac{\widetilde{\mathsf{L}}_T}{\varrho_T}\right) - f\left(\frac{\log|\mathsf{D}_T|^{-1}}{\varrho_T}\right)\right|\right) \leqslant 2C\,\mathbf{P}_T(A_T) + \frac{C}{2\varrho_T}.$$

Proposition 4.2.3 implies that $\mathbf{P}_T(A_T) \to 0$ (convergence to 1 of $\mathsf{Z}_T\mathsf{D}_T$ in L^2 implies convergence to 1 in probability, hence convergence to 0 in probability for the logarithm of the modulus) and therefore

$$\mathbf{E}_T\left(\left|f\left(\frac{\widetilde{\mathsf{L}}_T}{\varrho_T}\right) - f\left(\frac{\log|\mathsf{D}_T|^{-1}}{\varrho_T}\right)\right|\right) \to 0$$

as $T \to +\infty$.

We now come to the third term on the right-hand side of (4.7). Distinguishing according to the events

$$\mathrm{B_T} = \{|\log |\mathrm{D_T} \exp(\mathrm{P_T})|| > 1/2\}$$

and its complement, we get as before

$$\mathbf{E}_\mathrm{T}\left(\left|f\left(\frac{\log |\mathrm{D_T}|^{-1}}{\varrho_\mathrm{T}}\right) - f\left(\frac{\mathrm{Re}\,\mathrm{P_T}}{\varrho_\mathrm{T}}\right)\right|\right) \leqslant 2\mathrm{C}\,\mathbf{P}_\mathrm{T}(\mathrm{B_T}) + \frac{\mathrm{C}}{2\varrho_\mathrm{T}},$$

and this also tends to 0 as $\mathrm{T} \to +\infty$ by Proposition 4.2.4.

Finally, Proposition 4.2.5 implies that

$$\left|\mathbf{E}_\mathrm{T}\left(f\left(\frac{\mathrm{Re}\,\mathrm{P_T}}{\varrho_\mathrm{T}}\right)\right) - \mathbf{E}(f(\mathcal{N}))\right| \to 0$$

as $\mathrm{T} \to +\infty$, and hence we conclude the proof of the theorem, assuming the approximation statements. □

We now explain the proofs of these four propositions. We begin with the easiest part, which also happens to be where the transition to the pure probabilistic behavior happens. A key tool is the quantitative form of Proposition 3.2.5 contained in Lemma 3.2.6. More precisely, as in Section 3.2, let $\mathrm{X} = (\mathrm{X}_p)_p$ be a sequence of independent random variables uniformly distributed on $\mathbf{S}^1$. We define X_n for $n \geqslant 1$ by multiplicativity as in formula (3.7).

Lemma 4.2.6 *Let $(a(n))_{n\geqslant 1}$ be any sequence of complex numbers with bounded support. For any $\mathrm{T} \geqslant 2$ and $\sigma \geqslant 0$, we have*

$$\mathbf{E}_\mathrm{T}\left(\left|\sum_{n\geqslant 1}\frac{a(n)}{n^{\sigma+it}}\right|^2\right) = \sum_{n\geqslant 1}\frac{|a(n)|^2}{n^{2\sigma}} + \mathrm{O}\Big(\frac{1}{\mathrm{T}}\sum_{\substack{m,n\geqslant 1\\ m\neq 1}}\frac{|a(m)a(n)|}{(mn)^{\sigma-\frac{1}{2}}}\Big)$$

$$= \mathbf{E}\left(\left|\sum_{n\geqslant 1}\frac{\mathrm{X}_n}{n^{\sigma}}\right|^2\right) + \mathrm{O}\Big(\frac{1}{\mathrm{T}}\sum_{\substack{m,n\geqslant 1\\ m\neq 1}}\frac{|a(m)a(n)|}{(mn)^{\sigma-\frac{1}{2}}}\Big),$$

where the implied constant is absolute.

Proof We have

$$\mathbf{E}_\mathrm{T}\left(\left|\sum_{n\geqslant 1}\frac{a(n)}{n^{\sigma+it}}\right|^2\right) = \sum_m\sum_n\frac{a(m)\overline{a(n)}}{(mn)^{\sigma}}\mathbf{E}_\mathrm{T}\left(\left(\frac{n}{m}\right)^{it}\right).$$

We now apply Lemma 3.2.6 and separate the "diagonal" contribution where $m = n$ from the remainder. This leads to the first formula in the lemma, and the second then reflects the orthonormality of the sequence $(\mathrm{X}_n)_{n\geqslant 1}$. □

When applying this lemma, we call the first term the "diagonal" contribution and the second the "off-diagonal" one.

Proof of Proposition 4.2.5 We have $\mathsf{P}_T = \mathsf{Q}_T + \mathsf{R}_T$, where Q_T is the contribution of the primes and R_T the contribution of squares and higher powers of primes. We first claim that R_T is uniformly bounded in L^2 for all T. Indeed, using Lemma 4.2.6, we get

$$\begin{aligned}\mathbf{E}_T(|\mathsf{R}_T|^2) &= \mathbf{E}_T\left(\left|\sum_{k\geqslant 2}\sum_{p\leqslant X^{1/k}}\frac{1}{k}p^{-k\sigma_0}p^{-kit}\right|^2\right)\\ &= \sum_{\substack{p^k\leqslant X\\ k\geqslant 2}}\frac{1}{k^2}p^{-2k\sigma_0} + O\Big(\frac{1}{T}\sum_{k,l\geqslant 2}\sum_{\substack{p^k,q^l\leqslant X\\ p\neq q}}\frac{1}{kl}(pq)^{-2\sigma_0+\frac{1}{2}}\Big)\\ &\ll 1 + \frac{X^2\log X}{T} \ll 1\end{aligned}$$

since $X \ll T^{\varepsilon}$ for any $\varepsilon > 0$.

From this, it follows that it is enough to show that Q_T/ϱ_T converges in law to a standard complex Gaussian random variable $\mathcal{N}$. For this purpose, we use moments, that is, we compute

$$\mathbf{E}_T\left(\mathsf{Q}_T^k\overline{\mathsf{Q}_T^\ell}\right)$$

for integers $k, \ell \geqslant 0$, and we compare with the corresponding moment of the random variable

$$\mathrm{Q}_T = \sum_{p\leqslant X} p^{-\sigma_0}X_p.$$

After applying Lemma 3.2.6 again (as in the proof of Lemma 4.2.6), we find that

$$\mathbf{E}_T\left(\mathsf{Q}_T^k\overline{\mathsf{Q}_T^\ell}\right) = \mathbf{E}(\mathrm{Q}_T^k\overline{\mathrm{Q}}_T^\ell) + O\Big(\frac{1}{T}\sum_{m\neq n}(mn)^{-\sigma_0+1/2}\Big),$$

where the sum in the error term runs over integers m (resp. n) with at most k prime factors, counted with multiplicity, all of which are $\leqslant X$ (resp. at most ℓ prime factors, counted with multiplicity, all of which are $\leqslant X$). Hence this error term is

$$\ll \frac{1}{T}\Big(\sum_{p\leqslant X}1\Big)^{k+\ell} \ll \frac{X^{k+\ell}}{T}.$$

Next, we note that

$$\mathbf{V}(\mathrm{Q_T}) = \sum_{p\leqslant \mathrm{X}} p^{-2\sigma_0}\,\mathbf{V}(\mathrm{X}_p^2) = \frac{1}{2}\sum_{p\leqslant \mathrm{X}} p^{-2\sigma_0}.$$

We compute this sum by splitting in two ranges $p \leqslant \mathrm{Y}$ and $\mathrm{Y} < p \leqslant \mathrm{X}$ (recall that σ_0 depends on T). The second sum is

$$\ll \sum_{\mathrm{Y}<p\leqslant \mathrm{X}} \frac{1}{p} = \log\left(\frac{\log \mathrm{X}}{\log \mathrm{Y}}\right) + \mathrm{O}(1) \ll \log\log\log \mathrm{T}$$

by Proposition C.3.1 and (4.2). On the other hand, for $p \leqslant \mathrm{Y} = \mathrm{T}^{1/(\log\log \mathrm{T})^2}$, we have

$$p^{-2\sigma_0} = p^{-1}\exp\left(-2\frac{(\log p)}{(\log \mathrm{T})}\mathrm{W}\right) = p^{-1}\left(1+\mathrm{O}\left(\frac{\mathrm{W}}{(\log\log \mathrm{T})^2}\right)\right),$$

which, in view of (4.2), implies that $\mathbf{V}(\mathrm{Q_T}) \sim \frac{1}{2}\log\log \mathrm{T} = \varrho_{\mathrm{T}}^2$ as $\mathrm{T}\to+\infty$.

It is finally again a case of the Central Limit Theorem that $\mathrm{Q_T}/\sqrt{\mathbf{V}(\mathrm{Q_T})}$, and hence also $\mathrm{Q_T}/\varrho_{\mathrm{T}}$, converges in law to a standard complex Gaussian random variable, with convergence of the moments (Theorem B.7.2 and Theorem B.5.6 (2), Remark B.5.8), so the conclusion follows from the method of moments since $\mathrm{X}^{k+\ell}/\mathrm{T} \to 0$ as $\mathrm{T}\to+\infty$. □

The other propositions will now be proved in order. Some of the arithmetic results that we will used are only stated in Appendix C (with suitable references).

Proof of Proposition 4.2.1 We appeal to Hadamard's factorization of the Riemann zeta function (Proposition C.4.3) in the form of its corollary, Proposition C.4.4. Let $t \in \Omega_{\mathrm{T}}$ be such that there is no zero of zeta with ordinate t (this only excludes finitely many values of t for a given T). We have

$$\log|\zeta(\sigma_0+it)| - \log|\zeta(\tfrac{1}{2}+it)| = \mathrm{Re}\left(\int_{1/2}^{\sigma_0}\frac{\zeta'}{\zeta}(\sigma+it)\,d\sigma\right)$$
$$= \int_{1/2}^{\sigma_0}\mathrm{Re}\left(\frac{\zeta'}{\zeta}(\sigma+it)\right)d\sigma.$$

For any σ with $\frac{1}{2}\leqslant\sigma\leqslant\sigma_0$, we have

$$-\frac{\zeta'}{\zeta}(\sigma+it) = \sum_{|t-\mathrm{Im}(\varrho)|<1}\frac{1}{\sigma+it-\varrho} + \mathrm{O}(\log(2+|t|)),$$

by Proposition C.4.4, where the sum is over zeros ϱ of $\zeta(s)$, counted with multiplicity, such that $|\sigma+it-\varrho|<1$.

We fix $t_0 \in \Omega_T$ and integrate over t such that $|t - t_0| \leqslant 1$. This leads to

$$\int_{t_0-1}^{t_0+1} \left|\log|\zeta(\sigma_0+it)| - \log|\zeta(\tfrac{1}{2}+it)|\right|dt$$
$$\leqslant \sum_{|\operatorname{Im}(\varrho)-t_0|\leqslant 1} \int_{t_0-1}^{t_0+1}\int_{\frac{1}{2}}^{\sigma_0} \left|\operatorname{Re}\left(\frac{1}{\sigma+it-\varrho}\right)\right| dt d\sigma.$$

An elementary integral (!) gives

$$\int_{t_0-1}^{t_0+1} \left|\operatorname{Re}\left(\frac{1}{\sigma+it-\varrho}\right)\right| dt \leqslant \int_{\mathbf{R}} \left|\operatorname{Re}\left(\frac{1}{\sigma+it-\varrho}\right)\right| dt$$
$$= \int_{\mathbf{R}} \frac{|\sigma-\beta|}{(\sigma-\beta)^2+(t-\gamma)^2} dt = \pi$$

for all σ and ϱ. Hence we get

$$\frac{1}{\mathrm{T}}\int_{|t-t_0|\leqslant 1} \left|\log|\zeta(\tfrac{1}{2}+it-\varrho)| - \log|\zeta(\sigma_0+it-\varrho)|\right|dt \ll (\sigma_0-\tfrac{1}{2})\frac{m(t_0)}{\mathrm{T}},$$

where $m(t_0)$ is the number of zeros ϱ such that $|t_0 - \operatorname{Im}(\varrho)| \leqslant 1$. This is $\ll \log(2+|t_0|)$ by Proposition C.4.4 again. Finally, by summing the bound

$$\frac{1}{\mathrm{T}}\int_{|t-t_0|\leqslant 1} \left|\log\left|\zeta\left(\tfrac{1}{2}+it-\varrho\right)\right| - \log|\zeta(\sigma_0+it-\varrho)|\right|dt$$
$$\ll \left(\sigma_0-\tfrac{1}{2}\right)\frac{\log(2+|t_0|)}{\mathrm{T}}$$

over a partition of Ω_T in $\ll$ T intervals of length 2, we deduce that

$$\mathbf{E}_{\mathrm{T}}(|\mathsf{L}_{\mathrm{T}} - \widetilde{\mathsf{L}}_{\mathrm{T}}|) \ll (\sigma_0 - \tfrac{1}{2})\log \mathrm{T} = \mathrm{W}.$$

We have $\mathrm{W} = o(\varrho_{\mathrm{T}})$ (by a rather wide margin!), and the proposition follows. □

The last two propositions are more involved, and we present their proofs in separate sections.

4.3 Dirichlet Polynomial Approximation

We will prove Proposition 4.2.3 in this section, that is, we need to prove that

$$\mathbf{E}_{\mathrm{T}}(|1 - \mathsf{Z}_{\mathrm{T}}\mathsf{D}_{\mathrm{T}}|^2),$$

where $\mathsf{Z}_{\mathrm{T}}(t) = \zeta(\sigma_0 + it)$, tends to 0 as $\mathrm{T} \to +\infty$. This is arithmetically the most involved part of the proof.

First of all, we use the approximation formula

$$\zeta(\sigma_0+it)=\sum_{1\leqslant n\leqslant \mathrm{T}} n^{-\sigma_0-it}+\mathrm{O}\Big(\frac{\mathrm{T}^{1-\sigma_0}}{|t|+1}+\mathrm{T}^{-1/2}\Big)$$

for $t\in\Omega_{\mathrm{T}}$ (see Proposition C.4.5). Multiplying by D_{T}, we obtain

$$\begin{aligned}\mathbf{E}_{\mathrm{T}}(\mathsf{Z}_{\mathrm{T}}\mathsf{D}_{\mathrm{T}})&=\sum_{\substack{m\geqslant 1\\ n\leqslant \mathrm{T}}} a_{\mathrm{T}}(m)\mu(m)\,\mathbf{E}_{\mathrm{T}}((mn)^{-\sigma_0})\\&\quad+\mathrm{O}\Big(\mathrm{T}^{1/2}\sum_{m\geqslant 1}a_{\mathrm{T}}(m)\,\mathbf{E}_{\mathrm{T}}((|t|+1)^{-1})+\mathrm{T}^{-1/2}\sum_{m\geqslant 1}a_{\mathrm{T}}(m)m^{-\sigma_0}\Big).\end{aligned}$$

We recall that $|a_{\mathrm{T}}(n)|\leqslant 1$ for all n, and $a_{\mathrm{T}}(n)=0$ unless $n\ll \mathrm{T}^{\varepsilon}$, for any $\varepsilon>0$. Hence, by (3.4), this becomes

$$\begin{aligned}\mathbf{E}_{\mathrm{T}}(\mathsf{Z}_{\mathrm{T}}\mathsf{D}_{\mathrm{T}})&=1+\mathrm{O}\Big(\frac{1}{\mathrm{T}}\sum_{\substack{n\leqslant \mathrm{T}\\ m\neq n}}a_{\mathrm{T}}(m)(mn)^{-\sigma_0}(\log mn)\Big)+\mathrm{O}(\mathrm{T}^{-1/2+\varepsilon})\\&=1+\mathrm{O}(\mathrm{T}^{-1/2+\varepsilon})\end{aligned}$$

for any $\varepsilon>0$ (in the diagonal terms, only $m=n=1$ contributes, and in the off-diagonal terms $mn\neq 1$, we have $\mathbf{E}_{\mathrm{T}}((mn)^{-it})\ll \mathrm{T}^{-1}\log(mn)$). It follows that it suffices to prove that

$$\lim_{\mathrm{T}\to+\infty}\mathbf{E}_{\mathrm{T}}(|\mathsf{Z}_{\mathrm{T}}\mathsf{D}_{\mathrm{T}}|^2)=1.$$

We expand the mean-square using the formula for D_{T} and obtain

$$\mathbf{E}_{\mathrm{T}}(|\mathsf{Z}_{\mathrm{T}}\mathsf{D}_{\mathrm{T}}|^2)=\sum_{m,n}\frac{\mu(m)\mu(n)}{(mn)^{\sigma_0}}a_{\mathrm{T}}(m)a_{\mathrm{T}}(n)\,\mathbf{E}_{\mathrm{T}}\Big(\Big(\frac{m}{n}\Big)^{it}|\mathsf{Z}_{\mathrm{T}}|^2\Big).$$

Now the asymptotic formula of Proposition C.4.6 translates to a formula for $\mathbf{E}_{\mathrm{T}}\big((m/n)^{it}|\mathsf{Z}_{\mathrm{T}}|^2\big)$, namely,

$$\begin{aligned}\mathbf{E}_{\mathrm{T}}\Big(\Big(\frac{m}{n}\Big)^{it}|\mathsf{Z}_{\mathrm{T}}|^2\Big)&=\zeta(2\sigma_0)\Big(\frac{(m,n)^2}{mn}\Big)^{\sigma_0}\\&\quad+\zeta(2-2\sigma_0)\Big(\frac{(m,n)^2}{mn}\Big)^{1-\sigma_0}\mathbf{E}_{\mathrm{T}}\Big(\Big(\frac{|t|}{2\pi}\Big)^{1-2\sigma_0}\Big)\\&\quad+\mathrm{O}(\min(m,n)\mathrm{T}^{-\sigma_0+\varepsilon})\end{aligned}$$

for any $\varepsilon>0$, where the expectation is really the integral

$$\frac{1}{2\mathrm{T}}\int_{-\mathrm{T}}^{\mathrm{T}}\Big(\frac{|t|}{2\pi}\Big)^{1-2\sigma_0}dt,$$

and we recall that (m,n) denotes here the gcd of m and n.

Using the properties of $a_T(n)$, the error term is easily handled, since it is at most

$$\mathrm{T}^{-\sigma_0+\varepsilon}\sum_{m,n}(mn)^{-\sigma_0}a_T(m)a_T(n)\min(m,n)\leqslant \mathrm{T}^{-\sigma_0+\varepsilon}\Big(\sum_m m^{1/2}a_T(m)\Big)^2$$
$$\ll \mathrm{T}^{-\sigma_0+2\varepsilon}$$

for any $\varepsilon>0$. Thus we only need to handle the main terms, which we write as

$$\zeta(2\sigma_0)\mathrm{M}_1+\zeta(2-2\sigma_0)\,\mathbf{E}_\mathrm{T}\left(\left(\frac{|t|}{2\pi}\right)^{1-2\sigma_0}\right)\mathrm{M}_2, \qquad (4.8)$$

where

$$\mathrm{M}_1=\sum_{m,n}\frac{\mu(m)\mu(n)}{(mn)^{2\sigma_0}}a_T(m)a_T(n)(m,n)^{2\sigma_0}$$

and M_2 is the other term. Using the multiplicative structure of a_T, the first term factors in turn as $\mathrm{M}_1=\mathrm{M}_1'\mathrm{M}_1''$, where

$$\mathrm{M}_1'=\sum_{m,n}\frac{\mu(m)\mu(n)}{[m,n]^{2\sigma_0}}b_T(m)b_T(n),$$
$$\mathrm{M}_1''=\sum_{m,n}\frac{\mu(m)\mu(n)}{[m,n]^{2\sigma_0}}c_T(m)c_T(n).$$

We compare M_1' to the similar sum $\tilde{\mathrm{M}}_1'$ where $b_T(n)$ and $b_T(m)$ are replaced by characteristic functions of integers with all prime factors $\leqslant \mathrm{Y}$, forgetting only the requirement to have $\leqslant m_1$ prime factors. By Example C.1.7, we have

$$\tilde{\mathrm{M}}_1'=\prod_{p\leqslant \mathrm{Y}}\left(1-\frac{1}{p^{2\sigma_0}}\right).$$

The difference $\mathrm{M}_1'-\tilde{\mathrm{M}}_1'$ can be bounded from above by

$$2e^{-m_1}\sum_{m,n}\frac{|\mu(m)\mu(n)|}{[m,n]^{2\sigma_0}}e^{\Omega(m)},$$

where the sum runs over integers with all prime factors $\leqslant \mathrm{Y}$ (this step is a case of what is called "Rankin's trick": the condition $\Omega(m)>m_1$ is handled by bounding its characteristic function by the nonnegative function $e^{\Omega(m)-m_1}$). Again from Example C.1.7, this is at most

$$2(\log \mathrm{T})^{-100}\prod_{p\leqslant \mathrm{Y}}\left(1+\frac{1+2e}{p}\right)\ll(\log \mathrm{T})^{-90}$$

(by Proposition C.3.6). Thus

$$M_1' \sim \prod_{p \leqslant Y} (1 - p^{-2\sigma_0})$$

as $T \to +\infty$. One deals similarly with the second term M_1'', which turns out to satisfy

$$M_1'' \sim \prod_{Y < p \leqslant X} (1 - p^{-2\sigma_0}),$$

and hence

$$M_1 \sim \zeta(2\sigma_0) \prod_{p \leqslant X} (1 - p^{-2\sigma_0}) = \prod_{p > X} (1 - p^{-2\sigma_0}).$$

Now, by the choice of the parameters, we obtain from the Prime Number Theorem (Theorem C.3.3) the upper bound

$$\sum_{p > X} p^{-2\sigma_0} \ll \int_X^{+\infty} \frac{1}{t^{2\sigma_0}} \frac{dt}{\log t} \ll \frac{X^{1-2\sigma_0}}{(2\sigma_0 - 1)\log X} = \frac{X^{1-2\sigma_0}}{2\sqrt{W}} \leqslant \frac{1}{2\sqrt{W}}.$$

Since this tends to 0 as $T \to +\infty$, it follows that

$$\begin{aligned}\prod_{p > X} (1 - p^{-2\sigma_0}) &= \exp\Big(\sum_{p > X} \Big(\frac{1}{p^{2\sigma_0}} + O\Big(\frac{1}{p^{4\sigma_0}}\Big)\Big)\Big) \\ &= \exp\Big(- \sum_{p > X} p^{-2\sigma_0} \Big)(1 + o(1))\end{aligned}$$

converges to 1 as $T \to +\infty$.

There only remains to check that the second part M_2 of the main term (4.8 tends to 0 as $T \to +\infty$. We have

$$\begin{aligned} M_2 &= \sum_{m,n} \frac{\mu(m)\mu(n)}{mn} a_T(m) a_T(n) (m,n)^{2-2\sigma_0} \\ &= \sum_{m,n} \frac{\mu(m)\mu(n)}{[m,n]^{2-2\sigma_0}} a_T(m) a_T(n) (mn)^{1-2\sigma_0}. \end{aligned}$$

The procedure is very similar: we factor $M_2 = M_2' M_2''$, where M_2' has coefficients b_T instead of a_T, and M_2'' has c_T. Applying Example C.1.7 and Rankin's trick to both factors now leads to

$$\begin{aligned} M_2 &\sim \prod_{p \leqslant X} \Big(1 + \frac{1}{p^{2-2\sigma_0}} \Big(- \frac{1}{p^{2\sigma_0 - 1}} - \frac{1}{p^{2\sigma_0 - 1}} + \frac{1}{p^{4\sigma_0 - 2}} \Big)\Big) \\ &= \prod_{p \leqslant X} \Big(1 - \frac{2}{p} + \frac{1}{p^{2\sigma_0}}\Big). \end{aligned}$$

We deduce from this that the contribution of M_2 to (4.8) is

$$\sim \zeta(2-2\sigma_0)\,\mathrm{E}_{\mathrm{T}}\left(\left(\frac{|t|}{2\pi}\right)^{1-2\sigma_0}\right)\prod_{p\leqslant X}\left(1-\frac{2}{p}+\frac{1}{p^{2\sigma_0}}\right).$$

Since $\zeta(s)$ has a pole at $s = 1$ with residue 1, this last expression is

$$\ll \frac{T^{1-2\sigma_0}}{2\sigma_0-1}\prod_{p\leqslant X}\left(1-\frac{1}{p}\right) \ll \frac{T^{1-2\sigma_0}}{(2\sigma_0-1)\log X}.$$

In terms of the parameter W, since $2\sigma_0 - 1 = 2W/\log T$ and $X = T^{1/\sqrt{W}}$, the right-hand side is simply $\exp(-2W)W^{-1/2}$, and hence tends to 0 as $T \to +\infty$. This concludes the proof.

4.4 Euler Product Approximation

This section is devoted to the proof of Proposition 4.2.4. We need to prove that $\mathsf{D}_T \exp(\mathsf{P}_T)$ converges to 1 in probability. This involves some extra decomposition of P_T: we write

$$\mathsf{P}_T = \mathsf{Q}_T + \mathsf{R}_T,$$

where Q_T is the contribution to (4.1) of the prime powers $p^k \leqslant Y$.

In addition, for any integer $m \geqslant 0$, we denote by $\exp_m$ the Taylor polynomial of degree m of the exponential function at 0, that is,

$$\exp_m(z) = \sum_{j=0}^{m} \frac{z^j}{j!}.$$

We have an elementary lemma:

Lemma 4.4.1 *Let $z \in \mathbf{C}$ and $m \geqslant 0$. If $m \geqslant 100|z|$, then*

$$\exp_m(z) = e^z + O(\exp(-m)) = e^z(1 + O(\exp(-99|z|))).$$

Proof Indeed, since $j! \geqslant (j/e)^k$ for all $j \geqslant 0$ and $|z| \leqslant m/100$, the difference $e^z - \exp_m(z)$ is at most

$$\sum_{j>m} \frac{(m/100)^j}{j!} \leqslant \sum_{j>m}\left(\frac{em}{100j}\right)^j \ll \exp(-m).$$

□

We define

$$\mathsf{E}_T = \exp_{m_1}(-\mathsf{Q}_T) \quad \text{and} \quad \mathsf{F}_T = \exp_{m_2}(-\mathsf{R}_T),$$

where we recall that m_1 and m_2 are the parameters defined in (4.5). We have by definition $\mathsf{D}_T = \mathsf{B}_T\mathsf{C}_T$, with

$$\mathsf{B}_T(t) = \sum_{n\geqslant 1} b_T(n)\mu(n)n^{-\sigma_0-it} \quad \text{and} \quad \mathsf{C}_T(t) = \sum_{n\geqslant 1} c_T(n)\mu(n)n^{-\sigma_0-it},$$

where b_T and c_T are defined after the statement of Proposition 4.2.1, for example, $b_T(n)$ is the characteristic function of squarefree integers n such that n has $\leqslant m_1$ prime factors, all of which are $\leqslant Y$.

The idea of the proof is that, usually, Q_T (resp. R_T) is not too large, and then the random variable E_T is a good approximation to $\exp(-\mathsf{Q}_T)$. On the other hand, because of the shape of E_T (and the choice of the parameters), it will be possible to prove that E_T is close to B_T in L^2-norm, and similarly for F_T and C_T. Combining these facts will lead to the conclusion.

We first observe that, as in the beginning of the proof of Proposition 4.2.5, by the usual appeal to Lemma 3.2.6, we have

$$\mathbf{E}_T(|\mathsf{Q}_T|^2) \ll \varrho_T \quad \text{and} \quad \mathbf{E}_T(|\mathsf{R}_T|^2) \ll \log \varrho_T.$$

Markov's inequality implies that $\mathbf{P}_T(|\mathsf{Q}_T| > \varrho_T)$ tends to 0 as $T \to +\infty$. Now by Lemma 4.4.1, whenever $|\mathsf{Q}_T| \leqslant \varrho_T$, we have

$$\mathsf{E}_T = \exp(-\mathsf{Q}_T)\big(1 + O((\log T)^{-99})\big).$$

Similarly, the probability $\mathbf{P}_T(|\mathsf{R}_T| > \log \varrho_T)$ tends to 0, and whenever $|\mathsf{R}_T| \leqslant \log \varrho_T$, we have

$$\mathsf{F}_T = \exp(-\mathsf{R}_T)\big(1 + O((\log\log T)^{-99})\big).$$

For the next step, we claim that

$$\mathbf{E}_T\left(|\mathsf{E}_T - \mathsf{B}_T|^2\right) \ll (\log T)^{-60}, \tag{4.9}$$

$$\mathbf{E}_T\left(|\mathsf{F}_T - \mathsf{C}_T|^2\right) \ll (\log\log T)^{-60}. \tag{4.10}$$

We begin the proof of the first estimate with a lemma.

Lemma 4.4.2 *For $t \in \Omega_T$, we have*

$$\mathsf{E}_T(t) = \sum_{n\geqslant 1} \alpha(n)n^{-\sigma_0+it},$$

where the coefficients $\alpha(n)$ are zero unless $n \leqslant Y^{m_1}$ and n has only prime factors $\leqslant Y$. Moreover $|\alpha(n)| \leqslant 1$ for all n, and $\alpha(n) = \mu(n)b_T(n)$ if n has $\leqslant m_1$ prime factors, counted with multiplicity, and if there is no prime power p^k dividing n such that $p^k > Y$.

Proof Since

$$\mathsf{E}_{\mathrm{T}} = \exp_{m_1}(-\mathsf{Q}_{\mathrm{T}}) = \sum_{j=0}^{m_1} \frac{(-1)^j}{j!} \Big(\sum_{p^k \leqslant \mathrm{Y}} \frac{1}{kp^{k(\sigma_0+it)}} \Big)^j,$$

we obtain by expanding the jth power an expression of the desired kind, with coefficients

$$\alpha(n) = \sum_{0 \leqslant j \leqslant m_1} \frac{(-1)^j}{j!} \sum_{\substack{p_1^{k_1} \cdots p_j^{k_j} = n \\ p_i^{k_i} \leqslant \mathrm{Y}}} \frac{1}{k_1 \cdots k_j}.$$

We see from this expression that $\alpha(n)$ is 0 unless $n \leqslant \mathrm{Y}^{m_1}$ and n has only prime factors $\leqslant \mathrm{Y}$. Suppose now that n has $\leqslant m_1$ prime factors, counted with multiplicity, and that no prime power $p^k > \mathrm{Y}$ divides n. Then we may extend the sum defining $\alpha(n)$ to all $j \geqslant 0$, and remove the redundant conditions $p_i^{k_i} \leqslant \mathrm{Y}$, so that

$$\alpha(n) = \sum_{j \geqslant 0} \frac{(-1)^j}{j!} \sum_{p_1^{k_1} \cdots p_j^{k_j} = n} \frac{1}{k_1 \cdots k_j}.$$

But we recognize that this is the coefficient of n^{-s} in the expansion of

$$\exp\Big(- \sum_{k \geqslant 1} \frac{1}{k} p^{-ks} \Big) = \exp(-\log \zeta(s)) = \frac{1}{\zeta(s)} = \sum_{n \geqslant 1} \frac{\mu(n)}{n^s}$$

(viewed as a formal Dirichlet series, or by restricting to $\mathrm{Re}(s) > 1$). This means that, for such integers n, we have $\alpha(n) = \mu(n) = \mu(n) b_{\mathrm{T}}(n)$.

Finally, for any $n \geqslant 1$ now, we have

$$|\alpha(n)| \leqslant \sum_{j \geqslant 0} \frac{1}{j!} \sum_{p_1^{k_1} \cdots p_j^{k_j} = n} \frac{1}{k_1 \cdots k_j} = 1,$$

since the right-hand side is the coefficient of n^{-s} in $\exp(\log \zeta(s)) = \zeta(s)$. □

Now define $\delta(n) = \alpha(n) - \mu(n) b_{\mathrm{T}}(n)$ for all $n \geqslant 1$. We have

$$\mathbf{E}_{\mathrm{T}}\left(|\mathsf{E}_{\mathrm{T}} - \mathsf{B}_{\mathrm{T}}|^2\right) = \mathbf{E}_{\mathrm{T}}\Big(\Big| \sum_{n \geqslant 1} \frac{\delta(n)}{n^{\sigma_0+it}} \Big|^2 \Big),$$

which we estimate using Lemma 4.2.6. The contribution of the off-diagonal term is

$$\ll \frac{1}{\mathrm{T}} \sum_{m,n \leqslant \mathrm{Y}^{m_1}} |\delta(n)\delta(m)|(mn)^{\frac{1}{2}-\sigma_0} \leqslant \frac{4}{\mathrm{T}} \Big(\sum_{m \leqslant \mathrm{Y}^{m_1}} 1 \Big)^2 \ll \mathrm{T}^{-1+\varepsilon}$$

for any $\varepsilon > 0$, hence is negligible. The diagonal term is

$$\mathrm{M} = \sum_{n \geqslant 1} \frac{|\delta(n)|^2}{n^{2\sigma_0}} \leqslant \sum_{n \geqslant 1} \frac{|\delta(n)|^2}{n}.$$

By Lemma 4.4.2, we have $\delta(n) = 0$ unless either n has $> m_1$ prime factors, counted with multiplicity, or is divisible by a power p^k such that $p^k > \mathrm{Y}$ (and necessarily $p \leqslant \mathrm{Y}$ since δ is supported on integers only divisible by such primes). The contribution of the integers satisfying the first property is at most

$$\sum_{\substack{\Omega(n) > m_1 \\ p \mid n \Rightarrow p \leqslant \mathrm{Y}}} \frac{1}{n}.$$

We use Rankin's trick once more to bound this from above: for any fixed real number $\eta > 1$, we have

$$\sum_{\substack{\Omega(n) > m_1 \\ p \mid n \Rightarrow p \leqslant \mathrm{Y}}} \frac{1}{n} \leqslant \eta^{-m_1} \prod_{p \leqslant \mathrm{Y}} \Big(1 + \frac{\eta}{p} + \cdots\Big) \ll \eta^{-m_1} (\log \mathrm{Y})^{\eta}$$
$$\leqslant (\log \mathrm{T})^{-100 \log \eta + \eta}$$

by Proposition C.3.6. Selecting $\eta = e^{2/3} \leqslant 2$, for instance, this shows that this contribution is $\ll (\log \mathrm{T})^{-60}$.

The contribution of integers divisible by $p^k > \mathrm{Y}$ is at most

$$\Big(\sum_{\substack{p \leqslant \mathrm{Y} \\ p^k > \mathrm{Y}}} \frac{1}{p^k} \Big) \Big(\sum_{p \mid n \Rightarrow p \leqslant \mathrm{Y}} \frac{1}{n} \Big) \leqslant \frac{1}{\mathrm{Y}} \Big(\sum_{\substack{\sqrt{\mathrm{Y}} < p^k \leqslant \mathrm{Y} \\ k \geqslant 2}} 1 \Big) \prod_{p \leqslant \mathrm{Y}} \frac{1}{1 - p^{-1}}$$
$$\ll \mathrm{Y}^{-1/2} (\log \mathrm{Y}),$$

which is even smaller. This concludes the proof of (4.9).

The proof of the second estimate (4.10) is quite similar, with one extra consideration to handle. Indeed, arguing as in Lemma 4.4.2, we obtain the expression

$$\mathsf{F}_{\mathrm{T}}(t) = \sum_{n \geqslant 1} \beta(n) n^{-\sigma_0 + it},$$

for $t \in \Omega_T$, where the coefficients $\beta(n)$ are zero unless $n \leqslant X^{m_2}$ and n has only prime factors $\leqslant X$, satisfy $|\beta(n)| \leqslant 1$ for all n, and finally satisfy $\beta(n) = \mu(n)$ if n has $\leqslant m_2$ prime factors, counted with multiplicity, and if there is no prime power p^k dividing n with $Y < p^k \leqslant X$.

Using this, and defining now $\delta(n) = \beta(n) - \mu(n)c_T(n)$, we get from Lemma 4.2.6 the bound

$$\mathbf{E}_T\left(|\mathsf{F}_T - \mathsf{C}_T|^2\right) \ll \sum_{\substack{n \geqslant 1 \\ \delta(n) \neq 0}} \frac{1}{n^{2\sigma_0}} \leqslant \sum_{\substack{n \geqslant 1 \\ \delta(n) \neq 0}} \frac{1}{n}.$$

But the integers that satisfy $\delta(n) \neq 0$ must be of one of the following types:

(1) Those with $c_T(n) = 1$, which (by the previous discussion) must either have $\Omega(n) > m_2$ (and be divisible by primes $\leqslant X$ only), *or* must be divisible by a prime power p^k such that $p^k > X$ (the possibility that $p^k \leqslant Y$ is here excluded, because $c_T(n) = 1$ implies that n has no prime factor $< Y$). The contribution of these integers is handled as in the case of the bound (4.9) and is $\ll (\log\log T)^{-60}$.

(2) Those with $c_T(n) = 0$ and $\beta(n) \neq 0$; since

$$\beta(n) = \sum_{0 \leqslant j \leqslant m_2} \frac{(-1)^j}{j!} \sum_{\substack{p_1^{k_1} \cdots p_j^{k_j} = n \\ Y < p_i^{k_i} \leqslant X}} \frac{1}{k_1 \cdots k_j},$$

as in the beginning of the proof of Lemma 4.4.2, such an integer n has at least one factorization $n = p_1^{k_1} \cdots p_j^{k_j}$ for some $j \leqslant m_2$, where each prime power $p_i^{k_i}$ is between Y and X. Since $c_T(n) = 0$, either $\Omega(n) > m_2$, *or* n has a prime factor $p > X$, *or* n has a prime factor $p \leqslant Y$. The first two possibilities are again handled exactly like in the proof of (4.9), but the third is somewhat different. We proceed as follows to estimate its contribution, say, N. We have

$$N = \sum_{0 \leqslant j < m_2} N_j,$$

where

$$N_j = \sum_{p \leqslant Y} \sum_{\substack{n = p^k p_1^{k_1} \cdots p_j^{k_j} \\ Y < p_i^{k_i} \leqslant X}} \frac{1}{n}$$

is the contribution of integers with a factorization of length $j+1$ as a product of prime powers between Y and X. By multiplicativity, we get

$$\mathrm{N}_j \leqslant \Big(\sum_{p\leqslant \mathrm{Y}} \sum_{\mathrm{Y}<p^k\leqslant \mathrm{X}} \frac{1}{p^k}\Big)\Big(\sum_{p\leqslant \mathrm{X}} \sum_{\mathrm{Y}<p^k\leqslant \mathrm{X}} \frac{1}{p^k}\Big)^{j-1}.$$

Consider the first factor. For a given prime $p \leqslant \mathrm{Y}$, let l be the smallest integer such that $p^l > \mathrm{Y}$. The sum over k is then

$$\sum_{\mathrm{Y}<p^k\leqslant \mathrm{X}} \frac{1}{p^k} \leqslant \frac{1}{p^l} + \frac{1}{p^{l+1}} + \cdots \ll \frac{1}{p^l} \leqslant \frac{1}{\mathrm{Y}},$$

so that the first factor is $\ll \pi(\mathrm{Y})/\mathrm{Y} \ll (\log \mathrm{Y})^{-1}$. On the other hand, for the second factor, we have

$$\sum_{p\leqslant \mathrm{X}} \sum_{\mathrm{Y}<p^k\leqslant \mathrm{X}} \frac{1}{p^k} = \sum_{p\leqslant \mathrm{Y}} \sum_{\mathrm{Y}<p^k\leqslant \mathrm{X}} \frac{1}{p^k} + \sum_{\mathrm{Y}<p\leqslant \mathrm{X}} \sum_{\mathrm{Y}<p^k\leqslant \mathrm{X}} \frac{1}{p^k}$$

$$\ll \frac{\pi(\mathrm{Y})}{\mathrm{Y}} + \sum_{\mathrm{Y}<p\leqslant \mathrm{X}} \sum_{\mathrm{Y}<p^k\leqslant \mathrm{X}} \frac{1}{p^k},$$

where we used the bound arising from the first factor. For a given prime p with $\mathrm{Y} < p \leqslant \mathrm{X}$, the last sum over k is

$$\frac{1}{p} + \frac{1}{p^2} + \cdots \ll \frac{1}{p},$$

and the sum over p is therefore

$$\sum_{\mathrm{Y}<p\leqslant \mathrm{X}} \frac{1}{p} = \log\Big(\frac{\log \mathrm{X}}{\log \mathrm{Y}}\Big) + \mathrm{O}(1) = \log\log\log \mathrm{T} + \mathrm{O}(1),$$

using the values of X and Y and Proposition C.3.1. Hence the final estimate is

$$\mathrm{N} \ll \frac{1}{\log \mathrm{Y}} (\log\log\log \mathrm{T})^{m_2} \ll (\log\log \mathrm{T})(\log\log\log \mathrm{T})^{m_2} (\log \mathrm{T})^{-1} \to 0$$

as $\mathrm{T} \to +\infty$, from which we finally deduce that (4.10) holds.

With the mean-square estimates (4.9) and (4.10) in hand, we can now finish the proof of Proposition 4.2.5. Except on sets of measure tending to 0 as $\mathrm{T} \to +\infty$, we have

$$\mathsf{B}_{\mathrm{T}} = \mathsf{E}_{\mathrm{T}} + \mathrm{O}((\log \mathrm{T})^{-25}), \qquad \mathsf{E}_{\mathrm{T}} = \exp(-\mathsf{Q}_{\mathrm{T}})\big(1 + \mathrm{O}((\log \mathrm{T})^{-99})\big),$$

$$\frac{1}{\log \mathrm{T}} \leqslant \exp(-\mathsf{Q}_{\mathrm{T}}) \leqslant (\log \mathrm{T})$$

(where the first property follows from (4.9)), and hence

$$\mathsf{B}_{\mathrm{T}} = \exp(-\mathsf{Q}_{\mathrm{T}})\big(1 + \mathrm{O}((\log \mathrm{T})^{-20})\big),$$

again outside of a set of measure tending to 0. Similarly, using (4.10), we get

$$\mathsf{C}_{\mathrm{T}} = \exp(-\mathsf{R}_{\mathrm{T}})\big(1 + \mathrm{O}((\log\log \mathrm{T})^{-20})\big)$$

outside of a set of measure tending to 0. Multiplying the two equalities shows that

$$\mathsf{D}_{\mathrm{T}} = \exp(-\mathsf{P}_{\mathrm{T}})\big(1 + \mathrm{O}((\log\log \mathrm{T})^{-20})\big)$$

with probability tending to 1 as $\mathrm{T} \to +\infty$. This concludes the proof.

Exercise 4.4.3 Try to see what happens if one uses a single range $p^k \leqslant \mathrm{X}$, instead of having the distinction between $p^k \leqslant \mathrm{Y}$ and $\mathrm{Y} < p^k \leqslant \mathrm{X}$.

4.5 Further Topics

Generalizations of Selberg's Central Limit Theorem are much harder to come by than those of Bagchi's Theorem (which is another illustration of the fact that arithmetic L-functions have much more delicate properties on the critical line). There are very few other cases than that of the Riemann zeta function where such a statement is known (see the remarks in [95, §7] for references). For instance, consider the family of modular forms f that is described in Section 3.4. The natural question is now to consider the distribution (possibly with weights ω_f) of $\mathrm{L}(f, \frac{1}{2})$. First, it is a known fact (due to Waldspurger and Kohnen–Zagier) that $\mathrm{L}(f, \frac{1}{2}) \geqslant 0$ in that case. This property reflects a different type of expected distribution of the values $\mathrm{L}(f, \frac{1}{2})$, namely, one expects that the correct normalization is

$$f \mapsto \frac{\log \mathrm{L}(f, \frac{1}{2}) + \frac{1}{2}\log\log q}{\sqrt{\log\log q}},$$

in the sense that this defines a sequence of random variables on Ω_q that should converge in law to a standard (real) Gaussian random variable. Now observe that such a statement, if true, would immediately imply that the proportion of $f \in \Omega_q$ with $\mathrm{L}(f, \frac{1}{2}) = 0$ tends to 0 as $q \to +\infty$, and this is not currently known (this would indeed be a major result in the analytic theory of modular forms).

Nevertheless, there has been significant progress in this direction, for various families, in recent and ongoing work of Radziwiłł and Soundararajan. In [96], they prove *sub-Gaussian* upper bounds for the distribution of L-values in certain families similar to Ω_q (specifically, quadratic twists of a fixed modular form). In [97], they announce Gaussian lower bounds, but for families conditioned to have $\mathrm{L}(f, \frac{1}{2}) \neq 0$ (which, for a number of cases, is known to be a subfamily with positive density as the size tends to infinity).

In addition to these developments, it should be emphasized that Selberg's Theorem serves as a general guiding principle when studying any probabilistic question for the Riemann zeta function on the critical line, and the ideas in its proof are often the starting points toward other results. Indeed, some of the deepest works in probabilistic number theory in recent years have been devoted to studies of finer aspects of the distribution of the Riemann Zeta function on the critical line. A particular focus has been a conjecture of Fedorov, Hiary and Keating [39] that addresses the distribution of the maximum of $\zeta(1/2+it)$ when t varies over an interval of length 1 (and t is taken uniformly at random in $[-\mathrm{T},\mathrm{T}]$ or $[\mathrm{T},2\mathrm{T}]$ with $\mathrm{T}\to+\infty$). This leads to links with objects like log-correlated fields, branching random walks, or Gaussian multiplicative chaos. We refer to the Bourbaki seminar survey of Harper [54] for a discussion of the work of Najnudel [90] and Arguin–Belius–Bourgade–Radziwi–Soundararajan [1], and to Harper's recent preprint [55] for the latest developments in this direction.

One of the reasons that Central Limit Theorems are expected to hold is that they are known to follow from the widely believed moment conjectures for families of L-functions, which predict (with considerable evidence, theoretic, numerical and heuristic) the asymptotic behavior of the Laplace or Fourier transform of the logarithm of the special values of the L-functions. In other words, taking the example of the Riemann zeta function, these conjectures (due to Keating and Snaith [63]) predict the asymptotic behavior of

$$\mathbf{E}_{\mathrm{T}}(e^{s\log|\zeta(\frac{1}{2}+it)|})=\mathbf{E}_{\mathrm{T}}(|\zeta(\tfrac{1}{2}+it)|^s)=\frac{1}{2\mathrm{T}}\int_{-\mathrm{T}}^{\mathrm{T}}|\zeta(\tfrac{1}{2}+it)|^s dt$$

for suitable $s\in\mathbf{C}$. It is of considerable interest that, besides natural arithmetic factors (related to the independence of Proposition 3.2.5 or suitable analogues), these conjectures involve certain terms which originate in Random Matrix Theory. In addition to implying straightforwardly the Central Limit Theorem, note that the moment conjectures also immediately yield the generalization of (3.11) or (3.15), hence can be allowed to deduce general versions of Bagchi's Theorem and universality. Moreover, these moment conjectures (in suitably uniform versions) are also able to settle other interesting conjectures concerning the distribution of values of $\zeta(\frac{1}{2}+it)$. For instance, as shown by Kowalski and Nikeghbali [78], they are known to imply that the image of $t\mapsto\zeta(\frac{1}{2}+it)$, for $t\in\mathbf{R}$, is *dense* in $\mathbf{C}$ (a conjecture of Ramachandra).

[Further references: Katz–Sarnak [62], Blomer, Fouvry, Kowalski, Michel, Milićević, and Sawin [11].]

5

The Chebychev Bias

Probability tools	Arithmetic tools
Definition of convergence in law (§ B.3)	Primes in arithmetic progressions
Kronecker's Theorem (th. B.6.5)	Orthogonality of Dirichlet characters (prop. C.5.1)
Convergence in law using auxiliary parameters (prop. B.4.4)	Dirichlet L-functions (§ C.5)
Characteristic functions (§ B.5)	Generalized Riemann Hypothesis (conj. C.5.8)
Kolmogorov's Theorem for random series (th. B.10.1)	Explicit formula (th. C.5.6)
Method of moments (th. B.5.5)	Distribution of the zeros of L-functions (prop. C.5.3)
	Generalized Simplicity Hypothesis

5.1 Introduction

One of the most remarkable limit theorems in probabilistic number theory is related to a surprising feature of the distribution of prime numbers, which was first noticed by Chebychev [24] in 1853: there seemed to be many more primes p such that $p \equiv 3 \pmod 4$ than primes with $p \equiv 1 \pmod 4$ (any prime, except $p = 2$, must satisfy one of these two conditions). More precisely, he states:

> En cherchant l'expression limitative des fonctions qui déterminent la totalité des nombres premiers de la forme $4n + 1$ et de ceux de la forme $4n + 3$, pris au-dessous d'une limite très grande, je suis parvenu à reconnaître que ces deux

fonctions diffèrent notablement entre elles par leurs seconds termes, dont la valeur, pour les nombres $4n+3$, est plus grande que celle pour les nombres $4n+1$; ainsi, si de la totalité des nombres premiers de la forme $4n+3$, on retranche celle des nombres premiers de la forme $4n+1$, et que l'on divise ensuite cette différence par la quantité $\frac{\sqrt{x}}{\log x}$, on trouvera plusieurs valeurs de x telles, que ce quotient s'approchera de l'unité aussi près qu'on le voudra.[1]

It is unclear from Chebychev's very short note what exactly he had proved, or simply conjectured, and he did not publish anything more on this topic. It is definitely *not* the case that we have

$$\pi(x;4,3) > \pi(x;4,1)$$

for all $x \geqslant 2$, where (in general), for an integer $q \geqslant 1$ and an integer a, we write $\pi(x;q,a)$ for the number of primes $p \leqslant x$ such that $p \equiv a \pmod{q}$. Indeed, for $x = 26{,}861$, we have

$$\pi(x;4,3) = 1472 < 1473 = \pi(x;4,1)$$

(as discovered by Leech in 1957), and one can prove that there are infinitely many sign changes of the difference $\pi(x;4,3) - \pi(x;4,1)$.

In any case, by communicating his observations, Chebychev created a fascinating area of number theory. We will discuss some of the basic known results in this chapter, which put the question on a rigorous footing, and in particular confirm the existence of the *bias* toward the residue class of 3 modulo 4, in a precise sense (although this conclusion will depend on currently unproved conjectures). Because of this feature, the subject is called the study of the *Chebychev bias*.

5.2 The Rubinstein–Sarnak Distribution

In order to study the problem suggested by Chebychev, we consider for $\mathrm{X} \geqslant 1$ the probability space $\Omega_{\mathrm{X}} = [1, \mathrm{X}]$, with the probability measure

$$\mathbf{P}_{\mathrm{X}} = \frac{1}{\log \mathrm{X}} \frac{dx}{x}. \tag{5.1}$$

[1] English translation: "While searching for the limiting expression of the functions that determine the number of prime numbers of the form $4n+1$ and of those of the form $4n+3$, less than a very large limit, I have succeeded in recognizing that the second terms of these two functions differ notably from each other; its value [of this second term], for the numbers $4n+3$, is larger than that for the numbers $4n+1$; thus, if from the number of prime numbers of the form $4n+3$, we subtract that of the prime numbers of the form $4n+1$, and then divide this difference by the quantity $\frac{\sqrt{x}}{\log x}$, we will find several values of x such that this ratio will approach one as closely as we want."

Let $q \geqslant 1$ be an integer. We define a random variable on Ω_X, with values in the vector space $C_{\mathbf{R}}((\mathbf{Z}/q\mathbf{Z})^\times)$ of real-valued functions on the (fixed) finite group $(\mathbf{Z}/q\mathbf{Z})^\times$, by defining $\mathsf{N}_{X,q}(x)$, for $x \in \Omega_X$, to be the function such that

$$\mathsf{N}_{X,q}(x)(a) = \frac{\log x}{\sqrt{x}}\big(\varphi(q)\pi(x;q,a) - \pi(x)\big) \tag{5.2}$$

for $a \in (\mathbf{Z}/q\mathbf{Z})^\times$ (this could also, of course, be viewed as a random real vector with values in $\mathbf{R}^{|(\mathbf{Z}/q\mathbf{Z})^\times|}$, but the perspective of a function will be slightly more convenient).

We see that the knowledge of $\mathsf{N}_{X,q}$ allows us to compare the number of primes up to X in any family of invertible residue classes modulo q. It is therefore appropriate for the study of the questions suggested by Chebychev.

We observe that in the remainder of this chapter, we will consider q to be fixed (although there are interesting questions that one can ask about uniformity with respect to q). For this reason, we will often simplify the notation (especially during proofs) to write N_X instead of $\mathsf{N}_{X,q}$, and similarly dropping q in some other cases.

Remark 5.2.1 (1) If $q = 4$, then $(\mathbf{Z}/4\mathbf{Z})^\times = \{1,3\}$, and for $x \in \Omega_T$, the random function $\mathsf{N}_{X,4}(x)$ is given by

$$1 \mapsto \frac{\log x}{\sqrt{x}}(2\pi(x;4,1) - \pi(x)) \quad \text{and} \quad 3 \mapsto \frac{\log x}{\sqrt{x}}(2\pi(x;4,3) - \pi(x)).$$

(2) Recall that the fundamental theorem of Dirichlet, Hadamard and de la Vallée Poussin (Theorem C.3.7) shows that

$$\pi(x;q,a) \sim \frac{1}{\varphi(q)}\pi(x)$$

for all a coprime to q. Thus the random variables N_X are considering the correction term from the asymptotic behavior.

(3) The normalizing factor $(\log x)/\sqrt{x}$, which is the "correct one," is the same one that is suggested by Chebychev's quote.

The basic probabilistic result concerning these arithmetic quantities is the following:

Theorem 5.2.2 (Rubinstein–Sarnak) *Let $q \geqslant 1$. Assume the* Generalized Riemann Hypothesis *modulo q. Then the random functions $\mathsf{N}_{X,q}$ converge in law to a random function N_q. The support of N_q is contained in the hyperplane*

$$\mathrm{H}_q = \Big\{ f : (\mathbf{Z}/q\mathbf{Z})^\times \to \mathbf{R} \mid \sum_{a \in (\mathbf{Z}/q\mathbf{Z})^\times} f(a) = 0 \Big\}. \tag{5.3}$$

We call N_q the *Rubinstein–Sarnak distribution* modulo q.

Remark 5.2.3 One may wonder if the choice of the logarithmic weight in the probability measure $\mathbf{P}_X$ is necessary for such a statement of convergence in law: this is indeed the case, and we will say a few words to explain this in Remark 5.3.5.

The Generalized Riemann Hypothesis modulo q is originally a statement about the zeros of certain analytic functions, the *Dirichlet* L-*functions* modulo q. It has, however, a concrete formulation in terms of the distribution of prime numbers: it is equivalent to the statement that, for all integers a coprime with q and all $x \geqslant 2$, we have

$$\pi(x;q,a) = \frac{1}{\varphi(q)} \int_2^x \frac{dt}{\log t} + \mathrm{O}(x^{1/2}(\log qx)),$$

where the implied constant is absolute (see, e.g., [59, 5.14, 5.15] for this equivalence). The size of the (expected) error term, approximately $\sqrt{x}$, is related to the zeros of the Dirichlet L-functions, as we will see later; it explains that the normalization factor in (5.2) is the right one for the existence of a limit in law as in Theorem 5.2.2. Indeed, using the case $q = 1$, which is the formula

$$\pi(x) = \int_2^x \frac{dt}{\log t} + \mathrm{O}(x^{1/2}(\log x)),$$

we deduce that each value of the function N_X satisfies

$$\frac{\log x}{\sqrt{x}}(\varphi(q)\pi(x;q,a) - \pi(x)) = \mathrm{O}(\varphi(q)(\log qx)^2).$$

To see how Theorem 5.2.2 helps answer questions related to the Chebychev bias, we take $q = 4$. Then we expect that

$$\lim_{X\to+\infty} \mathbf{P}_X(\pi(x;4,3) > \pi(x;4,1)) = \mathbf{P}(\mathrm{N}_4 \in \mathrm{H}_4 \cap \mathrm{C}),$$

where $\mathrm{C} = \{(x_1,x_3) \mid x_3 > x_1\}$ (although whether this limit exists or not does not follow from Theorem 5.2.2, without further information concerning the properties of the limit N_4). Then Chebychev's basic observation could be considered to be confirmed if $\mathbf{P}(\mathrm{N}_4 \in \mathrm{H}_4 \cap \mathrm{C})$ is close to 1. But in the absence of any other information, it seems very hard to prove (or disprove) this last fact.

However, Rubinstein and Sarnak showed that one could go much further by making one extra assumption on the distribution of the zeros of Dirichlet L-functions. Indeed, one can then represent N_q explicitly as the sum of a series of independent random variables (and in particular compute explicitly the characteristic function of the random function N_q). We describe this random series in Section 5.4, since to do so at this point would lead to a statement that would appear highly unmotivated. The proof of Theorem 5.2.2 will lead us naturally to this next step (see Theorem 5.4.4 for the details).

Below, we write

$$\sideset{}{^*}\sum_{\chi \pmod q} (\ldots) \quad \text{and} \quad \sideset{}{^*}\prod_{\chi \pmod q} (\cdots)$$

for a sum or a product over nontrivial[2] Dirichlet characters modulo q; we recall that these are (completely) multiplicative functions on $\mathbf{Z}$ such that $\chi(n) = 0$ unless n is coprime to q, in which case we have $\chi(n) = \widetilde{\chi}(n)$ for some group homomorphism $\widetilde{\chi}\colon (\mathbf{Z}/q\mathbf{Z})^\times \to \mathbf{C}^\times$.

We define a function m_q on $(\mathbf{Z}/q\mathbf{Z})^\times$ by

$$m_q(a) = -\sideset{}{^*}\sum_{\substack{\chi \pmod q \\ \chi^2=1}} \overline{\chi(a)} \tag{5.4}$$

for $a \in (\mathbf{Z}/q\mathbf{Z})^\times$. This can also, using orthogonality of characters modulo q (see Proposition C.5.1), be expressed in the form

$$m_q(a) = 1 - \sum_{\substack{b \in (\mathbf{Z}/q\mathbf{Z})^\times \\ b^2 = a \pmod q}} 1,$$

from which we see that in fact we have simply two possible values, namely,

$$m_q(a) = \begin{cases} 1 & \text{if } a \text{ is not a square modulo } q, \\ 1 - \sigma_q & \text{otherwise,} \end{cases} \tag{5.5}$$

where

$$\sigma_q = |\{b \in (\mathbf{Z}/q\mathbf{Z})^\times \mid b^2 = 1\}| = |\{\chi \pmod q \mid \chi^2 = 1\}|$$

is also the index of the subgroup of squares in $(\mathbf{Z}/q\mathbf{Z})^\times$.

In the remaining sections of this chapter, we will explain the proof of Theorem 5.2.2, following Rubinstein and Sarnak. We will assume some familiarity with Dirichlet L-functions (in Section C.5, we recall the relevant definitions and standard facts). Readers who have not yet been exposed to these functions will probably find it easier to assume in what follows that $q = 4$. In this case, there is only one nontrivial Dirichlet L-function modulo 4, which is defined by

[2] We emphasize, for readers already familiar with analytic number theory, that this does not mean *primitive* characters.

$$L(\chi_4, s) = \sum_{k \geqslant 0} \frac{(-1)^k}{(2k+1)^s} = \sum_{n \geqslant 1} \chi_4(n) n^{-s},$$

corresponding to the character χ_4 such that

$$\chi_4(n) = \begin{cases} 0 & \text{if } n \text{ is even,} \\ (-1)^k & \text{if } n = 2k+1 \text{ is odd} \end{cases} \tag{5.6}$$

for $n \geqslant 1$. The arguments should then be reasonably transparent. In particular, any sum of the type

$$\sum_{\chi \pmod 4}^{*} (\cdots)$$

means that one only considers the expression on the right-hand side for the character χ_4 defined in (5.6).

5.3 Existence of the Rubinstein–Sarnak Distribution

The proof of Theorem 5.2.2 depends roughly on two ingredients:

- on the arithmetic side, we can represent the arithmetic random functions N_X as combinations of $x \mapsto x^{i\gamma}$, where the γ are ordinates of zeros of the L-functions modulo q;
- once this is done, we observe that Kronecker's Equidistribution Theorem (Theorem B.6.5) implies convergence in law for any function of this type.

There are some intermediate approximation steps involved, but the ideas are quite intuitive.

In this section, we always assume the validity of the Generalized Riemann Hypothesis modulo q, unless otherwise noted.

For a Dirichlet character χ modulo q, we define random variables ψ_χ on Ω_X by

$$\psi_\chi(x) = \frac{1}{\sqrt{x}} \sum_{n \leqslant x} \Lambda(n)\chi(n)$$

for $x \in \Omega_X$, where Λ is the von Mangoldt function (see Section C.4, especially (C.6), for the definition of this function).

The next lemma is a key step to express N_X in terms of Dirichlet characters. It looks first like standard harmonic analysis, but there is a subtle point in the proof that is crucial for the rest of the argument, and for the very existence of the Chebychev bias.

Lemma 5.3.1 *We have*

$$\mathsf{N}_{\mathrm{X},q} = m_q + \sum_{\chi \,(\mathrm{mod}\, q)}^{*} \psi_\chi \,\overline{\chi} + \mathsf{E}_{\mathrm{X},q},$$

where $\mathsf{E}_{\mathrm{X},q}$ *converges to* 0 *in probability as* $\mathrm{X} \to +\infty$.

Proof By orthogonality of the Dirichlet characters modulo q (see Proposition C.5.1), we have

$$\varphi(q)\pi(x;q,a) = \sum_{\chi \,(\mathrm{mod}\, q)} \overline{\chi(a)} \sum_{p \leqslant x} \chi(p),$$

hence

$$\frac{\log x}{\sqrt{x}}\big(\varphi(q)\pi(x;q,a) - \pi(x)\big) = \sum_{\chi \,(\mathrm{mod}\, q)}^{*} \overline{\chi(a)}\,\frac{\log x}{\sqrt{x}} \sum_{p \leqslant x} \chi(p) + \mathrm{O}\Big(\frac{\log x}{\sqrt{x}}\Big)$$

for $x \geqslant 2$, where the error term accounts for primes p dividing q (for which the trivial character takes the value 0 instead of 1); in particular, the implied constant depends on q.

We now need to connect the sum over primes, for a fixed character χ, to ψ_χ. Recall that the von Mangoldt functions differs little from the characteristic function of primes multiplied by the logarithm function. The sum of this simpler function is the random variable defined by

$$\theta_\chi(x) = \frac{1}{\sqrt{x}} \sum_{p \leqslant x} \chi(p) \log(p)$$

for $x \in \Omega_{\mathrm{X}}$. It is related to ψ_χ by

$$\theta_\chi(x) - \psi_\chi(x) = -\frac{1}{\sqrt{x}} \sum_{k \geqslant 2} \sum_{p^k \leqslant x} \chi(p^k) \log p = -\frac{1}{\sqrt{x}} \sum_{k \geqslant 2} \sum_{p^k \leqslant x} \chi(p)^k \log p.$$

We can immediately see that the contribution of $k \geqslant 3$ is very small: since the exponent k is at most of size $\log x$, and $|\chi(p)| \leqslant 1$ for all primes p, it is bounded by

$$\Big|\frac{1}{\sqrt{x}} \sum_{k \geqslant 2} \sum_{\substack{p^k \leqslant x \\ k \geqslant 3}} \chi(p)^k \log p\Big| \leqslant \frac{1}{\sqrt{x}} \sum_{3 \leqslant k \leqslant \log x} (\log x) x^{1/k} \ll \frac{(\log x)^2}{x^{1/6}},$$

where the implied constant is absolute.

For $k = 2$, there are two cases. If χ^2 is the trivial character, then

$$\frac{1}{\sqrt{x}} \sum_{p \leqslant \sqrt{x}} \chi(p)^2 \log p = \frac{1}{\sqrt{x}} \sum_{\substack{p \leqslant \sqrt{x} \\ p \nmid q}} \log p = 1 + \mathrm{O}\left(\frac{1}{\log x}\right)$$

by a simple form of the Prime Number Theorem in arithmetic progressions (the Generalized Riemann Hypothesis would of course give a much better error term, but this is not needed here). If χ^2 is nontrivial, then we have

$$\frac{1}{\sqrt{x}} \sum_{p \leqslant \sqrt{x}} \chi(p)^2 \log p \ll \frac{1}{\log x}$$

for the same reason. Thus we have

$$\theta_\chi(x) = \psi_\chi(x) - \delta_{\chi^2} + \mathrm{O}\left(\frac{1}{\log x}\right), \tag{5.7}$$

where δ_{χ^2} is 1 if χ^2 is trivial, and is zero otherwise.

By summation by parts, we have

$$\sum_{p \leqslant x} \chi(p) = \frac{1}{\log x} \sum_{p \leqslant x} \chi(p) \log p + \int_2^x \left(\sum_{p \leqslant t} \chi(p) \log p\right) \frac{dt}{t(\log t)^2}$$

for any Dirichlet character χ modulo q, so that

$$\begin{aligned} \frac{\log x}{\sqrt{x}}(\varphi(q)\pi(x;q,a) - \pi(x)) = & \sum_{\chi \,(\mathrm{mod}\, q)}^{*} \overline{\chi(a)}\,\theta_\chi(x) \\ & + \frac{\log x}{\sqrt{x}} \int_2^x \frac{\theta_\chi(t)}{t^{1/2}(\log t)^2} dt + \mathrm{O}\left(\frac{\log x}{\sqrt{x}}\right). \end{aligned} \tag{5.8}$$

We begin by handling the integral for a nontrivial character χ. We have $\theta_\chi(x) = \psi_\chi(x) + \mathrm{O}(1/\log x)$ if $\chi^2 \neq 1_q$, which implies

$$\int_2^x \frac{\theta_\chi(t)}{t^{1/2}(\log t)^2} dt = \int_2^x \frac{\psi_\chi(t)}{t^{1/2}(\log t)^2} dt + \mathrm{O}\left(\frac{x^{1/2}}{(\log x)^3}\right)$$

since

$$\int_2^x \frac{1}{t^{1/2}(\log t)^2} dt \ll \frac{x^{1/2}}{(\log x)^2}.$$

If χ^2 is trivial, we have an additional constant term $\theta_\chi(x) - \psi_\chi(x) = 1 + \mathrm{O}(1/\log x)$, and we get

$$\int_2^x \frac{\theta_\chi(t)}{t^{1/2}(\log t)^2} dt = \int_2^x \frac{\psi_\chi(t)}{t^{1/2}(\log t)^2} dt + \int_2^x \frac{1}{t^{1/2}(\log t)^2} dt + \mathrm{O}\left(\frac{x^{1/2}}{(\log x)^3}\right)$$
$$= \int_2^x \frac{\psi_\chi(t)}{t^{1/2}(\log t)^2} dt + \mathrm{O}\left(\frac{x^{1/2}}{(\log x)^2}\right).$$

Thus, in all cases, we get

$$\frac{\log x}{\sqrt{x}} \int_2^x \frac{\theta_\chi(t)}{t^{1/2}(\log t)^2} dt = \frac{\log x}{\sqrt{x}} \int_2^x \frac{\psi_\chi(t)}{t^{1/2}(\log t)^2} dt + \mathrm{O}\left(\frac{1}{\log x}\right).$$

Now comes the subtle point we previously mentioned. If we were to use the pointwise bound $\psi_\chi(t) \ll (\log t)^2$ (which is essentially the content of the Generalized Riemann Hypothesis) in the remaining integral, we would only get

$$\frac{\log x}{\sqrt{x}} \int_2^x \frac{\psi_\chi(t)}{t^{1/2}(\log t)^2} dt \ll \log x,$$

which is too big. So we need to use the integration process nontrivially. Precisely, by Corollary C.5.11, we have

$$\int_2^x \psi_\chi(t) dt \ll x$$

for all $x \geqslant 2$ (this reflects a “smoothing” effect due to the convergence of the series with terms $1/|\frac{1}{2} + i\gamma|^2$, where γ are the ordinates of zeros of $\mathrm{L}(s,\chi)$). Using integration by parts, we can then deduce that

$$\frac{\log x}{\sqrt{x}} \int_2^x \frac{\psi_\chi(t)}{t^{1/2}(\log t)^2} dt \ll \frac{\log x}{\sqrt{x}} \left(\frac{x}{x^{1/2}(\log x)^2} + \int_2^x \frac{t^{1/2} dt}{(\log t)^2}\right) \ll \frac{1}{\log x}.$$

Finally, we transform the first term of (5.8) to express it in terms of ψ_χ, again using (5.7). For any element $a \in (\mathbf{Z}/q\mathbf{Z})^\times$ and $x \in \Omega_X$, we have

$$\frac{\log x}{\sqrt{x}} \big(\varphi(q)\pi(x;q,a) - \pi(x)\big)$$
$$= -\sum_{\substack{\chi^2=1 \\ \chi\neq 1}} \overline{\chi(a)} + \sideset{}{^*}\sum_{\chi \,(\mathrm{mod}\, q)} \overline{\chi(a)}\psi_\chi(x) + \mathrm{O}\left(\frac{1}{\log x}\right)$$
$$= m_q(a) + \sideset{}{^*}\sum_{\chi \,(\mathrm{mod}\, q)} \overline{\chi(a)}\psi_\chi(x) + \mathrm{O}\left(\frac{1}{\log x}\right),$$

where the implied constant depends on q. Since the error term is $\ll (\log x)^{-1}$ for $x \in \Omega_X$, it converges to zero in probability, and this concludes the proof. □

We keep further on the notation of the lemma, except that we also sometimes write E_{X} for $\mathsf{E}_{\mathrm{X},q}$. Since E_{X} tends to 0 in probability and m_q is a fixed function on $(\mathbf{Z}/q\mathbf{Z})^\times$, Theorem 5.2.2 will follow (by Corollary B.4.2) from the convergence in law of the random functions

$$\mathsf{M}_{\mathrm{X},q} = \sum_{\chi \pmod q}^{*} \psi_\chi \overline{\chi}.$$

Now we express these functions in terms of zeros of L-functions. Here and later, a sum over zeros of a Dirichlet L-function always means implicitly that zeros are counted with their multiplicity.

We will denote by I_{X} the identity variable $x \mapsto x$ on Ω_{X}; thus, for a complex number s, the random variable $\mathsf{I}_{\mathrm{X}}^s$ is the function $x \mapsto x^s$ on Ω_{X}.

Below, when we have a random function X on $(\mathbf{Z}/q\mathbf{Z})^\times$, and a nonnegative random variable Y, the meaning of a statement of the form $\mathrm{X} = \mathrm{O}(\mathrm{Y})$ is that $\|\mathrm{X}\| = \mathrm{O}(\mathrm{Y})$, where the norm is the euclidean norm, that is, we have

$$\|\mathrm{X}\|^2 = \sum_{a \in (\mathbf{Z}/q\mathbf{Z})^\times} |\mathrm{X}(a)|^2.$$

Lemma 5.3.2 *We have*

$$\mathsf{M}_{\mathrm{X},q} = -\sum_{\chi \pmod q}^{*} \Bigl(\sum_{|\gamma|\leqslant \mathrm{X}} \frac{\mathsf{I}_{\mathrm{X}}^{i\gamma}}{\frac{1}{2}+i\gamma}\Bigr)\overline{\chi} + \mathrm{O}\Bigl(\frac{(\log \mathrm{X})^2}{\mathrm{X}^{1/2}}\Bigr),$$

where γ ranges over ordinates of zeros of $\mathrm{L}(s,\chi)$, *counted with multiplicity, and the implied constant depends on q.*

Proof The key ingredient is the (approximate) *explicit formula* of Prime Number Theory, which can be stated in the form

$$\psi_\chi = -\sum_{\substack{\mathrm{L}(\beta+i\gamma)=0\\ |\gamma|\leqslant \mathrm{X}}} \frac{\mathsf{I}_{\mathrm{X}}^{\beta-\frac{1}{2}+i\gamma}}{\beta+i\gamma} + \mathrm{O}\Bigl(\frac{\mathsf{I}_{\mathrm{X}}^{1/2}\log(\mathrm{X})^2}{\mathrm{X}}\Bigr),$$

where the sum is over zeros of the Dirichlet L-functions with $0 \leqslant \beta \leqslant 1$, counted with multiplicity (see Theorem C.5.6). Under the assumption of the Generalized Riemann Hypothesis modulo q, we always have $\beta = \frac{1}{2}$, and this formula implies

$$\psi_\chi = -\sum_{|\gamma|\leqslant \mathrm{X}} \frac{\mathsf{I}_{\mathrm{X}}^{i\gamma}}{\frac{1}{2}+i\gamma} + \mathrm{O}\Bigl(\frac{(\log \mathrm{X})^2}{\mathrm{X}^{1/2}}\Bigr).$$

Summing over the characters (the number of which is $\varphi(q) - 1 \leqslant q$), the formula follows. □

Probabilistically, we have now a finite linear combination (of length depending on X) of the random variables $\mathsf{l}_{\mathrm{X}}^{iy}$. The link with probability theory, and to the existence of the Rubinstein–Sarnak distribution, is then performed by the following theorem (quite similar to Proposition 3.2.5).

Proposition 5.3.3 *Let $k \geqslant 1$ be an integer. Let* F *be a finite set of real numbers, and let $(\alpha(t))_{t\in\mathrm{F}}$ be a family of elements in $\mathbf{C}^k$. The random vectors*

$$\sum_{t\in\mathrm{F}} \mathsf{l}_{\mathrm{X}}^{it}\,\alpha(t)$$

on Ω_{X} converge in law as $\mathrm{X} \to +\infty$.

Proof After a simple translation, this is a direct consequence of the Kronecker Equidistribution Theorem B.6.5. Indeed, consider the vector

$$z = \Big(\frac{t}{2\pi}\Big)_{t\in\mathrm{F}} \in \mathbf{R}^{\mathrm{F}}.$$

By Kronecker's Theorem, the probability measures μ_{Y} on $(\mathbf{R}/\mathbf{Z})^{\mathrm{F}}$ defined for $\mathrm{Y} > 0$ by

$$\mu_{\mathrm{Y}}(\mathrm{A}) = \frac{1}{\mathrm{Y}}|\{y \in [0,\mathrm{Y}] \mid yz \in \mathrm{A}\}|,$$

for any measurable set A, converge in law to the probability Haar measure μ on the subgroup T of $(\mathbf{R}/\mathbf{Z})^{\mathrm{F}}$ generated by the classes modulo $\mathbf{Z}^{\mathrm{F}}$ of the elements yz, where y ranges over $\mathbf{R}$.

We extend the isomorphism $\theta \mapsto e(\theta)$ from $\mathbf{R}/\mathbf{Z}$ to $\mathbf{S}^1$ componentwise to define an isomorphism of $(\mathbf{R}/\mathbf{Z})^{\mathrm{F}}$ to $(\mathbf{S}^1)^{\mathrm{F}}$. For any continuous function f on $(\mathbf{S}^1)^{\mathrm{F}}$, we observe that

$$\begin{aligned}\int_{(\mathbf{R}/\mathbf{Z})^{\mathrm{F}}} f(e(v))d\mu_{\mathrm{Y}}(v) &= \frac{1}{\mathrm{Y}}\int_0^{\mathrm{Y}} f(e(yz))dy \\ &= \frac{1}{\mathrm{Y}}\int_0^{\mathrm{Y}} f\big((e^{ity})_{t\in\mathrm{F}}\big)dy \\ &= \frac{1}{\mathrm{Y}}\int_1^{e^{\mathrm{Y}}} f((x^{it})_{t\in\mathrm{F}})\frac{dx}{x} = \mathbf{E}_{\mathrm{X}}\big(f((\mathsf{l}_{\mathrm{X}}^{it})_{t\in\mathrm{F}})\big)\end{aligned}$$

for $\mathrm{X} = e^{\mathrm{Y}}$, after the change of variable $x = e^y$. Hence the vector $(\mathsf{l}_{\mathrm{X}}^{it})_{t\in\mathrm{F}}$ converges in law as $\mathrm{X} \to +\infty$ to the image of μ by $v \mapsto e(v)$. Now we finish the proof of the proposition by composition with the continuous map from $(\mathbf{S}^1)^{\mathrm{F}}$ to $\mathbf{C}^k$ defined by

$$(z_t)_{t\in\mathrm{F}} \mapsto \sum_{t\in\mathrm{F}} z_t\alpha(t),$$

using Proposition B.3.2. □

From the proof, we see that we can make the result more precise:

Corollary 5.3.4 *With notation and assumptions as in Proposition 5.3.3, the random vectors*

$$\sum_{t\in\mathrm{F}} \mathsf{I}_{\mathrm{X}}^{it}\,\alpha(t)$$

on Ω_{X} converge in law as $\mathrm{X}\to+\infty$ to

$$\sum_{t\in\mathrm{F}} \mathrm{I}_t\,\alpha(t),$$

where $(\mathrm{I}_t)_{t\in\mathrm{F}}$ is a random variable with values in $(\mathbf{S}^1)^{\mathrm{F}}$ with law given by the probability Haar measure of the closure of the subgroup of $(\mathbf{S}^1)^{\mathrm{F}}$ generated by all elements $(x^{it})_{t\in\mathrm{F}}$ for $x\in\mathbf{R}$.

Remark 5.3.5 This proposition explains why the logarithmic weight in (5.1) is absolutely natural. It also hints that it is necessary. Indeed, the statement of the proposition becomes false if the probability measure $\mathbf{P}_{\mathrm{X}}$ on Ω_{X} is replaced by the uniform measure. This is already visible in the simplest case where $\mathrm{F}=\{t\}$ contains a single nonzero real number t; for instance, taking the test function f to be the identity, observe that with this other probability measure, the expectation of $x\mapsto x^{it}$ is

$$\frac{1}{\mathrm{X}-1}\int_1^{\mathrm{X}} x^{it}\,dx = \frac{1}{it+1}\,\frac{\mathrm{X}^{it+1}-1}{\mathrm{X}-1} \sim \frac{\mathrm{X}^{it}}{it+1},$$

which has no limit as $\mathrm{X}\to+\infty$.

Let $\mathrm{T}\geqslant 2$ be a parameter. It follows from Lemma 5.3.2 and Proposition 5.3.3 that for $\mathrm{X}\geqslant\mathrm{T}$, we have

$$\mathsf{M}_{\mathrm{X},q} = \mathsf{N}_{\mathrm{X},\mathrm{T},q} + \sideset{}{^*}\sum_{\chi \,(\mathrm{mod}\, q)} \Bigl(\sum_{\mathrm{T}\leqslant|\gamma|\leqslant\mathrm{X}} \frac{\mathsf{I}_{\mathrm{X}}^{i\gamma}}{\frac{1}{2}+i\gamma}\Bigr)\overline{\chi} + \mathrm{O}\Bigl(\frac{(\log\mathrm{X})^2}{\sqrt{\mathrm{X}}}\Bigr), \tag{5.9}$$

where

$$\mathsf{N}_{\mathrm{X},\mathrm{T},q} = -\sideset{}{^*}\sum_{\chi \,(\mathrm{mod}\, q)} \Bigl(\sum_{|\gamma|\leqslant\mathrm{T}} \frac{\mathsf{I}_{\mathrm{X}}^{i\gamma}}{\frac{1}{2}+i\gamma}\Bigr)\,\overline{\chi}$$

are random functions that converge in law as $X \to +\infty$ for any fixed $T \geqslant 2$. The next lemma will allow us to check that the remainder term in this approximation is small.

Lemma 5.3.6 *Let $k \geqslant 1$ be an integer. Let F be a countable set of real numbers, and let $(\alpha(t))_{t\in F}$ be a family of elements in $\mathbf{C}^k$. Assume that the following conditions hold for all $T \geqslant 2$ and all $t_0 \in \mathbf{R}$:*

$$\sum_{t\in F} \|\alpha(t)\|^2 |t|^{1/2} \log(1+|t|) < +\infty, \tag{5.10}$$

$$\sum_{\substack{t\in F\\ |t|\geqslant T}} \frac{\|\alpha(t)\|}{|t|^{1/4}} \ll \frac{(\log T)^2}{T^{1/4}}, \tag{5.11}$$

$$|\{t \in F \mid |t-t_0| \leqslant 1\}| \ll \log(1+|t_0|). \tag{5.12}$$

Then we have

$$\lim_{\substack{T\leqslant X\\ T\to+\infty}} \Big\| \sum_{\substack{t\in F\\ |t|\geqslant T}} \mathsf{l}_X^{it}\,\alpha(t) \Big\|_{L^2} = 0,$$

where the limit is over pairs (T, X) with $T \leqslant X$ and T tends to infinity.

In this statement, we use the Hilbert space $L^2(\Omega_X; \mathbf{R}^k)$ of $\mathbf{R}^k$-valued L^2-functions on Ω_X, with norm defined by

$$\|f\|_{L^2}^2 = \mathbf{E}_X(\|f\|^2)$$

for $f \in L^2(\Omega_X; \mathbf{R}^k)$.

Proof Note first that an explicit computation of the integral gives

$$\mathbf{E}_X(\mathsf{l}_X^{i(t_1-t_2)}) = \frac{1}{\log X}\frac{X^{i(t_1-t_2)}-1}{t_1-t_2}$$

for $t_1 \neq t_2$, hence the general bound

$$|\mathbf{E}_X(\mathsf{l}_X^{i(t_1-t_2)})| \leqslant \min\left(1, \frac{1}{\log X}\frac{2}{|t_1-t_2|}\right). \tag{5.13}$$

We will use this bound slightly wastefully (using the first estimate even when it is not the best of the two) to gain some flexibility.

All sums below involving t, t_1, t_2 are restricted to $t \in \mathrm{F}$. Assume $2^5 \leqslant \mathrm{T} \leqslant \mathrm{X}$. We have

$$\Bigl\| \sum_{\substack{t\in \mathrm{F}\\ |t|\geqslant \mathrm{T}}} \mathsf{l}_{\mathrm{X}}^{it}\alpha(t) \Bigr\|_{\mathrm{L}^2} = \mathbf{E}_{\mathrm{X}}\Bigl(\Bigl\| \sum_{\mathrm{T}\leqslant |t|\leqslant \mathrm{X}} \mathsf{l}_{\mathrm{X}}^{it}\alpha(t) \Bigr\|^2 \Bigr)$$

$$= \sum_{\mathrm{T}\leqslant |t_1|,|t_2|\leqslant \mathrm{X}} \alpha(t_1)\cdot\alpha(t_2)\,\mathbf{E}_{\mathrm{X}}\Bigl(\mathsf{l}_{\mathrm{X}}^{i(t_1-t_2)}\Bigr).$$

We write this double sum as $\mathrm{S}_1 + \mathrm{S}_2$, where S_1 is the contribution of the terms where $|t_1 - t_2| \leqslant |t_1 t_2|^{1/4}$, and S_2 is the remainder.

In the sum S_1, we first claim that if $\mathrm{T} \geqslant \sqrt{2}$, then the condition $|t_1 - t_2| \leqslant |t_1 t_2|^{1/4}$ implies $|t_2| \leqslant 2|t_1|$. Indeed, suppose that $|t_2| > 2|t_1|$. We have

$$|t_2| \leqslant |t_1 - t_2| + |t_1| \leqslant |t_1 t_2|^{1/4} + \tfrac{1}{2}|t_2|,$$

hence $|t_2| \leqslant 2|t_1 t_2|^{1/4}$, which implies $|t_2| \leqslant 2^{4/3}|t_1|^{1/3}$, and further

$$2|t_1| < |t_2| \leqslant 2^{4/3}|t_1|^{1/3},$$

which implies that $\mathrm{T} \leqslant |t_1| < \sqrt{2}$, reaching a contradiction.

Exchanging the roles of t_1 and t_2, we see also that $|t_1| \leqslant 2|t_2|$. In particular, it now follows that we also have

$$|t_2 - t_1| \leqslant |t_1 t_2|^{1/4} \leqslant 2|t_1|^{1/2} \quad \text{and} \quad |t_2 - t_1| \leqslant |t_1 t_2|^{1/4} \leqslant 2|t_2|^{1/2}.$$

Still for $\mathrm{T} \geqslant \sqrt{2}$, we get

$$\begin{aligned}
|\mathrm{S}_1| &\leqslant \sum_{\substack{\mathrm{T}\leqslant |t_1|,|t_2|\leqslant \mathrm{X}\\ |t_2-t_1|\leqslant |t_1t_2|^{1/4}}} |\alpha(t_1)\cdot\alpha(t_2)| \\
&\leqslant \frac{1}{2} \sum_{\substack{\mathrm{T}\leqslant |t_1|,|t_2|\leqslant \mathrm{X}\\ |t_2-t_1|\leqslant |t_1t_2|^{1/4}}} \bigl(\|\alpha(t_1)\|^2 + \|\alpha(t_2)\|^2\bigr) \\
&\leqslant \sum_{\mathrm{T}\leqslant |t_1|\leqslant \mathrm{X}} \|\alpha(t_1)\|^2 \sum_{\substack{\mathrm{T}\leqslant |t_2|\leqslant \mathrm{X}\\ |t_2-t_1|\leqslant 2|t_1|^{1/2}}} 1 + \sum_{\mathrm{T}\leqslant |t_2|\leqslant \mathrm{X}} \|\alpha(t_2)\|^2 \sum_{\substack{\mathrm{T}\leqslant |t_1|\leqslant \mathrm{X}\\ |t_2-t_1|\leqslant 2|t_2|^{1/2}}} 1 \\
&\ll \sum_{\mathrm{T}\leqslant |t|\leqslant \mathrm{X}} \|\alpha(t)\|^2 |t|^{1/2} \log(1+|t|)
\end{aligned}$$

by (5.12). This quantity tends to 0 as $\mathrm{T} \to +\infty$ since the series over all t converges by assumption (5.10).

For the sum S_2, we have

$$|S_2| \leqslant \frac{2}{\log X} \sum_{\substack{T \leqslant |t_1|, |t_2| \leqslant X \\ |t_2 - t_1| > |t_1 t_2|^{1/4}}} \frac{\|\alpha(t_1)\| \, \|\alpha(t_2)\|}{|t_1 - t_2|}$$

$$\leqslant \frac{2}{\log X} \sum_{\substack{T \leqslant |t_1|, |t_2| \leqslant X \\ |t_2 - t_1| > |t_1 t_2|^{1/4}}} \frac{\|\alpha(t_1)\| \, \|\alpha(t_2)\|}{|t_1 t_2|^{1/4}},$$

and therefore

$$|S_2| \leqslant \frac{2}{\log X} \sum_{T \leqslant |t_1| \leqslant X} \frac{\|\alpha(t_1)\|}{|t_1|^{1/4}} \sum_{T \leqslant |t_2| \leqslant X} \frac{\|\alpha(t_2)\|}{|t_2|^{1/4}} \ll \frac{1}{\log X} \frac{(\log T)^4}{T^{1/2}},$$

by (5.11). The lemma now follows. □

Remark 5.3.7 Although we have stated this lemma in some generality, it is far from the best that can be achieved along such lines.

The assumptions might look complicated, but note that (5.12) means that the density of F is roughly logarithmic; then (5.10) and (5.11) are certainly satisfied if the series with terms $\|\alpha(t)\|$ is convergent, and more generally when $\|\alpha(t)\|$ is comparable with $(1 + |t|)^{-\alpha}$ with $\alpha > 3/4$.

We will now finish the proof of Theorem 5.2.2. We apply Lemma 5.3.6 to the set F of ordinates γ of zeros of some $L(s, \chi)$, for χ a nontrivial character modulo q, and to

$$\alpha(\gamma) = \sum_{\substack{\chi \,(\mathrm{mod}\, q) \\ L(\frac{1}{2}+i\gamma)=0}}^{*} \frac{1}{\frac{1}{2} + i\gamma} \overline{\chi}$$

for $\gamma \in F$, viewing $\alpha(\gamma)$ as a vector in $\mathbf{C}^{(\mathbf{Z}/q\mathbf{Z})^\times}$, and taking into account the multiplicity of the zero $\frac{1}{2} + i\gamma$ for any character χ such that $L(\frac{1}{2} + i\gamma, \chi) = 0$. We need to check the three assumptions of the lemma.

From the asymptotic von Mangoldt formula (C.10), we first know that (5.12) holds for the zeros of a fixed L-function modulo q, with an implied constant depending on q, and hence it holds also for F.

We next have

$$\|\alpha(\gamma)\| \leqslant \sum_{\substack{\chi \,(\mathrm{mod}\, q) \\ L(\frac{1}{2}+i\gamma)=0}}^{*} \frac{1}{|\frac{1}{2} + i\gamma|} \|\overline{\chi}\| = \varphi(q)^{1/2} \sum_{\substack{\chi \,(\mathrm{mod}\, q) \\ L(\frac{1}{2}+i\gamma)=0}}^{*} \frac{1}{|\frac{1}{2} + i\gamma|} \leqslant \frac{\varphi(q)^{3/2}}{|\frac{1}{2} + i\gamma|} \tag{5.14}$$

by a trivial estimate of the number of characters of which $\frac{1}{2} + i\gamma$ can be a zero.

Condition (5.10) follows from (5.14), since we even have

$$\sum_{L(\frac{1}{2}+i\gamma,\chi)=0} \frac{1}{|\frac{1}{2}+i\gamma|^{1+\varepsilon}} < +\infty \tag{5.15}$$

for any fixed $\varepsilon > 0$ and any $\chi \pmod q$, and condition (5.11) is again an easy consequence of (5.15) and (5.14).

From (5.9), we conclude that for $X \geqslant T \geqslant 2$, we have

$$\mathsf{M}_X = \mathsf{N}_{X,T} + \mathsf{E}'_{X,T},$$

where

$$\mathsf{E}'_{X,T} = \sum_{\chi \pmod q}{}^{*} \Bigl(\sum_{T \leqslant |\gamma| \leqslant X} \frac{\mathsf{l}_X^{i\gamma}}{\frac{1}{2}+i\gamma} \Bigr) \overline{\chi} + O\Bigl(\frac{(\log X)^2}{\sqrt{X}} \Bigr).$$

These random functions converge to 0 in L^2, hence in L^1, by Lemma 5.3.6 as applied before. By Proposition B.4.4 (and Remark B.4.6), we conclude that the random functions M_X converge in law, and that their limit is the same as the limit as $T \to +\infty$ of the law of the limit of

$$-\sum_{\chi \pmod q}{}^{*} \Bigl(\sum_{|\gamma| \leqslant T} \frac{\mathsf{l}_X^{i\gamma}}{\frac{1}{2}+i\gamma} \Bigr) \overline{\chi}.$$

In the next section, we compute these limits, and hence the law of N_q, assuming that the zeros of the Dirichlet L-functions are "as independent as possible," so that Proposition 5.3.3 becomes explicit in the special case of interest.

To finish the proof of Theorem 5.2.2, we need to check the last assertion, namely, that the support of N_q is contained in the hyperplane (5.3). But note that

$$\begin{aligned}\sum_{a \in (\mathbf{Z}/q\mathbf{Z})^\times} \mathsf{N}_X(x)(a) &= \frac{\log x}{\sqrt{x}} \sum_{a \in (\mathbf{Z}/q\mathbf{Z})^\times} (\varphi(q)\pi(x;q,a) - \pi(x)) \\ &= \frac{\log x}{\sqrt{x}} \sum_{\substack{p \leqslant x \\ p \pmod q \notin (\mathbf{Z}/q\mathbf{Z})^\times}} 1 \ll \frac{\log x}{\sqrt{x}}\end{aligned}$$

for all $x \in \Omega_X$, since at most finitely many primes are not congruent to some $a \in (\mathbf{Z}/q\mathbf{Z})^\times$. Hence the random variables

$$\sum_{a \in (\mathbf{Z}/q\mathbf{Z})^\times} \mathsf{N}_X(a)$$

converge in probability to 0 as $X \to +\infty$, and by Corollary B.3.4, it follows that the support of N_q is contained in the zero set of the linear form

$$f \mapsto \sum_{a \in (\mathbf{Z}/q\mathbf{Z})^\times} f(a),$$

that is, in H_q.

5.4 The Generalized Simplicity Hypothesis

The proof of Theorem 5.2.2 now allows us to understand what is needed for the next step, which we take to be the explicit determination of the random variable N_q. Indeed, the proof tells us that N_q is the limit, as $T \to +\infty$, of the random variables that are themselves the limits in law as $X \to +\infty$ of the random function given by the finite sum

$$m_q - \sum_{\chi \pmod q}^{*} \sum_{\substack{L(\frac{1}{2}+i\gamma,\chi)=0 \\ |\gamma| \leqslant T}} \frac{\mathsf{l}_X^{i\gamma}}{\frac{1}{2}+i\gamma} \overline{\chi(a)},$$

which converge by Proposition 5.3.3. The *proof* of that proposition shows how this limit $N_{q,T}$ can be computed in principle. Precisely, let X_T be the set of pairs (χ, γ), where χ runs over nontrivial Dirichlet characters modulo q and γ runs over the ordinates of the nontrivial zeros of $L(s, \chi)$ with $|\gamma| \leqslant T$. Then, by Corollary 5.3.4, we have

$$N_{q,T} = m_q - \sum_{\chi \pmod q}^{*} \sum_{\substack{L(\frac{1}{2}+i\gamma,\chi)=0 \\ |\gamma| \leqslant T}} \frac{I_{\chi,\gamma}}{\frac{1}{2}+i\gamma} \overline{\chi(a)}, \tag{5.16}$$

where $(I_{\chi,\gamma})$ is distributed on $(\mathbf{S}^1)^{X_T}$ according to the probability Haar measure of the closure S_T of the subgroup generated by the elements $(x^{i\gamma})_{(\chi,\gamma) \in X_T}$ for $x \in \mathbf{R}$.

Thus, *to compute* N_q *explicitly*, we "simply" need to know what the subgroup $S_{q,T}$ is. If (hypothetically) this subgroup was *equal* to $(\mathbf{S}^1)^{X_{q,T}}$, then the $(I_{\chi,\gamma})$ would simply be independent and uniformly distributed on $\mathbf{S}^1$, and we would immediately obtain a formula for N_q from (5.16) as a sum of a series of independent terms.

This hypothesis is however too optimistic. Indeed, there is an "obvious" type of dependency among the ordinates γ, which amount to restrictions on the subgroup S_T in $(\mathbf{S}^1)^{X_T}$. Beyond these relations, there are none that are immediately apparent. The *Generalized Simplicity Hypothesis* modulo q is then the statement that, in fact, these obvious relations should exhaust all possible constraints satisfied by S_T.[3]

These systematic relations between the elements of X_T are simply the following: a complex number $\frac{1}{2} + i\gamma$ is a zero of $L(s, \chi)$ if and only if the conjugate $\frac{1}{2} - i\gamma$ is a zero of $L(s, \overline{\chi})$, simply because $\overline{L(\bar{s}, \chi)} = L(s, \overline{\chi})$ as holomorphic functions; hence (χ, γ) belongs to X_T if and only if $(\overline{\chi}, -\gamma)$ does.

We are therefore led to the so-called Generalized Simplicity Hypothesis modulo q.

Definition 5.4.1 Let $q \geqslant 1$ be an integer. The *Generalized Simplicity Hypothesis* holds modulo q if the family of *nonnegative* ordinates γ of the nontrivial zeros of all nontrivial Dirichlet L-functions modulo q, with multiplicity taken into account, is linearly independent over $\mathbf{Q}$.

We emphasize that we are looking at the family of the ordinates, not just the set of values. In particular, the Generalized Simplicity Hypothesis modulo q implies that

- for a given $\gamma \geqslant 0$, there is at most one primitive Dirichlet character χ modulo q such that $L(\frac{1}{2} + i\gamma, \chi) = 0$;
- all nontrivial zeros are of multiplicity 1;
- we have $L(\frac{1}{2}, \chi) \neq 0$ for any nontrivial character χ.

All these statements are highly nontrivial conjectures!

Lemma 5.4.2 *Under the assumption of the Generalized Simplicity Hypothesis modulo q, the subgroup S_T is given by*

$$S_T = \{(z_{\chi,\gamma}) \in (\mathbf{S}^1)^{X_T} \mid z_{\overline{\chi},-\gamma} = \overline{z_{\chi,\gamma}} \text{ for all } (\chi, \gamma) \in X_T\}, \tag{5.17}$$

for all $T \geqslant 2$. In particular, denoting by X_T^+ the set of pairs (χ, γ) in X_T with $\gamma \geqslant 0$, the projection

$$(z_{\chi,\gamma}) \mapsto (z_{\chi,\gamma})_{(\chi,\gamma) \in X_T^+} \tag{5.18}$$

from S_T to $(\mathbf{S}^1)^{X_T^+}$ is surjective.

Proof Indeed, S_T is contained in the subgroup $\widetilde{S}_T$ in the right-hand side of (5.17), because each vector $(x^{i\gamma})_{(\chi,\gamma) \in X_T}$ has this property for $x \in \mathbf{R}$, by the relation between zeros of the L-functions of χ and $\overline{\chi}$.

[3] In other words, it is an application of Occam's Razor.

To show that S_T is not a proper subgroup of $\widetilde{S}_T$, it is enough to prove the last assertion, since an element of $\widetilde{S}_T$ is uniquely determined by the value of the projection (5.18). But if that projection is not surjective, then there exists a nonzero family of integers $(m_{\chi,\gamma})_{(\chi,\gamma)\in X_T^+}$ such that

$$\prod_{(\chi,\gamma)\in X_T^+} x^{im_{\chi,\gamma}\gamma} = 1$$

for all $x \in \mathbf{R}$, and this implies

$$\sum_{\chi \pmod q}^{*} \sum_{\gamma\geqslant 0} m_{\chi,\gamma}\gamma = 0,$$

which contradicts the Generalized Simplicity Hypothesis modulo q. □

Remark 5.4.3 If we were also considering problems involving the comparison of the number of primes in arithmetic progressions with different moduli, say, modulo q_1 and q_2, then there would be another systematic source of relations between the zeros of the L-functions modulo q_1 and q_2. Precisely, if d is a common divisor of q_1 and q_2, and χ_0 a Dirichlet character modulo d, corresponding to a character χ_0 of $(\mathbf{Z}/d\mathbf{Z})^\times$, then there is a Dirichlet character χ_i modulo q_i, for $i = 1, 2$, corresponding to the composition

$$(\mathbf{Z}/q_i\mathbf{Z})^\times \to (\mathbf{Z}/d\mathbf{Z})^\times \xrightarrow{\chi_0} \mathbf{C}^\times,$$

and we have

$$L(s,\chi_i) = \prod_{p\mid q_i/d} (1 - \chi_0(p)p^{-s})L(s,\chi_0),$$

which shows that the ordinates of the nontrivial zeros of $L(s,\chi_1)$ and $L(s,\chi_2)$ are the same.

Because of this, the correct formulation of the Generalized Simplicity Hypothesis, without reference to a single modulus q, is that *the nonnegative ordinates of zeros of the* L*-functions of all primitive Dirichlet characters are* **Q***-linearly independent*; this is the statement as formulated in [105].

We can now state precisely the computation of the law of the random function N_q under the assumption of the Generalized Simplicity Hypothesis modulo q.

To do this, let X^+ be the set of all pairs (χ,γ) where χ is a nontrivial Dirichlet character modulo q and $\gamma \geqslant 0$ is a *nonnegative* ordinate of a nontrivial zero of $L(s,\chi)$, that is, we have $L(\frac{1}{2}+i\gamma,\chi) = 0$. Let $(I_{\chi,\gamma})_{(\chi,\gamma)\in X^+}$ be a family of independent random variables all uniformly distributed over the circle $\mathbf{S}^1$.

Define further

$$\mathrm{I}_{\overline{\chi},-\gamma} = \overline{\mathrm{I}}_{\chi,\gamma} \tag{5.19}$$

for all ordinates $\gamma \geqslant 0$ of a zero of $\mathrm{L}(s,\chi)$. We have then defined random variables $\mathrm{I}_{\chi,\gamma}$ for all ordinates of a zero of $\mathrm{L}(s,\chi)$.

Theorem 5.4.4 (Rubinstein–Sarnak) *Let $q \geqslant 1$. In addition to the Generalized Riemann Hypothesis, assume the Generalized Simplicity Hypothesis modulo q. Then the law of N_q is the law of the series*

$$m_q - \sum_{\chi \,(\mathrm{mod}\, q)}^{*} \Big(\sum_{\substack{\gamma \\ \mathrm{L}(\frac{1}{2}+i\gamma,\chi)=0}} \frac{\mathrm{I}_{\chi,\gamma}}{\frac{1}{2}+i\gamma} \Big) \overline{\chi}, \tag{5.20}$$

where the series converges almost surely and in L^2 as the limit of partial sums

$$\lim_{\mathrm{T}\to+\infty} \sum_{\chi \,(\mathrm{mod}\, q)}^{*} \sum_{\substack{\mathrm{L}(\frac{1}{2}+i\gamma,\chi)=0 \\ |\gamma|\leqslant \mathrm{T}}} \frac{\mathrm{I}_{\chi,\gamma}}{\frac{1}{2}+i\gamma}. \tag{5.21}$$

In these formulas, for each Dirichlet character χ modulo q, the sum runs over the ordinates of zeros of $\mathrm{L}(s,\chi)$.

Remark 5.4.5 (1) Since the Generalized Simplicity Hypothesis modulo q implies that each zero has multiplicity one (even as we vary χ modulo q), there is no need to worry about this issue when defining the series over the zeros.

(2) This result shows that the random function N_q is probabilistically quite subtle. It is somewhat analogue to Bagchi's measure, or to one of its Bohr–Jessen specializations (see Theorem 3.2.1), with a sum (or a product) of rather simple individual independent random variables, but it retains important arithmetic features because the sum and the coefficients involve the zeros of Dirichlet L-functions (instead of the primes that occur in Bagchi's random Euler product).

One important contrasting feature, in comparison with either Theorem 3.2.1 (or Selberg's Theorem) is that the series defining N_q is not far from being absolutely convergent, which is not the case at all of the series

$$\sum_p \frac{\mathrm{X}_p}{p^s}$$

that occurs in Bagchi's Theorem when $\frac{1}{2} < \mathrm{Re}(s) < 1$.

Before giving the proof, we can draw some simple conclusions from Theorem 5.4.4, in the direction of confirming the existence of a bias for certain residue classes.

Under the assumptions of Theorem 5.4.4, we have $\mathbf{E}(\mathrm{N}_q) = m_q$, since the convergence also holds in L^2, and $\mathbf{E}(\mathrm{I}_{\chi,\gamma}) = 0$ for all (χ, γ). Using either (5.4) or (5.5), we know that

$$\frac{1}{\varphi(q)} \sum_{a \in (\mathbf{Z}/q\mathbf{Z})^\times} m_q(a) = 0 \quad \text{and} \quad \frac{1}{\varphi(q)} \sum_{a \in (\mathbf{Z}/q\mathbf{Z})^\times} m_q(a)^2 = \sigma_q = \sum_{\chi^2=1}^{*} 1.$$

It is natural to say that "not all residue classes modulo q are equal," as far as representing primes is concerned, if the average function m_q of N_q is not constant (assuming that Theorem 5.4.4 is applicable). This is equivalent (by (5.5)) to the existence of at least one $b \neq 1$ such that $b^2 = 1$, and therefore holds whenever $q \neq 2$, since one can always take $b = -1$.

This statement can be considered to be the simplest general confirmation of the Chebychev bias; note that $q = 2$ is of course an exception, since all primes (with one exception) are odd.

Remark 5.4.6 (1) The mean-square σ_q of m_q is also the size of the quotient group

$$(\mathbf{Z}/q\mathbf{Z})^\times / ((\mathbf{Z}/q\mathbf{Z})^\times)^2$$

of invertible residues modulo quadratic residues, minus 1. Using the Chinese Remainder Theorem, this expression can be computed in terms of the factorization of q, namely, if we write

$$q = \prod_p p^{n_p},$$

then we obtain

$$\sigma_q = 2^{\min(n_2 - 1, 2)} \prod_{\substack{p \mid q \\ p \geqslant 3}} 2 - 1$$

(because for p odd, the group of squares is of index 2 in $(\mathbf{Z}/p^{n_p}\mathbf{Z})^\times$ if $n_p \geqslant 1$, whereas for $p = 2$, it is trivial if $n_p = 1$ or $n_p = 2$, and of index 4 if $n_2 \geqslant 3$).

(2) Consider once more the case $q = 4$. Then $m_4(1) = -1$ and $m_4(3) = 1$, and in particular we certainly expect to have, in general, more primes congruent to 3 modulo 4 than there are congruent to 1 modulo 4.

In fact, using Theorem 5.4.4 and numerical tables of zeros of the Dirichlet L-functions modulo 4 up to some bound T, one can get approximations to the distribution of N_4 (e.g., through the characteristic function of N_4, and approximate Fourier inversion). Rubinstein and Sarnak [105, §4] established in this manner that

$$\mathbf{P}(\mathrm{N}_4 \in \mathrm{H}_4 \cap \mathrm{C}) = 0.9959\ldots$$

(under the assumptions of Theorem 5.4.4 modulo 4). This confirms a very strong bias for primes to be $\equiv 3$ modulo 4, but also shows that one has sometimes $\pi(x;4,1) > \pi(x;4,3)$ (in fact, in the sense of the probability measure $\mathbf{P}_{\mathrm{X}}$, this happens with probability about $1/250$, and we have already mentioned that the first occurrence of this reverse inequality is for $\mathrm{X} = 26861$).

We now give the proof of Theorem 5.4.4. We first check that the series (5.20) converges almost surely and in L^2 in the sense of the limit (5.21).[4]

It suffices to prove that each value $\mathrm{N}_q(a)$ of the random function N_q converges almost surely and in L^2. To check this, we first observe that for any $\mathrm{T} \geqslant 2$, we have

$$\begin{aligned}\sum_{\chi \,(\mathrm{mod}\, q)}^{*} \sum_{\substack{\mathrm{L}(\frac{1}{2}+i\gamma,\chi)=0 \\ |\gamma|\leqslant \mathrm{T}}} \frac{\mathrm{I}_{\chi,\gamma}}{\frac{1}{2}+i\gamma}\overline{\chi(a)} &= \sum_{\chi \,(\mathrm{mod}\, q)}^{*} \sum_{\substack{\mathrm{L}(\frac{1}{2}+i\gamma,\chi)=0 \\ 0<\gamma\leqslant \mathrm{T}}} \left(\frac{\mathrm{I}_{\chi,\gamma}}{\frac{1}{2}+i\gamma}\overline{\chi(a)} + \frac{\mathrm{I}_{\overline{\chi},-\gamma}}{\frac{1}{2}-i\gamma}\chi(a)\right) \\ &= 2\sum_{\chi \,(\mathrm{mod}\, q)}^{*} \sum_{\substack{\mathrm{L}(\frac{1}{2}+i\gamma,\chi)=0 \\ 0<\gamma\leqslant \mathrm{T}}} \mathrm{Re}\left(\frac{\mathrm{I}_{\chi,\gamma}}{\frac{1}{2}+i\gamma}\overline{\chi(a)}\right) \qquad (5.22)\end{aligned}$$

according to the definition (5.19) of $\mathrm{I}_{\chi,\gamma}$ for negative γ (we use here the fact that, under the Generalized Simplicity Hypothesis, no zero has ordinate $\gamma = 0$).

The right-hand side of (5.22) is the partial sum of a series of independent random variables, and we can apply Kolmogorov's Theorem, B.10.1. Indeed, we have

$$\mathbf{E}\left(\mathrm{Re}\left(\frac{\mathrm{I}_{\chi,\gamma}}{\frac{1}{2}+i\gamma}\overline{\chi(a)}\right)\right) = 0$$

[4] This convergence could be proved without any condition, not even the Generalized Riemann Hypothesis, but the series has no arithmetic meaning without such assumptions.

for any pair (χ, γ), and

$$\sum_{\chi \,(\mathrm{mod}\, q)}^{*} \sum_{\gamma>0} \mathbf{V}\left(\mathrm{Re}\left(\frac{\mathrm{I}_{\chi,\gamma}}{\frac{1}{2}+i\gamma}\overline{\chi(a)}\right)\right) \leqslant \sum_{\chi \,(\mathrm{mod}\, q)}^{*} \sum_{\gamma>0} \mathbf{E}\left(\left|\frac{\mathrm{I}_{\chi,\gamma}}{\frac{1}{2}+i\gamma}\right|^{2}\right)$$
$$= \sum_{\chi \,(\mathrm{mod}\, q)}^{*} \sum_{\gamma>0} \frac{1}{\frac{1}{4}+\gamma^{2}} < +\infty$$

by Proposition C.5.3 (2), so that the series converges almost surely and in L^2, by Kolmogorov's Theorem, as claimed.

Now we need only go through the steps described above when motivating Definition 5.4.1. The random function N_q is the limit as $\mathrm{T} \to +\infty$ of

$$\mathrm{N}_{q,\mathrm{T}} = m_q - \lim_{\mathrm{X}\to+\infty}\left(\sum_{\chi \,(\mathrm{mod}\, q)}^{*} \sum_{\substack{\mathrm{L}(\frac{1}{2}+i\gamma,\chi)=0 \\ |\gamma|\leqslant \mathrm{T}}} \frac{\mathrm{X}^{i\gamma}}{\frac{1}{2}+i\gamma}\overline{\chi(a)}\right).$$

We write once more

$$m_q - \sum_{\chi \,(\mathrm{mod}\, q)}^{*} \sum_{\substack{\mathrm{L}(\frac{1}{2}+i\gamma,\chi)=0 \\ |\gamma|\leqslant \mathrm{T}}} \frac{\mathrm{X}^{i\gamma}}{\frac{1}{2}+i\gamma}\overline{\chi(a)}$$
$$= m_q - 2\sum_{\chi \,(\mathrm{mod}\, q)}^{*} \sum_{\substack{\mathrm{L}(\frac{1}{2}+i\gamma,\chi)=0 \\ 0<\gamma\leqslant \mathrm{T}}} \mathrm{Re}\left(\frac{\mathrm{X}^{i\gamma}}{\frac{1}{2}+i\gamma}\overline{\chi(a)}\right).$$

By Proposition 5.3.3, or Corollary 5.3.4, as explained above, and the Generalized Simplicity Hypothesis modulo q (precisely through Lemma 5.4.2), the limit as $\mathrm{X} \to +\infty$ of these random functions is simply

$$m_q - 2\sum_{\chi \,(\mathrm{mod}\, q)}^{*} \sum_{\substack{\mathrm{L}(\frac{1}{2}+i\gamma,\chi)=0 \\ 0<\gamma\leqslant \mathrm{T}}} \mathrm{Re}\left(\frac{\mathrm{I}_{\chi,\gamma}}{\frac{1}{2}+i\gamma}\overline{\chi(a)}\right),$$

which in turn converge to the random function N_q as $\mathrm{T} \to +\infty$ by definition. This concludes the proof of Theorem 5.4.4.

Theorem 5.4.4 is equivalent to the computation of the characteristic function of N_q, viewed as a random vector, that is, of the function

$$t \mapsto \mathbf{E}(e^{it\cdot \mathrm{N}_q})$$

for $t \in \mathbf{R}^{(\mathbf{Z}/q\mathbf{Z})^\times}$, where

$$t \cdot f = \sum_{a \in (\mathbf{Z}/q\mathbf{Z})^\times} t_a f(a)$$

for $t = (t_a) \in \mathbf{R}^{(\mathbf{Z}/q\mathbf{Z})^\times}$ and $f\colon (\mathbf{Z}/q\mathbf{Z})^\times \to \mathbf{R}$. (Indeed, this is how the result is presented in [105, §3.1].)

To state the formula for the characteristic function, define the Bessel function J_0 on $\mathbf{R}$ by

$$\mathrm{J}_0(x) = \frac{1}{2\pi} \int_0^{2\pi} e^{ix\cos(t)}\,dt.$$

It is elementary that J_0 is a real-valued and even function of x.

Corollary 5.4.7 *Let $q \geqslant 2$ be an integer. Assume the Generalized Riemann Hypothesis and the Generalized Simplicity Hypothesis modulo q. The characteristic function of the law of the Rubinstein–Sarnak distribution N_q modulo q is given by*

$$\mathbf{E}(e^{it\cdot \mathrm{N}_q}) = \exp(it \cdot m_q) \prod_{\chi \,(\mathrm{mod}\, q)}^{*} \prod_{\substack{\gamma > 0 \\ \mathrm{L}(\frac{1}{2}+i\gamma,\chi)=0}} \mathrm{J}_0\left(\frac{2\,|t \cdot \overline{\chi}|}{(\frac{1}{4}+\gamma^2)^{1/2}}\right)$$

for $t \in \mathbf{R}^{(\mathbf{Z}/q\mathbf{Z})^\times}$, where, for each Dirichlet character χ modulo q, the product runs over the positive ordinates of zeros of $\mathrm{L}(s,\chi)$.

Proof Using the previous argument, we write the series defining N_q in the form

$$\begin{aligned} m_q &- \sum_{\chi\,(\mathrm{mod}\, q)}^{*} \sum_{\gamma>0} \left(\frac{\mathrm{I}_{\chi,\gamma}}{\frac{1}{2}+i\gamma}\,\overline{\chi} + \frac{\mathrm{I}_{\overline{\chi},-\gamma}}{\frac{1}{2}-i\gamma}\,\chi \right) \\ &= m_q - 2\,\mathrm{Re}\left(\sum_{\chi\,(\mathrm{mod}\, q)}^{*} \sum_{\gamma>0} \frac{\mathrm{I}_{\chi,\gamma}}{\frac{1}{2}+i\gamma}\,\overline{\chi} \right). \end{aligned}$$

Since the characteristic function of a limit in law is the pointwise limit of the characteristic functions of the sequence involved, we obtain using the independence of the random variables $(\mathrm{I}_{\chi,\gamma})$ the convergent product formula

$$\begin{aligned}\mathbf{E}(e^{it\cdot \mathrm{N}_q}) &= e^{it\cdot m_q} \prod_{\chi \pmod q}^{*} \prod_{\gamma>0} \mathbf{E}\left(e^{-2it\cdot \mathrm{Re}\left(\frac{\mathrm{I}_{\chi,\gamma}}{1/2+i\gamma}\overline{\chi}\right)}\right)\\ &= e^{it\cdot m_q} \prod_{\chi \pmod q}^{*} \prod_{\gamma>0} \varphi\left(\frac{t\cdot\overline{\chi}}{\frac{1}{2}+i\gamma}\right),\end{aligned}$$

where, for $z \in \mathbf{C}$, we defined

$$\varphi(z) = \mathbf{E}\left(e^{-2i\,\mathrm{Re}(z\mathrm{I})}\right)$$

for a random variable I uniformly distributed over the unit circle. By invariance of the law of I under rotation (i.e., the law of $ze^{i\theta}\mathrm{I}$ is the same as that of $z\mathrm{I}$ for any $\theta \in \mathbf{R}$), applied to the angle θ such that $ze^{i\theta} = |z|$, we have

$$\varphi(z) = \mathbf{E}(e^{-2i\,\mathrm{Re}(|z|\mathrm{I})}) = \mathbf{E}(e^{-2i|z|\,\mathrm{Re}(\mathrm{I})}) = \frac{1}{2\pi}\int_0^{2\pi} e^{-2i|z|\cos(t)}dt = \mathrm{J}_0(2|z|).$$

Hence we obtain

$$\mathbf{E}(e^{it\cdot \mathrm{N}_q}) = e^{it\cdot m_q} \prod_{\chi \pmod q}^{*} \prod_{\gamma>0} \mathrm{J}_0\left(2\frac{|t\cdot\overline{\chi}|}{|\frac{1}{2}+i\gamma|}\right),$$

as claimed. □

Another consequence of Theorem 5.4.4 is an estimate for the probability that N_q takes large values.

Corollary 5.4.8 *There exists a constant $c_q > 0$ such that, for $\mathrm{A} > 0$, we have*

$$\begin{aligned}c_q^{-1}\exp(-\exp(c_q\mathrm{A}^{1/2})) &\leqslant \liminf_{\mathrm{X}\to+\infty} \mathbf{P}_{\mathrm{X}}(\|\mathsf{N}_{\mathrm{X},q}\| \geqslant \mathrm{A})\\ &\leqslant \limsup_{\mathrm{X}\to+\infty} \mathbf{P}_{\mathrm{X}}(\|\mathsf{N}_{\mathrm{X},q}\| > \mathrm{A}) \leqslant c_q\exp(-\exp(c_q^{-1}\mathrm{A}^{1/2})).\end{aligned}$$

Proof We view N_q as a random variable with values in the complex finite-dimensional Banach space of complex-valued functions on $(\mathbf{Z}/q\mathbf{Z})^\times$. We have the series representation

$$\mathrm{N}_q = m_q - 2\sum_{\chi \pmod q}^{*} \sum_{\substack{\gamma>0\\ \mathrm{L}(\frac{1}{2}+i\gamma,\chi)=0}} \mathrm{Re}\left(\frac{\mathrm{I}_{\chi,\gamma}}{\frac{1}{2}+i\gamma}\overline{\chi}\right).$$

This series converges almost surely, the terms are independent and the random variables $\mathrm{I}_{\chi,\gamma}$ are bounded by 1 in modulus. Moreover,

$$\mathbf{P}(\|\mathrm{N}_q\| > \mathrm{A}) \leqslant \mathbf{P}(\|\widetilde{\mathrm{N}}_q\| > \mathrm{A}),$$

where

$$\widetilde{\mathrm{N}}_q = m_q - 2 \sideset{}{^*}\sum_{\chi \,(\mathrm{mod}\, q)} \sum_{\substack{\gamma>0 \\ \mathrm{L}(\frac{1}{2}+i\gamma,\chi)=0}} \frac{\mathrm{I}_{\chi,\gamma}}{\frac{1}{2}+i\gamma} \overline{\chi},$$

since $\|\mathrm{N}_q\| \leqslant \|\widetilde{\mathrm{N}}_q\|$. By Corollary C.5.5, the functions

$$-\frac{2}{\frac{1}{2}+i\gamma}\overline{\chi}$$

satisfy the bounds described in Remark B.11.14 (2), namely,

$$\sideset{}{^*}\sum_{\chi \,(\mathrm{mod}\, q)} \sum_{\substack{0<\gamma<\mathrm{T} \\ \mathrm{L}(\frac{1}{2}+i\gamma,\chi)=0}} \left\| \frac{1}{\frac{1}{2}+i\gamma} \overline{\chi} \right\| \gg (\log \mathrm{T})^2$$

and

$$\sideset{}{^*}\sum_{\chi \,(\mathrm{mod}\, q)} \sum_{\substack{\gamma>\mathrm{T} \\ \mathrm{L}(\frac{1}{2}+i\gamma,\chi)=0}} \left\| \frac{1}{\frac{1}{2}+i\gamma} \overline{\chi} \right\|^2 \ll \frac{\log \mathrm{T}}{\mathrm{T}}$$

for $\mathrm{T} \geqslant 1$. Thus by Remark B.11.14 (2), and the convergence in law of $\mathsf{N}_{\mathrm{X},q}$ to N_q, we deduce the upper bound

$$\limsup_{\mathrm{X}\to+\infty} \mathbf{P}_{\mathrm{X}}(\|\mathsf{N}_{\mathrm{X},q}\| > \mathrm{A}) \leqslant \mathbf{P}(\|\mathrm{N}_q\| > \mathrm{A}) \leqslant c \exp(-\exp(c^{-1}\mathrm{A}^{1/2}))$$

for some real number $c > 0$.

In the case of the lower bound, it suffices to prove it for $\mathrm{N}_q(a)$, where a is any fixed element of $(\mathbf{Z}/q\mathbf{Z})^{\times}$. Since the series expressing $\mathrm{N}_q(a)$ is not exactly of the form required for the lower bound in Remark B.11.14 (2) (and in Proposition B.11.13), we first transform it a bit. We have

$$\mathrm{Re}\left(\frac{\mathrm{I}_{\chi,\gamma}}{\frac{1}{2}+i\gamma}\overline{\chi(a)}\right) = \frac{1}{2(\frac{1}{4}+\gamma^2)}\mathrm{Re}(\mathrm{I}_{\chi,\gamma}\overline{\chi(a)}) + \frac{\gamma}{\frac{1}{4}+\gamma^2}\mathrm{Im}(\mathrm{I}_{\chi,\gamma}\overline{\chi(a)})$$

for any pair (χ,γ), which implies that

$$\mathrm{N}_q(a) = m_q(a) + e_q(a) - 2 \sideset{}{^*}\sum_{\chi \,(\mathrm{mod}\, q)} \sum_{\substack{\gamma>0 \\ \mathrm{L}(\frac{1}{2}+i\gamma,\chi)=0}} \frac{\gamma}{\frac{1}{4}+\gamma^2}\mathrm{Im}(\mathrm{I}_{\chi,\gamma}\overline{\chi(a)}),$$

where the random variable $e_q(a)$ (arising from the sum of the first terms in the previous expression) is uniformly bounded (by Proposition C.5.3 (2)). Now we can apply the lower bound in Remark B.11.14 (2) to the last series: the random variables $\mathrm{Im}(\mathrm{I}_{\chi,\gamma}\overline{\chi(a)})$ are independent, symmetric and bounded by 1, and the assumptions on the size of the coefficients are provided by Corollary C.5.5 again. □

5.5 Further Results

In recent years, the Chebychev bias has been a popular topic in analytic number theory; besides further studies of the original setting that we have discussed, it has also been generalized in many ways. We only indicate a few examples here, without any attempt to completeness.

In the first direction, there have been many studies of the properties of the Rubinstein–Sarnak measures, and of the consequences concerning various "races" between primes (see, for instance, the papers of Granville and Martin [50] and Harper and Lamzouri [56]). In parallel, attempts have been made to weaken the assumptions used by Rubinstein and Sarnak to establish properties of their measures (recall that the *existence* of the measure does not require the Generalized Simplicity Hypothesis). Among these, we refer in particular to the work of Devin [26], who found a much weaker condition that ensures that the Rubinstein–Sarnak measure is absolutely continuous.

Among generalizations, it seems worth mentioning the discussion by Sarnak [107] of a bias related to elliptic curves, as well as the recent extensive work of Fiorilli and Jouve [38] concerning Artin L-functions. In another direction, Kowalski [68] and later Cha–Fiorilli–Jouve [23] have considered analogue questions over finite fields, where the main difference is that relations between zeros of the analogues of the Dirichlet L-functions may well exist (although they are rare), leading to new phenomena.

6

The Shape of Exponential Sums

Probability tools	Arithmetic tools
Definition of convergence in law (§ B.3)	Kloosterman sums (§ C.6)
Kolmogorov's Theorem for random series (th. B.10.1)	Riemann Hypothesis over finite fields (th. C.6.4)
Convergence of finite distributions (def. B.11.2)	Average Sato–Tate Theorem
Kolmogorov's Criterion for tightness (prop. B.11.10)	Weyl criterion (§ B.6)
Fourier coefficients criterion (prop. B.11.8)	Deligne's Equidistribution Theorem
Sub-Gaussian random variables (§ B.8)	
Talagrand's inequality (th. B.11.12)	
Support of a random series (prop. B.10.8)	

6.1 Introduction

We consider in this chapter a rather different type of arithmetic objects: exponential sums and their partial sums. Although the ideas that we will present apply to very general situations, we consider as usual only an important special case: the partial sums of *Kloosterman sums* modulo primes. In Section C.6, we give some motivation for the type of sums (and questions) discussed in this chapter.

Thus let p be a prime number. For any pair (a,b) of invertible elements in the finite field $\mathbf{F}_p = \mathbf{Z}/p\mathbf{Z}$, the (normalized) Kloosterman sum $\mathrm{Kl}(a,b;p)$ is defined by the formula

$$\mathrm{Kl}(a,b;p) = \frac{1}{\sqrt{p}} \sum_{x\in\mathbf{F}_p^\times} e\left(\frac{ax+b\bar{x}}{p}\right),$$

where we recall that we denote by $e(z)$ the 1-periodic function defined by $e(z) = e^{2i\pi z}$, and that $\bar{x}$ is the inverse of x modulo p.

These are finite sums, and they are of great importance in many areas of number theory, especially in relation with automorphic and modular forms and with analytic number theory (see [66] for a survey of the origin of these sums and of their applications, due to Poincaré, Kloosterman, Linnik, Iwaniec, and others). Among their remarkable properties is the following estimate for the modulus of $\mathrm{Kl}(a,b;p)$, due to A. Weil: for any $(a,b) \in \mathbf{F}_p^\times \times \mathbf{F}_p^\times$, we have

$$|\,\mathrm{Kl}(a,b;p)| \leqslant 2. \tag{6.1}$$

This is a very strong result if one considers that $\mathrm{Kl}(a,b;p)$ is, up to dividing by $\sqrt{p}$, the sum of $p-1$ roots of unity, so that the "trivial" estimate is that $|\,\mathrm{Kl}(a,b;p)| \leqslant (p-1)/\sqrt{p}$. What this reveals is that the arguments of the summands $e((ax+b\bar{x})/p)$ in $\mathbf{C}$ vary in a very complicated manner that leads to this remarkable cancellation property. This is due essentially to the very "random" behavior of the map $x \mapsto \bar{x}$ when seen at the level of representatives of x and $\bar{x}$ in the interval $\{0,\ldots,p-1\}$.

From a probabilistic point of view, the order of magnitude $\sqrt{p}$ of the sum (before normalization) is not unexpected. If we simply heuristically model an exponential sum as above by a random walk with independent summands uniformly distributed on the unit circle, say,

$$\mathrm{S_N} = \mathrm{X_1} + \cdots + \mathrm{X_N},$$

where the random variables (X_n) are independent and uniform on the unit circle, then the Central Limit Theorem implies a convergence in law of $\mathrm{X_N}/\sqrt{\mathrm{N}}$ to a standard complex Gaussian random variable, which shows that $\sqrt{\mathrm{N}}$ is the "right" order of magnitude. Note however that probabilistic analogies of this type would also suggest that $\mathrm{S_N}$ is sometimes (although rarely) larger than $\sqrt{\mathrm{N}}$ (the law of the iterated logarithm suggests that it should almost surely reach values as large as $\sqrt{\mathrm{N}}(\log\log \mathrm{N})$; see, e.g., [9, Th. 9.5]). Hence Weil's bound (6.1) indicates that the summands defining the Kloosterman sum have very special properties.

This probabilistic analogy and the study of random walks (or sheer curiosity) suggests to look at the partial sums of Kloosterman sums, and the way they move in the complex plane. This requires some ordering of the sum defining

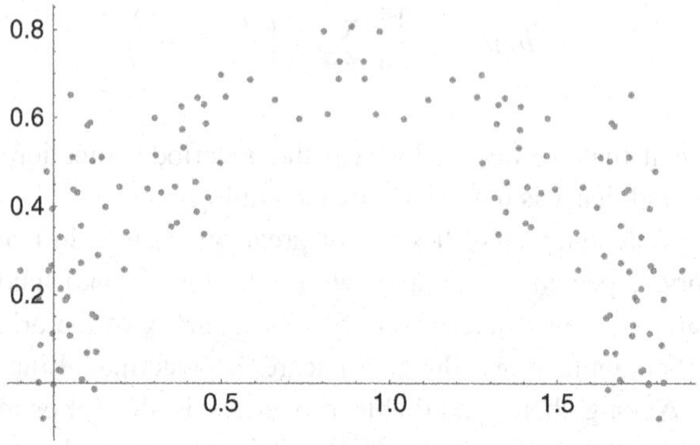

Figure 6.1 The partial sums of Kl(1, 1; 139).

$\mathrm{Kl}(a,b;p)$, which we simply achieve by summing over $1 \leqslant x \leqslant p-1$ in increasing order. Thus we will consider the $p-1$ points

$$z_j = \frac{1}{\sqrt{p}} \sum_{1 \leqslant x \leqslant j} e\left(\frac{ax + b\bar{x}}{p}\right)$$

for $1 \leqslant j \leqslant p-1$. We illustrate this for the sum Kl(1, 1; 139) in Figure 6.1.

Because this cloud of points is not particularly enlightening, we refine the construction by joining the successive points with line segments. This gives the result in Figure 6.2 for Kl(1, 1; 139). If we change the values of a and b, we observe that the figures change in apparently random and unpredictable way, although some basic features remain (the final point is on the real axis, which reflects the easily proven fact that $\mathrm{Kl}(a,b;p) \in \mathbf{R}$, and there is a reflection symmetry with respect to the line $x = \frac{1}{2}\,\mathrm{Kl}(a,b;p)$). For instance, Figure 6.3 shows the curves corresponding to Kl(2, 1; 139), Kl(3, 1; 139) and Kl(4, 1; 139); see [71] for many more pictures.

We then ask whether there is a definite statistical behavior for these *Kloosterman paths* as $p \to +\infty$, when we pick $(a,b) \in \mathbf{F}_p^\times \times \mathbf{F}_p^\times$ uniformly at random. As we will see, this is indeed the case!

To state the precise result, we introduce some further notation. Thus, for p prime and $(a,b) \in \mathbf{F}_p^\times \times \mathbf{F}_p^\times$, we denote by $\mathsf{K}_p(a,b)$ the function

$$[0,1] \longrightarrow \mathbf{C}$$

such that, for $0 \leqslant j \leqslant p-2$, the value at a real number t such that

$$\frac{j}{p-1} \leqslant t < \frac{j+1}{p-1}$$

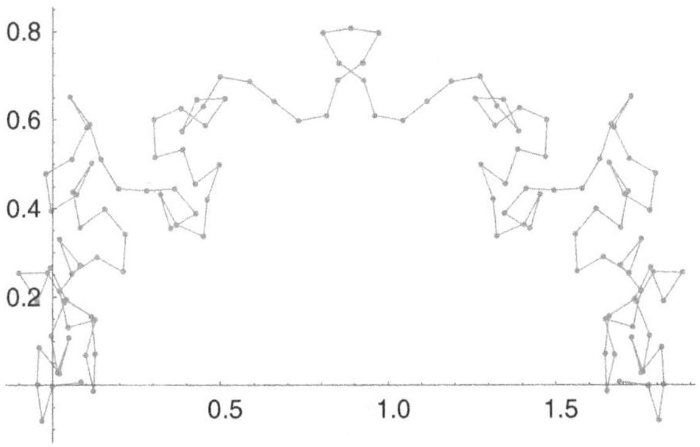

Figure 6.2 The partial sums of $\mathrm{Kl}(1,1;139)$, joined by line segments.

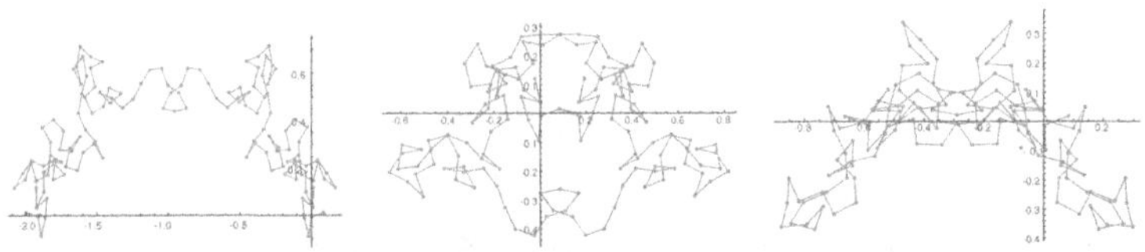

Figure 6.3 The partial sums of $\mathrm{Kl}(a,1;139)$ for $a = 2,3,4$.

is obtained by interpolating linearly between the consecutive partial sums

$$z_j = \frac{1}{\sqrt{p}} \sum_{1\leqslant x\leqslant j} e\left(\frac{ax+b\bar{x}}{p}\right) \quad \text{and} \quad z_{j+1} = \frac{1}{\sqrt{p}} \sum_{1\leqslant x\leqslant j+1} e\left(\frac{ax+b\bar{x}}{p}\right).$$

The path $t \mapsto \mathsf{K}_p(a,b)(t)$ is the polygonal path described above; for $t = 0$, we have $\mathsf{K}_p(a,b)(0) = 0$, and for $t = 1$, we obtain $\mathsf{K}_p(a,b)(1) = \mathrm{Kl}(a,b;p)$.

Let $\Omega_p = \mathbf{F}_p^\times \times \mathbf{F}_p^\times$. We view K_p as a random variable

$$\Omega_p \longrightarrow \mathrm{C}([0,1]),$$

where $\mathrm{C}([0,1])$ is the Banach space of continuous functions $\varphi\colon [0,1] \to \mathbf{C}$ with the supremum norm $\|\varphi\|_\infty = \sup|\varphi(t)|$. Alternatively, we may think of the *family* of random variables $(\mathsf{K}_p(t))_{t\in[0,1]}$ such that

$$(a,b) \mapsto \mathsf{K}_p(a,b)(t)$$

and view it as a "stochastic process" with t playing the role of "time."

Here is the theorem that gives the limiting behavior of these arithmetically defined random variables, proved by Kowalski and Sawin in [79].

Theorem 6.1.1 *Let* $(\mathrm{ST}_h)_{h\in\mathbf{Z}}$ *be a sequence of independent random variables, all distributed according to the Sato–Tate measure*

$$\mu_{\mathrm{ST}} = \frac{1}{\pi}\sqrt{1-\frac{x^2}{4}}dx$$

on $[-2,2]$.

(1) *The random Fourier series*

$$\mathrm{K}(t) = t\mathrm{ST}_0 + \sum_{\substack{h\in\mathbf{Z}\\ h\neq 0}} \frac{e(ht)-1}{2i\pi h}\mathrm{ST}_h$$

defined for $t \in [0,1]$ *converges uniformly almost surely, in the sense of symmetric partial sums*

$$\mathrm{K}(t) = t\mathrm{ST}_0 + \lim_{\mathrm{H}\to+\infty} \sum_{\substack{h\in\mathbf{Z}\\ 1\leqslant |h|<\mathrm{H}}} \frac{e(ht)-1}{2i\pi h}\mathrm{ST}_h.$$

This random Fourier series defines a $\mathrm{C}([0,1])$*-valued random variable* K.

(2) *As* $p \to +\infty$*, the random variables* K_p *converge in law to* K*, in the sense of* $\mathrm{C}([0,1])$*-valued variables.*

The Sato–Tate measure is better known in probability as a semi-circle law, but its appearance in Theorem 6.1.1 is really due to the group-theoretic interpretation that often arises in number theory, and reflects the choice of name. Namely, we recall (see Example B.6.1 (3)) that μ_{ST} is the direct image under the trace map of the probability Haar measure on the compact group $\mathrm{SU}_2(\mathbf{C})$.

Note in particular that the theorem implies, by taking $t = 1$, that the Kloosterman sums $\mathrm{Kl}(a,b;p) = \mathsf{K}_p(a,b)(1)$, viewed as random variables on Ω_p, become asymptotically distributed like $\mathrm{K}(1) = \mathrm{ST}_0$, that is, that Kloosterman sums are Sato–Tate distributed in the sense that for any real numbers $-2 \leqslant \alpha < \beta \leqslant 2$, we have

$$\frac{1}{(p-1)^2}|\{(a,b) \in \mathbf{F}_p^\times \times \mathbf{F}_p^\times \mid \alpha < \mathrm{Kl}(a,b;p) < \beta\}| \longrightarrow \int_\alpha^\beta d\mu_{\mathrm{ST}}(t).$$

This result is a famous theorem of N. Katz [61]. In some sense, Theorem 6.1.1 is a "functional" extension of this equidistribution theorem. In fact, the key arithmetic ingredient in the proof is an extension of the results and methods developed by Katz to prove many similar statements.

Remark 6.1.2 Although we do not require this, we mention a few regularity properties of the random series $\mathrm{K}(t)$: it is almost surely nowhere differentiable, but almost surely Hölder-continuous of order α for any $\alpha < 1/2$ (see the references in [79, Prop. 2.1]; these follow from general results of Kahane).

6.2 Proof of the Distribution Theorem

We will explain the proof of the theorem. We use a slightly different approach than the original article, bypassing the method of moments, and exploiting some simplifications that arise from the consideration of this single example.

The proof will be complete from a probabilistic point of view, but it relies on an extremely deep arithmetic result that we will only be able to view as a black box in this book. The crucial underlying result is the very general form of the Riemann Hypothesis over finite fields, and the formalism that is attached to it. This is due to Deligne, and the particular application we use relies extensively on the additional work of Katz. All of this builds on the algebraic-geometric foundations of Grothendieck and his school (see [59, Ch. 11] for an introduction).

In outline, the proof has three steps:

- **Step 1**: Show that the random Fourier series K exists, as a $C([0,1])$-valued random variable.
- **Step 2**: Prove that (a small variant of) the sequence of Fourier coefficients of K_p converges in law to the sequence of Fourier coefficients of K.
- **Step 3**: Prove that the sequence $(\mathsf{K}_p)_p$ is tight (Definition B.3.6), using Kolmogorov's Tightness Criterion (Proposition B.11.10).

Once this is done, a simple probabilistic statement (Proposition B.11.8, which is a variant of Prokhorov's Theorem, B.11.4) shows that the combination of (2) and (3) implies that K_p converges to K. Both steps (2) and (3) involve nontrivial arithmetic information; indeed, the main input in (2) is exceptionally deep, as we will explain soon.

We denote by $\mathbf{P}_p$ and $\mathbf{E}_p$ the probability and expectation with respect to the uniform measure on $\Omega_p = \mathbf{F}_p^\times \times \mathbf{F}_p^\times$. Before we begin the proof in earnest, it is useful to see *why* the limit arises, and why it is precisely this random Fourier series. The idea is to use discrete Fourier analysis to represent the partial sums of Kloosterman sums.

Lemma 6.2.1 *Let $p \geqslant 3$ be a prime and $a, b \in \mathbf{F}_p^\times$. Let $t \in [0,1]$. Then we have*

$$\frac{1}{\sqrt{p}} \sum_{1 \leqslant n \leqslant (p-1)t} e\left(\frac{an + b\bar{n}}{p}\right) = \sum_{|h|<p/2} \alpha_p(h,t)\,\mathrm{Kl}(a-h,b;p), \qquad (6.2)$$

where

$$\alpha_p(h,t) = \frac{1}{p} \sum_{1 \leqslant n \leqslant (p-1)t} e\left(\frac{nh}{p}\right).$$

Proof This is a case of the discrete Plancherel formula, applied to the characteristic (indicator) function of the discrete interval of summation; to check it quickly, insert the definitions of $\alpha_p(h,t)$ and of $\mathrm{Kl}(a-h,b;p)$ in the right-hand side of (6.2). This shows that it is equal to

$$\begin{aligned}
&\sum_{|h|<p/2} \alpha_p(h,t)\,\mathrm{Kl}(a-h,b;p) \\
&= \frac{1}{p^{3/2}} \sum_{|h|<p/2} \sum_{1 \leqslant n \leqslant (p-1)t} \sum_{m \in \mathbf{F}_p} e\left(\frac{nh}{p}\right) e\left(\frac{(a-h)m + b\bar{m}}{p}\right) \\
&= \frac{1}{\sqrt{p}} \sum_{1 \leqslant n \leqslant (p-1)t} \sum_{m \in \mathbf{F}_p} e\left(\frac{am + b\bar{m}}{p}\right) \frac{1}{p} \sum_{h \in \mathbf{F}_p} e\left(\frac{h(n-m)}{p}\right) \\
&= \frac{1}{\sqrt{p}} \sum_{1 \leqslant n \leqslant (p-1)t} e\left(\frac{an + b\bar{n}}{p}\right),
\end{aligned}$$

as claimed, since by the orthogonality of characters we have

$$\frac{1}{p} \sum_{h \in \mathbf{F}_p} e\left(\frac{h(n-m)}{p}\right) = \delta(n,m)$$

for any n, $m \in \mathbf{F}_p$, where $\delta(n,m) = 1$ if $n = m$ modulo p, and otherwise $\delta(n,m) = 0$. □

If we observe that $\alpha_p(h,t)$ is essentially a Riemann sum for the integral

$$\int_0^t e(ht)dt = \frac{e(ht)-1}{2i\pi h}$$

for all $h \neq 0$, and that $\alpha_p(0,t) \to t$ as $p \to +\infty$, we see that the right-hand side of (6.2) looks like a Fourier series of the same type as $\mathrm{K}(t)$, with coefficients given by shifted Kloosterman sums $\mathrm{Kl}(a-h,b;p)$ instead of ST_h. Now the crucial arithmetic information is contained in the following very deep theorem:

Theorem 6.2.2 (Katz; Deligne) *Fix an integer $b \neq 0$. For p prime not dividing b, consider the random variable*

$$\mathsf{S}_p \colon a \mapsto (\mathrm{Kl}(a-h,b;p))_{h \in \mathbf{Z}}$$

on $\mathbf{F}_p^\times$ *with uniform probability measure, taking values in the compact topological space*

$$\hat{\mathrm{T}} = \prod_{h \in \mathbf{Z}} [-2, 2].$$

Then S_p *converges in law to the product probability measure*

$$\bigotimes_{h \in \mathbf{Z}} \mu_{\mathrm{ST}}.$$

In other words, the sequence of random variables $a \mapsto \mathrm{Kl}(a - h, b; p)$ *converges in law to a sequence* $(\mathrm{ST}_h)_{h \in \mathbf{Z}}$ *of independent Sato–Tate distributed random variables.*

Because of this theorem, the formula (6.2) suggests that $\mathsf{K}_p(t)$ converges in law to the random series

$$t\mathrm{ST}_0 + \sum_{\substack{h \in \mathbf{Z} \\ h \neq 0}} \frac{e(ht) - 1}{2i\pi h} \mathrm{ST}_h,$$

which is exactly $\mathrm{K}(t)$. We now proceed to the implementation of the three steps above, which will use this deep arithmetic ingredient.

Remark 6.2.3 There is a subtlety in the argument: although Theorem 6.2.2 holds for any fixed b, when averaging only over a, we cannot at the current time prove the analogue of Theorem 6.1.1 for fixed b, because the proof of tightness in the last step uses crucially both averages.

Step 1. (Existence and properties of the random Fourier series)
We can write the series $\mathrm{K}(t)$ as

$$\mathrm{K}(t) = t\mathrm{ST}_0 + \sum_{h \geqslant 1} \left(\frac{e(ht) - 1}{2i\pi h} \mathrm{ST}_h - \frac{e(-ht) - 1}{2i\pi h} \mathrm{ST}_{-h} \right).$$

The summands here, namely,

$$\mathrm{X}_h = \frac{e(ht) - 1}{2i\pi h} \mathrm{ST}_h - \frac{e(-ht) - 1}{2i\pi h} \mathrm{ST}_{-h}$$

for $h \geqslant 1$, are independent and have expectation 0 since $\mathbf{E}(\mathrm{ST}_h) = 0$ (see (B.8)). Furthermore, since ST_h is independent of ST_{-h}, and they have variance 1, we have

$$\sum_{h \geqslant 1} \mathbf{V}(\mathrm{X}_h) = \sum_{h \geqslant 1} \left(\left| \frac{e(ht) - 1}{2i\pi h} \right|^2 + \left| \frac{e(-ht) - 1}{2i\pi h} \right|^2 \right) \leqslant \sum_{h \geqslant 1} \frac{1}{h^2} < +\infty$$

for any $t \in [0,1]$. From Kolmogorov's criterion for almost sure convergence of random series with finite variance (Theorem B.10.1), it follows that for any $t \in [0,1]$, the series $\mathrm{K}(t)$ converges almost surely and in L^2 to a complex-valued random variable.

To prove convergence in $C([0,1])$, we will use convergence of finite distributions combined with Kolmogorov's Tightness Criterion. Consider the partial sums

$$\mathrm{K}_{\mathrm{H}}(t) = t\mathrm{ST}_0 + \sum_{1\leqslant|h|\leqslant \mathrm{H}} \frac{e(ht)-1}{2i\pi h}\mathrm{ST}_h$$

for $\mathrm{H} \geqslant 1$. These are $C([0,1])$-valued random variables. The convergence of $\mathrm{K}_{\mathrm{H}}(t)$ to $\mathrm{K}(t)$ in L^1, for any $t \in [0,1]$, implies (see Lemma B.11.3) that the sequence $(\mathrm{K}_{\mathrm{H}})_{\mathrm{H}\geqslant 1}$ converges to K in the sense of finite distributions. Therefore, by Proposition B.11.10, the sequence converges in the sense of $C([0,1])$-valued random variables if there exist constants $\mathrm{C} \geqslant 0$, $\alpha > 0$ and $\delta > 0$ such that for any $\mathrm{H} \geqslant 1$, and real numbers $0 \leqslant s < t \leqslant 1$, we have

$$\mathbf{E}(|\mathrm{K}_{\mathrm{H}}(t) - \mathrm{K}_{\mathrm{H}}(s)|^{\alpha}) \leqslant \mathrm{C}|t-s|^{1+\delta}. \tag{6.3}$$

We will take $\alpha = 4$. We have

$$\mathrm{K}_{\mathrm{H}}(t) - \mathrm{K}_{\mathrm{H}}(s) = (t-s)\mathrm{ST}_0 + \sum_{1\leqslant|h|\leqslant \mathrm{H}} \frac{e(ht)-e(hs)}{2i\pi h}\mathrm{ST}_h.$$

This is a sum of independent, centered and bounded random variables, so that by Proposition B.8.2 (1) and (2), it is σ_{H}^2-sub-Gaussian with

$$\sigma_{\mathrm{H}}^2 = |t-s|^2 + \sum_{1\leqslant|h|\leqslant \mathrm{H}} \left|\frac{e(ht)-e(hs)}{2i\pi h}\right|^2 \leqslant |t-s|^2 + \sum_{h\neq 0} \left|\frac{e(ht)-e(hs)}{2i\pi h}\right|^2.$$

By Parseval's formula for ordinary Fourier series, we have

$$|t-s|^2 + \sum_{h\neq 0} \left|\frac{e(ht)-e(hs)}{2i\pi h}\right|^2 = \int_0^1 |\varphi_{s,t}(x)|^2 dx,$$

where $\varphi_{s,t}$ is the characteristic function of the interval $[s,t]$. Therefore $\sigma_{\mathrm{H}}^2 \leqslant |t-s|$. By the properties of sub-Gaussian random variables (see Proposition B.8.3 in Section B.8), we deduce that there exists $\mathrm{C} \geqslant 0$ such that

$$\mathbf{E}(|\mathrm{K}_{\mathrm{H}}(t) - \mathrm{K}_{\mathrm{H}}(s)|^4) \leqslant \mathrm{C}\sigma_{\mathrm{H}}^4 \leqslant \mathrm{C}|t-s|^2,$$

which establishes (6.3).

Step 2. (Computation of Fourier coefficients)

As in Section B.11, we will denote by $C_0([0,1])$ the subspace of functions $f \in C([0,1])$ such that $f(0) = 0$. For $f \in C_0([0,1])$, the sequence $\mathrm{FT}(f) = (\widetilde{f}(h))_{h\in\mathbf{Z}}$ is defined by $\widetilde{f}(0) = f(1)$ and

$$\widetilde{f}(h) = \int_0^1 (f(t) - tf(1))e(-ht)dt$$

for $h \neq 0$. The map FT is a continuous linear map from $C_0([0,1])$ to $C_0(\mathbf{Z})$, the Banach space of functions $\mathbf{Z} \to \mathbf{C}$ that tend to zero at infinity.

Lemma 6.2.4 *The "Fourier coefficients"* $\mathrm{FT}(\mathsf{K}_p)$ *converge in law to* $\mathrm{FT}(\mathrm{K})$, *in the sense of convergence of finite distribution.*

We begin by computing the Fourier coefficients of a polygonal path. Let z_0 and z_1 be complex numbers, and $t_0 < t_1$ real numbers. We define $\Delta = t_1 - t_0$ and $f \in C([0,1])$ by

$$f(t) = \begin{cases} \frac{1}{\Delta}(z_1(t - t_0) + z_0(t_1 - t)) & \text{if } t_0 \leqslant t \leqslant t_1, \\ 0 & \text{otherwise,} \end{cases}$$

which parameterizes the segment from z_0 to z_1 over the interval $[t_0, t_1]$.

Let $h \neq 0$ be an integer. By direct computation, we find

$$\begin{aligned} \int_0^1 f(t)e(-ht)dt &= -\frac{1}{2i\pi h}(z_1 e(-ht_1) - z_0 e(-ht_0)) \\ &\quad + \frac{1}{2i\pi h}(z_1 - z_0)e(-ht_0)\frac{1}{\Delta}\left(\int_0^\Delta e(-hu)du\right) \\ &= -\frac{1}{2i\pi h}(z_1 e(-ht_1) - z_0 e(-ht_0)) \\ &\quad + \frac{1}{2i\pi h}(z_1 - z_0)\frac{\sin(\pi h\Delta)}{\pi h\Delta} e\left(-h\left(t_0 + \frac{\Delta}{2}\right)\right). \end{aligned} \tag{6.4}$$

Consider now an integer $n \geqslant 1$ and a family $(z_0, \ldots, z_n)$ of complex numbers. For $0 \leqslant j \leqslant n-1$, let f_j be the function as above relative to the points (z_j, z_{j+1}) and the interval $[j/n, (j+1)/n]$, and define

$$f = \sum_{j=0}^{n-1} f_j$$

so that f parameterizes the polygonal path joining z_0 to z_1 to $\ldots$ to z_n, each over time intervals of equal length $1/n$.

For $h \neq 0$, we obtain by summing (6.4), using a telescoping sum and the relations $z_0 = f(0)$, $z_n = f(1)$, the formula

$$\int_0^1 f(t)e(-ht)dt = -\frac{1}{2i\pi h}(f(1) - f(0)) + \frac{1}{2i\pi h}\frac{\sin(\pi h/n)}{\pi h/n}\sum_{j=0}^{n-1}(z_{j+1} - z_j)e\left(-\frac{h(j+\frac{1}{2})}{n}\right). \tag{6.5}$$

We specialize this general formula to Kloosterman paths. Let p be a prime, $(a,b) \in \mathbf{F}_p^\times \times \mathbf{F}_p^\times$, and apply the formula above to $n = p - 1$ and the points

$$z_j = \frac{1}{\sqrt{p}}\sum_{1 \leqslant x \leqslant j} e\left(\frac{ax + b\bar{x}}{p}\right), \qquad 0 \leqslant j \leqslant p - 1.$$

For $h \neq 0$, the hth Fourier coefficient of $\mathrm{K}_p - t\mathrm{K}_p(1)$ is the random variable on Ω_p that maps (a,b) to

$$\frac{1}{2i\pi h}\frac{\sin(\pi h/(p-1))}{\pi h/(p-1)}e\left(-\frac{h}{2(p-1)}\right)\frac{1}{\sqrt{p}}\sum_{x=1}^{p-1} e\left(\frac{ax + b\bar{x}}{p}\right)e\left(-\frac{hx}{p-1}\right).$$

Note that for fixed h, we have

$$e\left(-\frac{hx}{p-1}\right) = e\left(-\frac{hx}{p}\right)e\left(-\frac{hx}{p(p-1)}\right) = e\left(-\frac{hx}{p}\right)(1 + \mathrm{O}(p^{-1}))$$

for all p and all x such that $1 \leqslant x \leqslant p - 1$, hence

$$\frac{1}{\sqrt{p}}\sum_{x=1}^{p-1} e\left(\frac{ax + b\bar{x}}{p}\right)e\left(-\frac{hx}{p-1}\right) = \mathrm{Kl}(a - h, b; p) + \mathrm{O}\left(\frac{1}{\sqrt{p}}\right),$$

where the implied constant depends on h. Let

$$\beta_p(h) = \frac{\sin(\pi h/(p-1))}{\pi h/(p-1)}e\left(-\frac{h}{2(p-1)}\right).$$

Note that $|\beta_p(h)| \leqslant 1$, so we can express the hth Fourier coefficient as

$$\frac{1}{2i\pi h}\mathrm{Kl}(a - h, b; p)\beta_p(h) + \mathrm{O}(p^{-1/2}),$$

where the implied constant depends on h.

Note further that the 0th component of $\mathrm{FT}(\mathrm{K}_p)$ is $\mathrm{Kl}(a,b;p)$. Since $\beta_p(h) \to 1$ as $p \to +\infty$ for each fixed h, we deduce from Katz's equidistribution theorem (Theorem 6.2.2) and from Lemma B.4.3 (applied to the vectors of

Fourier coefficients at $h_1, \ldots, h_m$ for arbitrary $m \geqslant 1$) that $\mathrm{FT}(\mathsf{K}_p)$ converges in law to FT(K) in the sense of finite distributions.

Step 3. (Tightness of the Kloosterman paths)

We now come to the second main step of the proof of Theorem 6.1.1: the fact that the sequence $(\mathsf{K}_p)_p$ is tight. According to Kolmogorov's Criterion (Proposition B.11.10), it is enough to find constants $\mathrm{C} \geqslant 0$, $\alpha > 0$ and $\delta > 0$ such that, for all primes $p \geqslant 3$ and all t and s with $0 \leqslant s < t \leqslant 1$, we have

$$\mathbf{E}_p(|\mathsf{K}_p(t) - \mathsf{K}_p(s)|^{\alpha}) \leqslant \mathrm{C}|t-s|^{1+\delta}. \tag{6.6}$$

We denote by $\gamma \geqslant 0$ the real number such that

$$|t-s| = (p-1)^{-\gamma}.$$

So γ is larger when t and s are closer. The proof of (6.6) involves two different ranges.

Assume first that $\gamma > 1$ (that is, that $|t-s| < 1/(p-1)$). In that range, we use the polygonal nature of the paths $x \mapsto \mathsf{K}_p(x)$, which implies that

$$|\mathsf{K}_p(t) - \mathsf{K}_p(s)| \leqslant \sqrt{p-1}|t-s| \leqslant \sqrt{|t-s|}$$

(since the "velocity" of the path is $(p-1)/\sqrt{p} \leqslant \sqrt{p-1}$). Consequently, for any $\alpha > 0$, we have

$$\mathbf{E}_p(|\mathsf{K}_p(t) - \mathsf{K}_p(s)|^{\alpha}) \leqslant |t-s|^{\alpha/2}. \tag{6.7}$$

In the remaining range $\gamma \leqslant 1$, we will use the discontinuous partial sums $\tilde{\mathsf{K}}_p(t)$ instead of $\mathsf{K}_p(t)$. To check that this is legitimate, note that

$$|\tilde{\mathsf{K}}_p(t) - \mathsf{K}_p(t)| \leqslant \frac{1}{\sqrt{p}}$$

for all primes $p \geqslant 3$ and all t. Hence, using Hölder's inequality, we derive for $\alpha \geqslant 1$ the relation

$$\begin{aligned}\mathbf{E}_p(|\mathsf{K}_p(t) - \mathsf{K}_p(s)|^{\alpha}) &= \mathbf{E}_p(|\tilde{\mathsf{K}}_p(t) - \tilde{\mathsf{K}}_p(s)|^{\alpha}) + \mathrm{O}(p^{-\alpha/2}) \\ &= \mathbf{E}_p(|\tilde{\mathsf{K}}_p(t) - \tilde{\mathsf{K}}_p(s)|^{\alpha}) + \mathrm{O}(|t-s|^{\alpha/2}),\end{aligned} \tag{6.8}$$

where the implied constant depends only on α.

We take $\alpha = 4$. The following computation of the fourth moment is an idea that goes back to Kloosterman's very first nontrivial estimate for individual Kloosterman sums.

We have

$$\tilde{\mathsf{K}}_p(t) - \tilde{\mathsf{K}}_p(s) = \frac{1}{\sqrt{p}} \sum_{n \in \mathrm{I}} e\left(\frac{an + b\bar{n}}{p}\right),$$

where I is the discrete interval

$$(p-1)s < n \leqslant (p-1)t$$

of summation. The length of I is

$$\lfloor (p-1)t \rfloor - \lceil (p-1)s \rceil \leqslant 2(p-1)|t-s|$$

since $(p-1)|t-s| \geqslant 1$.

By expanding the fourth power, we get

$$\begin{aligned}
&\mathbf{E}_p(|\tilde{\mathsf{K}}_p(t) - \tilde{\mathsf{K}}_p(s)|^4) \\
&\quad = \frac{1}{(p-1)^2} \sum_{(a,b)\in \mathbf{F}_p^\times \times \mathbf{F}_p^\times} \left| \frac{1}{\sqrt{p}} \sum_{n\in \mathrm{I}} e\left(\frac{an+b\bar{n}}{p}\right)\right|^4 \\
&\quad = \frac{1}{p^2(p-1)^2} \sum_{a,b} \sum_{n_1,\ldots,n_4\in \mathrm{I}} e\left(\frac{a(n_1+n_2-n_3-n_4)}{p}\right) \\
&\qquad \times e\left(\frac{b(\bar{n}_1+\bar{n}_2-\bar{n}_3-\bar{n}_4)}{p}\right).
\end{aligned}$$

After exchanging the order of the sums, which "separates" the two variables a and b, we get

$$\begin{aligned}
\frac{1}{p^2(p-1)^2} \sum_{n_1,\ldots,n_4\in \mathrm{I}} &\left(\sum_{a\in \mathbf{F}_p^\times} e\left(\frac{a(n_1+n_2-n_3-n_4)}{p}\right)\right) \\
&\times \left(\sum_{b\in \mathbf{F}_p^\times} e\left(\frac{b(\bar{n}_1+\bar{n}_2-\bar{n}_3-\bar{n}_4)}{p}\right)\right).
\end{aligned}$$

The orthogonality relations for additive character (namely, the relation

$$\frac{1}{p} \sum_{a\in \mathbf{F}_p^\times} e\left(\frac{ah}{p}\right) = \delta(h,0) - \frac{1}{p}$$

for any $h \in \mathbf{F}_p$) imply that

$$\mathbf{E}_p(|\tilde{\mathsf{K}}_p(t) - \tilde{\mathsf{K}}_p(s)|^4) = \frac{1}{(p-1)^2} \sum_{\substack{n_1,\ldots,n_4\in \mathrm{I} \\ n_1+n_2=n_3+n_4 \\ \bar{n}_1+\bar{n}_2=\bar{n}_3+\bar{n}_4}} 1 + \mathrm{O}(|\mathrm{I}|^3(p-1)^{-3}). \quad (6.9)$$

Fix first n_1 and n_2 in I with $n_1 + n_2 \neq 0$. Then if (n_3, n_4) satisfy

$$n_1 + n_2 = n_3 + n_4 \quad \text{and} \quad \bar{n}_1 + \bar{n}_2 = \bar{n}_3 + \bar{n}_4,$$

the value of $n_3 + n_4$ is fixed, and $\bar{n}_1 + \bar{n}_2$ is nonzero, so

$$n_3 n_4 = \frac{n_3 + n_4}{\bar{n}_1 + \bar{n}_2}$$

(in $\mathbf{F}_p^\times$) is also fixed. Hence there are at most two pairs (n_3, n_4) that satisfy the equations for these given (n_1, n_2). This means that the contribution of these n_1, n_2 to (6.9) is $\leqslant 2|\mathrm{I}|^2(p-1)^{-2}$. Similarly, if $n_1 + n_2 = 0$, the equations imply that $n_3 + n_4 = 0$, and hence the solutions are determined uniquely by (n_1, n_3). Hence the contribution is then $\leqslant |\mathrm{I}|^2(p-1)^2$, and we get

$$\mathbf{E}_p(|\tilde{\mathsf{K}}_p(t) - \tilde{\mathsf{K}}_p(s)|^4) \ll |\mathrm{I}|^2(p-1)^{-2} + |\mathrm{I}|^3(p-1)^{-3} \ll |t-s|^2,$$

where the implied constants are absolute. Using (6.8), this gives

$$\mathbf{E}_p(|\mathsf{K}_p(t) - \mathsf{K}_p(s)|^4) \ll |t-s|^2 \tag{6.10}$$

with an absolute implied constant. Combined with (6.7) with $\alpha = 4$ in the former range, this completes the proof of tightness.

Final Step. (Proof of Theorem 6.1.1) In view of Proposition B.11.8, the theorem follows directly from the results of Steps 2 and 3.

Remark 6.2.5 The proof of tightness uses crucially that we average over both a and b to reduce the problem to counting the number of solutions of certain equations over $\mathbf{F}_p$ (see (6.9)), which turn out to be accessible. Since $\mathrm{Kl}(a,b;p) = \mathrm{Kl}(ab;1,p)$ for all a and b in $\mathbf{F}_p^\times$, it seems natural to try to prove an analogue of Theorem 6.1.1 when averaging only over a, with $b = 1$ fixed. The convergence of finite distributions extends to that setting (since Theorem 6.2.2 holds for any fixed b), but a proof of tightness is not currently known for fixed b. Using moment estimates (derived from Deligne's Riemann Hypothesis) and the trivial bound

$$\left|\tilde{\mathsf{K}}_p(t) - \tilde{\mathsf{K}}_p(s)\right| \leqslant |\mathrm{I}|p^{-1/2},$$

one can check that it is enough to prove a suitable estimate for the average over a in the restricted range where

$$\frac{1}{2} - \eta \leqslant \gamma \leqslant \frac{1}{2} + \eta$$

for some fixed but arbitrarily small value of $\eta > 0$ (see [79, §3]). The next exercise illustrates this point.

Exercise 6.2.6 Assume p is odd. Let $\Omega'_p = \mathbf{F}_p^\times \times (\mathbf{F}_p^\times)^2$, where $(\mathbf{F}_p^\times)^2$ is the set of nonzero squares in $\mathbf{F}_p^\times$. We denote by $\mathsf{K}'_p(t)$ the random variable $\mathsf{K}_p(t)$

restricted to Ω'_p, with the uniform probability measure, for which $\mathbf{P}'_p(\cdot)$ and $\mathbf{E}'_p(\cdot)$ denote probability and expectation.

(1) Prove that $\mathrm{FT}(\mathsf{K}'_p)$ converges to $\mathrm{FT}(\mathsf{K})$ in the sense of finite distributions.

(2) For $n \in \mathbf{F}_p$, prove that

$$\sum_{b \in (\mathbf{F}_p^\times)^2} e\left(\frac{bn}{p}\right) = \frac{p-1}{2}\delta(n,0) + O(\sqrt{p}),$$

where the implied constant is absolute. [**Hint**: Show that if $n \in \mathbf{F}_p^\times$, we have

$$\left|\sum_{b \in \mathbf{F}_p^\times} e\left(\frac{nb^2}{p}\right)\right| = \sqrt{p},$$

where the left-hand sum is known as a *quadratic Gauss sum*; see Example C.6.2 (1) and Exercise C.6.5.]

(3) Deduce that if $|t-s| \geqslant 1/p$, then

$$\mathbf{E}'_p(|\mathsf{K}'_p(t) - \mathsf{K}'_p(s)|^4) \ll \sqrt{p}|t-s|^3 + |t-s|^2,$$

where the implied constant is absolute.

(3) Using notation as in the proof of tightness for K_p, prove that if $\eta > 0$, $\alpha \geqslant 1$ and

$$\frac{1}{2} + \eta \leqslant \gamma \leqslant 1,$$

then

$$\mathbf{E}'_p(|\mathsf{K}'_p(t) - \mathsf{K}'_p(s)|^\alpha) \ll |t-s|^{\alpha\eta} + |t-s|^{\alpha/2},$$

where the implied constant depends only on α.

(4) Prove that if $\eta > 0$ and

$$0 \leqslant \gamma \leqslant \frac{1}{2} - \eta,$$

then there exists $\delta > 0$ such that

$$\mathbf{E}'_p(|\mathsf{K}'_p(t) - \mathsf{K}'_p(s)|^4) \ll |t-s|^{1+\delta},$$

where the implied constant depends only on η.

(5) Conclude that (K'_p) converges in law to K in $C([0,1])$. [**Hint**: It may be convenient to use the variant of Kolmogorov's tightness Criterion in Proposition B.11.11.]

6.3 Applications

We can use Theorem 6.1.1 to gain information on partial sums of Kloosterman sums. We will give two examples, one concerning large values of the partial sums, and the other dealing with the support of the Kloosterman paths, following [12].

Theorem 6.3.1 *For p prime and $\mathrm{A} > 0$, let $\mathrm{M}_p(\mathrm{A})$ and $\mathrm{N}_p(\mathrm{A})$ be the events*

$$\mathrm{M}_p(\mathrm{A}) = \Big\{(a,b) \in \mathbf{F}_p^\times \times \mathbf{F}_p^\times \mid \max_{1 \leqslant j \leqslant p-1} \frac{1}{\sqrt{p}} \Big| \sum_{1 \leqslant n \leqslant j} e\Big(\frac{an + b\bar{n}}{p}\Big)\Big| > \mathrm{A}\Big\},$$

$$\mathrm{N}_p(\mathrm{A}) = \Big\{(a,b) \in \mathbf{F}_p^\times \times \mathbf{F}_p^\times \mid \max_{1 \leqslant j \leqslant p-1} \frac{1}{\sqrt{p}} \Big| \sum_{1 \leqslant n \leqslant j} e\Big(\frac{an + b\bar{n}}{p}\Big)\Big| \geqslant \mathrm{A}\Big\}.$$

There exists a positive constant $c > 0$ such that, for any $\mathrm{A} > 0$, we have

$$\begin{aligned} c^{-1}\exp(-\exp(c\mathrm{A})) &\leqslant \liminf_{p \to +\infty} \mathbf{P}_p(\mathrm{N}_p(\mathrm{A})) \\ &\leqslant \limsup_{p \to +\infty} \mathbf{P}_p(\mathrm{M}_p(\mathrm{A})) \leqslant c\exp(-\exp(c^{-1}\mathrm{A})). \end{aligned}$$

In particular, partial sums of normalized Kloosterman sums are unbounded (whereas the full normalized Kloosterman sums are always of modulus at most 2), but large values of partial sums are extremely rare.

Proof The functions $t \mapsto \mathsf{K}_p(a,b)(t)$ describe polygonal paths in the complex plane. Since the maximum modulus of a point on such a path is achieved at one of the vertices, it follows that

$$\max_{1 \leqslant j \leqslant p-1} \frac{1}{\sqrt{p}} \Big| \sum_{1 \leqslant n \leqslant j} e\Big(\frac{an + b\bar{n}}{p}\Big)\Big| = \|\mathsf{K}_p(a,b)\|_\infty,$$

so that the event $\mathrm{M}_p(\mathrm{A})$ is the same as $\{\|\mathsf{K}_p\|_\infty > \mathrm{A}\}$, and $\mathrm{N}_p(\mathrm{A})$ is the same as $\{\|\mathsf{K}_p\|_\infty \geqslant \mathrm{A}\}$.

By Theorem 6.1.1 and composition with the norm map (Proposition B.3.2), the real-valued random variables $\|\mathsf{K}_p\|_\infty$ converge in law to the random variable $\|\mathrm{K}\|_\infty$, the norm of the random Fourier series K. By elementary properties of convergence in law, we have therefore

$$\mathbf{P}(\|\mathrm{K}\|_\infty > \mathrm{A}) \leqslant \liminf_{p \to +\infty} \mathbf{P}_p(\mathrm{N}_p(\mathrm{A})) \leqslant \limsup_{p \to +\infty} \mathbf{P}_p(\mathrm{M}_p(\mathrm{A})) \leqslant \mathbf{P}(\|\mathrm{K}\|_\infty \geqslant \mathrm{A}).$$

So the problem is reduced to questions about the limiting random Fourier series.

We first consider the upper bound. Here it suffices to prove the existence of a constant $c > 0$ such that

$$\mathbf{P}(\|\operatorname{Im}(\mathrm{K})\|_\infty > \mathrm{A}) \leqslant c\exp(-\exp(c^{-1}\mathrm{A})),$$
$$\mathbf{P}(\|\operatorname{Re}(\mathrm{K})\|_\infty > \mathrm{A}) \leqslant c\exp(-\exp(c^{-1}\mathrm{A})).$$

We will do this for the real part, since the imaginary part is very similar and can be left as an exercise. The random variable $\mathrm{R} = \operatorname{Re}(\mathrm{K})$ takes values in the separable real Banach space $C_{\mathbf{R}}([0,1])$ of real-valued continuous functions on $[0,1]$. It is almost surely the sum of the random Fourier series

$$\mathrm{R} = \sum_{h \geqslant 0} \varphi_h \mathrm{Y}_h,$$

where $\varphi_h \in C_{\mathbf{R}}([0,1])$ and the random variables Y_h are defined by

$$\varphi_0(t) = 2t, \qquad \mathrm{Y}_0 = \tfrac{1}{2}\mathrm{ST}_0,$$
$$\varphi_h(t) = \frac{\sin(2\pi h t)}{8\pi h}, \quad \mathrm{Y}_h = \tfrac{1}{4}(\mathrm{ST}_h + \mathrm{ST}_{-h}) \text{ for } h \geqslant 1.$$

We note that the random variables (Y_h) are independent and that $|\mathrm{Y}_h| \leqslant 1$ (almost surely) for all h. We can then apply the bound of Proposition B.11.13 (1) to conclude.

We now prove the lower bound. It suffices to prove that there exists $c > 0$ such that

$$\mathbf{P}(|\operatorname{Im}(\mathrm{K}(1/2))| > \mathrm{A}) \geqslant c^{-1}\exp(-\exp(c\mathrm{A})), \tag{6.11}$$

since this implies that

$$\mathbf{P}(\|\mathrm{K}\|_\infty > \mathrm{A}) \geqslant c^{-1}\exp(-\exp(c\mathrm{A})).$$

We have

$$\operatorname{Im}(\mathrm{K}(1/2)) = -\frac{1}{2\pi}\sum_{h \neq 0} \frac{\cos(\pi h) - 1}{h}\mathrm{ST}_h = \frac{1}{\pi}\sum_{h \geqslant 1} \frac{1}{h}\mathrm{ST}_h,$$

which is a series that converges almost surely in $\mathbf{R}$ with independent terms, and where $\frac{1}{\pi}\mathrm{ST}_h$ is symmetric and $\leqslant 1$ in absolute value for all h. Thus the bound

$$\mathbf{P}(|\operatorname{Im}(\mathrm{K}(1/2))| > \mathrm{A}) \geqslant c^{-1}\exp(-\exp(c\mathrm{A}))$$

for some $c > 0$ follows immediately from Proposition B.11.13 (2). □

Remark 6.3.2 In the lower bound, the point $1/2$ could be replaced by any $t \in]0,1[$ for the imaginary part, and one could also use the real part and any t such

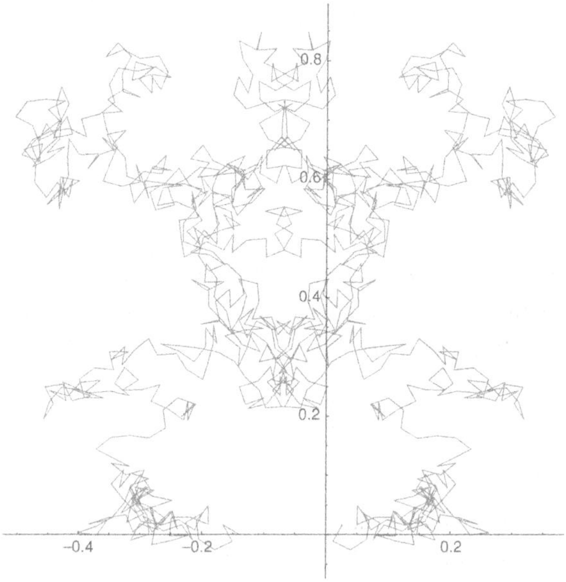

Figure 6.4 The partial sums of Kl(88, 1; 1021).

that $t \notin \{0, 1/2, 1\}$; the symmetry of the Kloosterman paths with respect to the line $x = \frac{1}{2}\,\mathrm{Kl}(a, b; p)$ shows that the real part of $\mathsf{K}_p(a, b)(1/2)$ is $\frac{1}{2}\,\mathrm{Kl}(a, b; p)$, and this is a real number in $[-1, 1]$.

For our second application, we compute the support of the random Fourier series K.

Theorem 6.3.3 *The support of the law of* K *is the set of all* $f \in C_0([0, 1])$ *such that*

(1) *we have* $f(1) \in [-2, 2]$;
(2) *for all* $h \neq 0$, *we have* $\widetilde{f}(h) \in i\mathbf{R}$ *and*

$$|\widetilde{f}(h)| \leqslant \frac{1}{\pi |h|}.$$

Proof Denote by $\mathcal{S}$ the set described in the statement. Then $\mathcal{S}$ is closed in $C([0, 1])$, since it is the intersection of closed sets. By Theorem 6.1.1, a sample function $f \in C([0, 1])$ of the random process K is almost surely given by a series

$$f(t) = \alpha_0 t + \sum_{h \neq 0} \frac{e(ht) - 1}{2\pi i h} \alpha_h$$

that is uniformly convergent in the sense of symmetric partial sums, for some real numbers α_h such that $|\alpha_h| \leqslant 2$. We have $\widetilde{f}(0) = f(1) \in [-2,2]$, and the uniform convergence implies that for $h \neq 0$, we have

$$\widetilde{f}(h) = \frac{\alpha_h}{2i\pi h},$$

so that f certainly belongs to $\mathcal{S}$. Consequently, the support of K is contained in $\mathcal{S}$.

We now prove the converse inclusion. By Lemma B.3.3, the support of K contains the set of continuous functions with uniformly convergent (symmetric) expansions

$$t\alpha_0 + \sum_{h\neq 0} \frac{e(ht)-1}{2\pi ih}\alpha_h,$$

where $\alpha_h \in [-2,2]$ for all $h \in \mathbf{Z}$. In particular, since 0 belongs to the support of the Sato–Tate measure, $\mathcal{S}$ contains all finite sums of this type.

Let $f \in \mathcal{S}$ and put $g(t) = f(t) - tf(1)$. We have

$$f(t) - tf(1) = \lim_{\mathrm{N}\to+\infty} \sum_{|h|\leqslant \mathrm{N}} \widehat{g}(h)e(ht)\left(1 - \frac{|h|}{\mathrm{N}}\right)$$

in $\mathrm{C}_0([0,1])$, by the uniform convergence of Cesàro means of the Fourier series of a continuous periodic function (see, e.g., [121, III, th. 3.4]). Evaluating at 0 and subtracting yields

$$\begin{aligned} f(t) &= tf(1) + \lim_{\mathrm{N}\to+\infty} \sum_{\substack{|h|\leqslant \mathrm{N}\\ h\neq 0}} \widetilde{f}(h)(e(ht)-1)\left(1 - \frac{|h|}{\mathrm{N}}\right) \\ &= tf(1) + \lim_{\mathrm{N}\to+\infty} \sum_{\substack{|h|\leqslant \mathrm{N}\\ h\neq 0}} \frac{\alpha_h}{2i\pi h}(e(ht)-1)\left(1 - \frac{|h|}{\mathrm{N}}\right) \end{aligned}$$

in $\mathrm{C}([0,1])$, where $\alpha_h = 2i\pi h\widetilde{f}(h)$ for $h \neq 0$. Then $\alpha_h \in \mathbf{R}$ and $|\alpha_h| \leqslant 2$ by the assumption that $f \in \mathcal{S}$, so each function

$$tf(1) + \sum_{\substack{|h|\leqslant \mathrm{N}\\ h\neq 0}} \frac{e(ht)-1}{2\pi ih}\alpha_h\left(1 - \frac{|h|}{\mathrm{N}}\right),$$

belongs to the support of K. Since the support is closed, we conclude that f also belongs to the support of K. □

The support of K is an interesting set of functions. Testing whether a function $f \in \mathrm{C}_0([0,1])$ belongs to it, or not, is straightforward if the Fourier coefficients of f are known, and a positive or negative answer has interesting arithmetic consequences, by Lemma B.3.3. In particular, since 0 clearly belongs to the support of K, we get:

Corollary 6.3.4 *For any $\varepsilon > 0$, we have*

$$\liminf_{p\to+\infty} \frac{1}{(p-1)^2}\left|\left\{(a,b) \in \mathbf{F}_p^\times \times \mathbf{F}_p^\times \mid \max_{0\leqslant j\leqslant p-1}\left|\frac{1}{\sqrt{p}}\sum_{1\leqslant x\leqslant j} e\left(\frac{ax+b\bar{x}}{p}\right)\right| < \varepsilon\right\}\right| > 0.$$

We refer to [12] for further examples of functions belonging (or not) to the support of K and mention only a remarkable result of J. Bober: the support of K contains space-filling curves, that is, functions f such that the image of f has nonempty interior.

6.4 Generalizations

The method of Kowalski and Sawin can be extended to study the "shape" of many other exponential sums. On the other hand, natural generalizations require different tools, when the Riemann Hypothesis is not applicable anymore. This was achieved by Ricotta and Royer [101] for Kloosterman sums modulo p^n when $n \geqslant 2$ is fixed and $p \to +\infty$, and later, they succeeded with Shparlinski [102] in obtaining convergence in law in that setting with a single variable a. If p is fixed and $n \to +\infty$, the corresponding study was done by Milićević and Zhang [87], where tools related to p-adic analysis are crucial. In the three cases, the limit random Fourier series are similar, but have coefficients that have distributions different from the Sato–Tate distribution.

Related developments concern quantitative versions of Theorem 6.3.1: how large (and how often) can one make a partial sum of Kloosterman sums? Results of this kind have been proved by Lamzouri [82] and Bonolis [14], and in great generality by Autissier, Bonolis and Lamzouri [3].

Finally, in another direction, Cellarosi and Marklof [22] have established beautiful functional limit theorems for other types of exponential sums closer to the Weyl sums that arise in the circle method, and especially for quadratic Weyl sums. The tools as well as the limiting functions are completely different.

[Further references: Iwaniec and Kowalski [59, Ch. 11].**]**

7

Further Topics

We explained at the beginning of this book that we would restrict our focus on a certain special type of results in probabilistic number theory: convergence in law of arithmetically defined sequences of random variables. In this chapter, we will quickly survey (with some references) some important and beautiful results that either do not exactly fit our precise setting, or require rather deeper tools than we wished to assume, or could develop from scratch.

7.1 Equidistribution Modulo 1

We have begun this book with the motivating "founding" example of the Erdős–Kac Theorem, which is usually interpreted as the first result in probabilistic number theory. However, one could arguably say that at the time when this was first proved, there already existed a substantial theory that is really part of probabilistic number theory in our sense, namely, the theory of *equidistribution modulo* 1, due especially to Weyl [120]. Indeed, this concerns originally the study of the fractional parts of various sequences $(x_n)_{n\geqslant 1}$ of real numbers, and the fact that in many cases, including many when x_n has some arithmetic meaning, the fractional parts become *equidistributed* in $[0,1]$ with respect to the Lebesgue measure.

We now make this more precise in probabilistic terms. For a real number x, we will denote (as in Chapter 3) by $\langle x\rangle$ the fractional part of x, namely, the unique real number in $[0,1[$ such that $x-\langle x\rangle \in \mathbf{Z}$. We can identify this value with the point $e(x)=e^{2i\pi x}$ on the unit circle, or with its image in $\mathbf{R}/\mathbf{Z}$, either of which might be more convenient. Given a sequence $(x_n)_{n\geqslant 1}$ of real numbers, we define random variables $\mathrm{S_N}$ on $\Omega_{\mathrm{N}}=\{1,\ldots,\mathrm{N}\}$ (with uniform probability measure) by

$$\mathrm{S_N}(n)=\langle x_n\rangle.$$

Then the sequence $(x_n)_{n\geqslant 1}$ is said to be *equidistributed modulo* 1 if the random variables S_N converge in law to the uniform probability measure dx on $[0, 1]$, as $N \to +\infty$.

Among other things, Weyl proved the following results:

Theorem 7.1.1 (1) *Let* $P \in \mathbf{R}[X]$ *be a polynomial of degree* $d \geqslant 1$ *with leading term* ξX^d *where* $\xi \notin \mathbf{Q}$*. Then the sequence* $(P(n))_{n\geqslant 1}$ *is equidistributed modulo* 1.

(2) *Let* $k \geqslant 1$ *be an integer, and let* $\xi = (\xi_1, \ldots, \xi_d) \in (\mathbf{R}/\mathbf{Z})^d$*. The closure* T *of the set* $\{n\xi \mid n \in \mathbf{Z}\} \subset (\mathbf{R}/\mathbf{Z})^d$ *is a compact subgroup of* $(\mathbf{R}/\mathbf{Z})^d$ *and the* T*-valued random variables on* Ω_N *defined by*

$$\mathsf{K}_N(n) = n\xi$$

converge in law as $N \to +\infty$ *to the probability Haar measure on* T.

The second part of this theorem is the same as Theorem B.6.5, (1). We sketch partial proofs of the first property, which is surprisingly elementary, given the Weyl Criterion (Theorem B.6.3).

We proceed by induction on the degree $d \geqslant 1$ of the polynomial $P \in \mathbf{R}[X]$, using a rather clever trick for this purpose. We may assume that $P(0) = 0$ (as the reader should check). If $d = 1$, then $P = \xi X$ for some real numbers ξ and $P(n) = n\xi$; the assumption is that ξ is irrational, and the result then follows from the 1-dimensional case of the second part, as explained in Example B.6.6.

Suppose that $d = \deg(P) \geqslant 2$ and that the statement is known for polynomials of smaller degree. We use the following:

Lemma 7.1.2 *Let* $(x_n)_{n\geqslant 1}$ *be a sequence of real numbers. Suppose that for any integer* $h \neq 0$*, the sequence* $(x_{n+h} - x_n)_n$ *is equidistributed modulo* 1*. Then* (x_n) *is equidistributed modulo* 1.

Sketch of the proof We leave this as an exercise to the reader; the key step is to use the following very useful inequality of van der Corput: for any integer $N \geqslant 1$, for any family $(a_n)_{1\leqslant n\leqslant N}$ of complex numbers, and for any integer $H \geqslant 1$, we have

$$\left|\sum_{n=1}^{N} a_n\right|^2 \leqslant \left(1 + \frac{N+1}{H}\right) \sum_{|h|<H} \left(1 - \frac{|h|}{H}\right) \sum_{\substack{1\leqslant n\leqslant N\\ 1\leqslant n+h\leqslant N}} a_{n+h}\bar{a}_n.$$

We also leave the proof of this inequality as an exercise... ☐

In the special case of $K_N(n) = \langle P(n) \rangle$, this means that we have to consider auxiliary sequences $K'_N(n) = \langle P(n+h) - P(n) \rangle$, which corresponds to the same problem for the polynomials

$$P(X+h) - P(X) = \xi(X+h)^d - \xi X^d + \cdots = d\xi X^{d-1} + \cdots.$$

Since these polynomials have degree $d-1$, and leading coefficient $d\xi \notin \mathbf{Q}$, the induction hypothesis applies to prove that the random variables K'_N converge to the Lebesgue measure. By the lemma, so does K_N.

Remark 7.1.3 The reader might ask what happens in Theorem B.6.3 if we replace the integers $n \leqslant N$ by primes taken uniformly from those that are $\leqslant N$. The answer is that the same properties hold – for both assertions, we have the same limit in law, under the same conditions on the polynomial for the first one. The proofs are quite a bit more involved however, and depend on Vinogradov's fundamental insight on the "bilinear" nature of the prime numbers. We refer to [59, 13.5, 21.2] for an introduction.

Exercise 7.1.4 Suppose that $0 < \alpha < 1$. Prove that the sequence $(\langle n^\alpha \rangle)_{n \geqslant 1}$ is equidistributed modulo 1.

Even in situations where equidistribution modulo 1 holds, there remain many fascinating and widely open questions when one attempts to go "beyond" equidistribution to understand fluctuations and variations that lie deeper. One of the best known problem in this area is that of the distribution of the *gaps* in a sequence that is equidistributed modulo 1.

Thus let $(x_n)_{n \geqslant 1}$ be a sequence in $\mathbf{R}/\mathbf{Z}$ that is equidistributed modulo 1. For $N \geqslant 1$, consider the set of the N first values

$$\{x_1, \ldots, x_N\}$$

of the sequence. The complement in $\mathbf{R}/\mathbf{Z}$ of these points is a disjoint union of "intervals" (in $[0, 1[$, all but one of them are literally subintervals, and the last one "wraps-around"). The number of these intervals is $\leqslant N$ (there might indeed be less than N, since some of the values x_i might coincide). The question that arises is: what is the distribution of the lengths of these gaps? Stated in a different way, the intervals in question are the connected components of $\mathbf{R}/\mathbf{Z} - \{x_1, \ldots, x_N\}$, and we are interested in the Lebesgue measure of these connected components.

Let Ω_N be the set of the intervals in $\mathbf{R}/\mathbf{Z} - \{x_1, \ldots, x_N\}$, with uniform probability measure. We define random variables by

$$G_N(I) = N \operatorname{length}(I)$$

for $I \in \Omega_N$. Note that the average gap is going to be about $1/N$, so that the multiplication by N leads to a natural normalization where the average of G_N is about 1.

In the case of purely random points located in $\mathbf{S}^1$ independently at random, a classical probabilistic result is that the analogue random variables converge in law to an *exponential random variable* E on $[0, +\infty[$, that is, a random variable such that

$$\mathbf{P}(a < E < b) = \int_a^b e^{-x} dx$$

for any nonnegative real numbers $a < b$. This is also called the "Poisson" behavior. For any (deterministic, for instance, arithmetic) sequence (x_n) that is equidistributed modulo 1, one can then ask whether a similar distribution will arise.

Already the special case of the sequence $(\langle n\xi \rangle)$, for a fixed irrational number ξ, leads to a particularly nice and remarkable answer, the "Three Gaps Theorem" (conjectured by Steinhaus and first proved by Sós [113]). This says that there are *at most three* distinct gaps between the fractional parts $\langle n\xi \rangle$ for $1 \leqslant n \leqslant N$, independently of N and $\xi \notin \mathbf{Q}$.

Although this is in some sense unrelated to our main interests (there is no probabilistic limit theorem here!) we will indicate in Exercise 7.1.5 the steps that lead to a recent proof due to Marklof and Strömbergsson [86]. It is rather modern in spirit, as it depends on the use of lattices in $\mathbf{R}^2$, and especially on the space of lattices.

Very little is known in other cases, but numerical experiments are often easy to perform and lead at least to various conjectural statements. For instance, let $0 < \alpha < 1$ be fixed and put $x_n = \langle n^\alpha \rangle$. By Exercise 7.1.4, the sequence $(x_n)_{n \geqslant 1}$ is equidistributed modulo 1. In this case, it is expected that G_N should have the exponential limiting behavior for all α *except* for $\alpha = \frac{1}{2}$. Remarkably, this exceptional case is the only one where the answer is known! This is a result of Elkies and McMullen that we will discuss below in Section 7.5.

Exercise 7.1.5 Throughout this exercise, we fix an irrational number $\xi \notin \mathbf{Q}$.

(1) For $g \in SL_2(\mathbf{R})$ and $0 \leqslant t < 1$, show that

$$\varphi(g,t) = \inf\{y > 0 \mid \text{there exists } x \text{ such that } -t < x \leqslant 1 - t \text{ and } (x, y) \in \mathbf{Z}^2 g\}$$

exists. Show that the function φ that it defines satisfies $\varphi(\gamma g, t) = \varphi(g, t)$ for all $\gamma \in SL_2(\mathbf{Z})$.

(2) Let $N \geqslant 1$ and $1 \leqslant n \leqslant N$. Prove that the gap between $\langle n\xi \rangle$ and the "next" element of the set

$$\{ \langle \xi \rangle, \ldots, \langle N\xi \rangle \}$$

(i.e., the next one in "clockwise order") is equal to

$$\frac{1}{N}\varphi\left(g_N, \frac{n}{N}\right),$$

where

$$g_N = \begin{pmatrix} 1 & \xi \\ 0 & 1 \end{pmatrix} \begin{pmatrix} N^{-1} & 0 \\ 0 & N \end{pmatrix} \in SL_2(\mathbf{R}).$$

(3) Let $g \in SL_2(\mathbf{R})$ be fixed. Consider the set

$$A_g = g\mathbf{Z}^2 \cap \Big( ]-1, 1[\times]0, +\infty[\Big).$$

Show that there exists $a = (x_1, y_1) \in A_g$ with $y_1 > 0$ minimal.

(4) Show that either there exists $b = (x_2, y_2) \in A_g$, not proportional to a, with y_2 minimal, or $\varphi(g,t) = y_1$ for all t.

(5) Assume that $y_2 > y_1$. Show that (a, b) is a basis of the lattice $g\mathbf{Z}^2 \subset \mathbf{R}^2$, and that x_1 and x_2 have opposite signs. Let

$$I_1 =]0, 1] \cap]-x_1, 1 - x_1] \quad \text{and} \quad I_2 =]0, 1] \cap]-x_2, 1 - x_2].$$

Prove that

$$\varphi(g,t) = \begin{cases} y_2 & \text{if } t \in I_1, \\ y_1 & \text{if } t \in I_2,\ t \notin I_1, \\ y_1 + y_2 & \text{otherwise.} \end{cases}$$

(6) If $y_2 = y_1$, show that $t \mapsto \varphi(g,t)$ takes at most three values by considering similarly $a' = (x_1', y_1') \in A_g$ with $x_1' \geqslant 0$ minimal, and $b' = (x_2', y_2')$ with $x_2' < 0$ maximal.

7.2 Roots of Polynomial Congruences and the Chinese Remainder Theorem

One case of equidistribution modulo 1 deserves mention since it involves some interesting philosophical points, and has been the subject of a number of important works.

Let f be a fixed integral monic polynomial of degree $d \geqslant 1$. For any integer $q \geqslant 1$, the number $\varrho_f(q)$ of roots of f modulo q is finite, and the function ϱ_f is multiplicative (by the Chinese Remainder Theorem); moreover it is elementary that the set M_f of integers $q \geqslant 1$ such that $\varrho_f(q) \geqslant 1$ is infinite. On the other hand, we always have $\varrho_f(p) \leqslant d$ for p prime, so $\varrho_f(q) \leqslant d^{\omega(q)}$ at least when q is squarefree.

Exercise 7.2.1 Prove that M_f is infinite. [**Hint**: It suffices to check that the set of primes p such that $\varrho_f(p) \geqslant 1$ is infinite; assuming that it is not, show that the set of values $f(n)$ for $n \geqslant 1$ would be "too small."]

The question is then: is it true that the fractional parts $\langle a/q \rangle$ of the roots $a \in \mathbf{Z}/q\mathbf{Z}$ of f modulo q, when $\varrho_f(q) \geqslant 1$, become equidistributed modulo 1?

This problem admits a number of variants, and the deepest is undoubtedly the case of equidistribution of $\langle a/p \rangle$ when the modulus p is restricted to be a prime number. Indeed, it is only when $d = 2$ and f is irreducible that the equidistribution of roots modulo primes has been proven, first by Duke–Friedlander–Iwaniec [29] for quadratic polynomials with negative discriminant, and by Toth [118] for quadratic polynomials with positive discriminant, that is, with two real roots.

When all moduli q are taken into account, on the other hand, one can prove equidistribution for any irreducible polynomial, as was first done by Hooley [57]. However, although one might think that this provides evidence for the stronger statement modulo primes, it turns out that this result has in fact almost nothing to do with roots of polynomials!

More precisely, Kowalski and Soundararajan [80] show that equidistribution holds for the fractional parts of elements of sets modulo q obtained by the Chinese Remainder Theorem, starting from subsets A_{p^ν} of $\mathbf{Z}/p^\nu\mathbf{Z}$, under the sole condition that A_p should have at least two elements for a positive proportion of the primes.

In other words, for p prime and $\nu \geqslant 1$, let $\mathrm{A}_{p^\nu} \subset \mathbf{Z}/p^\nu\mathbf{Z}$ be an arbitrary subset of residue classes, and for $q \geqslant 1$, define $\mathrm{A}_q \subset \mathbf{Z}/q\mathbf{Z}$ to be the set of $x \pmod q$ such that, for all primes p dividing q, with exact exponent ν, we have $x \pmod{p^\nu} \in \mathrm{A}_{p^\nu}$. Define $\varrho(q) = |\mathrm{A}_q|$, which is a multiplicative function, and let Ω be the set of all $q \geqslant 1$ such that A_q is not empty. For any $q \in \Omega$, let Δ_q be the probability measure on $\mathbf{R}/\mathbf{Z}$ given by

$$\Delta_q = \frac{1}{\varrho(q)} \sum_{x \in \mathrm{A}_q} \delta_{\langle \frac{a}{q} \rangle},$$

where δ_x denotes a Dirac mass at x (the measure Δ_q is the image of the uniform probability measure on A_q by the map $a \mapsto \langle \frac{a}{q} \rangle$). Then [80, Th. 1.1] implies the following:

Theorem 7.2.2 *Suppose that there exists $\alpha > 0$ such that*

$$\sum_{\substack{p \leqslant Q \\ \varrho(p) \geqslant 2}} 1 \geqslant \alpha \pi(Q)$$

for all Q large enough. Let N(Q) be the number of $q \leqslant Q$ such that A_q is not empty. Then the probability measures

$$\frac{1}{N(Q)} \sum_{\substack{q \leqslant Q \\ q \in \Omega}} \Delta_q$$

converge to the Lebesgue measure on $\mathbf{R}/\mathbf{Z}$.

Example 7.2.3 Let $f \in \mathbf{Z}[X]$ be monic and without repeated roots. If $\deg(f) \geqslant 2$, then this theorem applies to the case where A_{p^ν} is the set of roots of f modulo p^ν, because a basic theorem of algebraic number theory (the Chebotarev Density Theorem; see, for instance, [91, Th. 13.4]) implies that there is a positive proportion of primes p for which f has $\deg(f) \geqslant 2$ distinct roots in $\mathbf{Z}/p\mathbf{Z}$. However, the theorem shows that we can replace A_{p^ν} by any other subset A'_{p^ν} of $\mathbf{Z}/p^\nu\mathbf{Z}$ with the same cardinality, without changing the conclusion concerning the fractional parts modulo all q, whereas (of course) we could select A'_p in such a way that there is no equidistribution modulo primes, in the sense that the measures

$$\frac{1}{P(Q)} \sum_{\substack{p \leqslant Q \\ p \in \Omega}} \Delta_p,$$

where P(Q) is the number of primes $p \leqslant Q$ in Ω, do not converge to the Lebesgue measure.

Remark 7.2.4 Theorem 7.2.2 does not correspond exactly to the setting considered in [57], which concerns (implicitly) the slightly different probability measures

$$\frac{1}{M(Q)} \sum_{\substack{q \leqslant Q \\ q \in \Omega}} \varrho(q) \Delta_q, \tag{7.1}$$

where

$$M(Q) = \sum_{q \leqslant Q} \varrho(q).$$

Interestingly, these two ways of making precise the idea of equidistribution modulo q are *not* equivalent: it is shown in [80, Prop. 2.8] that there exist choices of subsets (A_p) to which Theorem 7.2.2 applies, but for which the measures (7.1) *do not* converge to the uniform measure.

7.3 Gaps between Primes

The Prime Number Theorem

$$\pi(x) \sim \frac{x}{\log x}$$

indicates that the average gap between successive prime numbers of size x is about $\log x$. A natural problem, especially in view of the many conjectures that exist concerning the distribution of primes (such as the Twin Prime conjecture), is to understand the distribution of these gaps.

One way to do this, which is consistent with our general framework, is the following. For any integer $N \geqslant 1$, we define the probability space Ω_N to be the set of integers n such that $1 \leqslant n \leqslant N$ (as in Chapter 2), with the uniform probability measure. Fix $\lambda > 0$. We then define the random variables

$$G_{\lambda,N}(n) = \pi(n + \lambda \log n) - \pi(n),$$

which measures how many primes exist in the interval starting at n of length equal to λ times the average gap.

A precise conjecture exists concerning the limiting behavior of $G_{\lambda,N}$ as $N \to +\infty$:

Conjecture 7.3.1 *The sequence $(G_{\lambda,N})_N$ converges in law as $N \to +\infty$ to a Poisson random variable with parameter λ, that is, for any integer $r \geqslant 0$, we have*

$$\mathbf{P}_N(G_{\lambda,N} = r) \to e^{-\lambda} \frac{\lambda^r}{r!}.$$

To the author's knowledge, this conjecture first appears in the work of Gallagher [45], who in fact proved that it would follow from a suitably uniform version of the famous Hardy-Littlewood k-tuple conjecture. (Interestingly, the same assumption would imply also a generalization of Conjecture 7.3.1 where one considers suitably normalized gaps between simultaneous prime values of a family of polynomials, e.g., between twin primes; see [73], where Gallagher's argument is presented in a probabilistic manner very much in the style of this book.)

Part of the interest of Conjecture 7.3.1 is that the distribution obtained for the gaps is exactly what one expects from "purely random" sets (see the discussion by Feller in [37, I.3, I.4]).

7.4 Cohen–Lenstra Heuristics

In this section, we will assume some basic knowledge concerning algebraic number theory. We refer, for instance, to the book [58] of Ireland and Rosen for an elementary introduction to this subject, in particular to [58, Ch. 12], and to the book [91] of Neukirch for a complete account.

Beginning with a famous paper of Cohen and Lenstra [25], there is by now an impressive body of work concerning the limiting behavior of certain arithmetic measures of a rather different nature than all those we have described up to now. For these, the underlying arithmetic objects are families of number fields of certain kinds, and the random variables of interest are given by the *ideal class groups* of the number fields, or some invariants of the ideal class groups, such as their p-primary subgroups (recall that, as a finite abelian group, the ideal class group C of a number field K can be represented as a direct product of groups of order a power of p, which are zero for all but finitely many p).

The basic idea of Cohen and Lenstra is that the ideal class groups, in suitable families, should behave (in general) in such a way that a given finite abelian group C appears as an ideal class group with "probability" proportional to the inverse $1/\mathrm{Aut}(\mathrm{C})$ of the order of the automorphism group of C, so that, for instance, obtaining a group of order p^2 of the form $\mathbf{Z}/p\mathbf{Z} \times \mathbf{Z}/p\mathbf{Z}$, with automorphism group of size about p^4, is much more unlikely than obtaining the cyclic group $\mathbf{Z}/p^2\mathbf{Z}$, which has automorphism group of size $p^2 - p$.

Imaginary quadratic fields provide a first basic (and still very open!) special case. Using our way of presenting probabilistic number theory, one could define the finite probability spaces Ω_{D} of negative "fundamental discriminants" $-d$ (that is, either $-d$ is a squarefree integer congruent to 3 modulo 4, or $-d = 4\delta$ where δ is squarefree and congruent to 1 or 2 modulo 4) with $1 \leqslant d \leqslant \mathrm{D}$ and the uniform probability measure, and one would define for each D and each prime p a random variable $\mathsf{P}_{p,\mathrm{D}}$ taking values in the set A_p of isomorphism classes of finite abelian groups of order a power of p, such that $\mathsf{P}_{p,\mathrm{D}}(-d)$ is the p-part of the class group of $\mathbf{Q}(\sqrt{-d})$. One of the conjectures ("heuristics") of Cohen and Lenstra is that if $p \geqslant 3$, then $\mathsf{P}_{p,\mathrm{D}}$ should converge in law as $\mathrm{D} \to +\infty$ to the probability measure μ_p on A_p such that

$$\mu_p(\mathrm{A}) = \frac{1}{\mathrm{Z}_p} \frac{1}{|\operatorname{Aut}(\mathrm{A})|}$$

for any group $\mathrm{A} \in \mathrm{A}_p$, where Z_p is the constant required to make the measure thus defined a probability measure (the existence of this measure – in other words, the convergence of the series defining Z_p – is something that of course requires a proof).

Very few unconditional results are known toward these conjectures, and progress often requires significant ideas. There has however been striking advances by Ellenberg, Venkatesh and Westerland [32] in some analogue problems for quadratic extensions of polynomial rings over finite fields, where geometric methods make the problem more accessible, and in fact allow the use of essentially topological ideas (see the Bourbaki report [98] of O. Randal-Williams).

7.5 Ratner Theory

Although all the results that we have described up to now are beautiful and important, maybe the most remarkably versatile tool that can be considered to lie within our chosen context is *Ratner theory*, named after the fundamental work of M. Ratner [99]. We lack the expertise to present anything more than a few selected statements of applications of this theory; we refer to the survey of É. Ghys [47] and to the book of Morris [89] for an introduction (Section 1.4 of that book lists more applications of Ratner Theory), and to that of Einsiedler and Ward [30] for background results on ergodic theory and dynamical systems (some of which also have remarkable applications in number theory).

We illustrate the remarkable power of this theory with the beautiful result of Elkies and McMullen [31] which was already mentioned in Section 7.1. We consider the sequence of fractional parts of $\sqrt{n}$ for $n \geqslant 1$ (viewed as elements of $\mathbf{R}/\mathbf{Z}$). As in the previous section, for any integer $\mathrm{N} \geqslant 1$, we define the space Ω_N to be the set of connected components of $\mathbf{R}/\mathbf{Z} - \{\langle 1\rangle, \ldots, \langle\sqrt{\mathrm{N}}\rangle\}$, with uniform probability measure, and we define random variables on Ω_N by

$$\mathsf{G}_\mathrm{N}(\mathrm{I}) = \mathrm{N} \operatorname{length}(\mathrm{I}).$$

Elkies and McMullen found the limiting distribution of G_N as $\mathrm{N} \to +\infty$. It is a very nongeneric probability measure on $\mathbf{R}$!

Theorem 7.5.1 (Elkies–McMullen) *As* $\mathrm{N} \to +\infty$, *the random variables* G_N *converge in law to a random variable on* $[0, +\infty[$ *with probability law* $\mu_{\mathrm{EM}} = \frac{6}{\pi^2} f(x)dx$, *where* f *is continuous, analytic on the intervals* $[0, 1/2]$, $[1/2, 2]$ *and* $[2, +\infty[$, *is not of class* C^3, *and satisfies* $f(x) = 1$ *if* $0 \leqslant x \leqslant 1/2$.

This is [31, Th. 1.1]. The restriction of the density f to the two intervals $[1/2, 2]$ and $[2, +\infty[$ can be written down explicitly and it is an "elementary" function. For instance, if $1/2 \leqslant x \leqslant 2$, then let $r = \frac{1}{2}x^{-1}$ and

$$\psi(r) = \arctan\left(\frac{2r-1}{\sqrt{4r-1}}\right) - \arctan\left(\frac{1}{\sqrt{4r-1}}\right);$$

we then have

$$f(x) = \frac{2}{3}(4r-1)^{3/2}\psi(r) + (1-6r)\log r + 2r - 1$$

(see [31, (3.53)]).

We give the barest outline of the proof, in order to simply point out what kind of results are meant by Ratner Theory. The paper of Elkies and McMullen also gives a detailed and highly readable introduction to this area.

The proof studies the gap distribution by means of the function L_N defined for $x \in \mathbf{R}/\mathbf{Z}$ so that $\mathrm{L}_\mathrm{N}(x)$ is the measure of the gap interval containing x (with $\mathrm{L}_\mathrm{N}(x) = 0$ if x is one of the boundary points of the gap intervals for $\langle\sqrt{1}\rangle, \ldots, \langle\sqrt{\mathrm{N}}\rangle$). We can then check that for $t \in \mathbf{R}$, the total measure in $\mathbf{R}/\mathbf{Z}$ of the points lying in a gap interval of length $< t$, which is equal to the Lebesgue measure

$$\mu(\{x \in \mathbf{R}/\mathbf{Z} \mid \mathrm{L}_\mathrm{N}(x) < t\}),$$

is given by

$$\int_0^t t d(\mathbf{P}_\mathrm{N}(\mathsf{G}_\mathrm{N} < t)) = t\,\mathbf{P}_\mathrm{N}(\mathsf{G}_\mathrm{N} < t) - \int_0^t \mathbf{P}_\mathrm{N}(\mathsf{G}_\mathrm{N} < t)dt.$$

Concretely, this means that it is enough to understand the limiting behavior of L_N in order to understand the limit gap distribution. Note that there is nothing special about the specific sequence considered in that part of the argument.

Fix $t \geqslant 0$. The key insight that leads to questions involving Ratner theory is that if N is a square of an integer, then the probability

$$\mu(\{x \in \mathbf{R}/\mathbf{Z} \mid \mathrm{L}_\mathrm{N}(x) < t\})$$

can be shown (asymptotically as $\mathrm{N} \to +\infty$) to be very close to the probability that a certain affine lattice $\Lambda_{\mathrm{N},x}$ in $\mathbf{R}^2$ intersects the triangle Δ_t with vertices

$(0,0)$, $(1,0)$ and $(0,2t)$ (with area t). The lattice has the form $\Lambda_{N,x} = g_{N,x} \cdot \mathbf{Z}^2$ for some (fairly explicit) affine transformation $g_{N,t}$.

Let $\mathrm{ASL}_2(\mathbf{R})$ be the group of affine transformations

$$z \mapsto z_0 + g(z)$$

of $\mathbf{R}^2$ whose linear part $g \in \mathrm{GL}_2(\mathbf{R})$ has determinant 1, and $\mathrm{ASL}_2(\mathbf{Z})$ the subgroup of those affine transformations of determinant 1 where both the translation term z_0 and the linear part have coefficients in $\mathbf{Z}$. Then the lattices $\Lambda_{N,x}$ can be interpreted as elements of the quotient space

$$\mathrm{M} = \mathrm{ASL}_2(\mathbf{Z}) \backslash \mathrm{ASL}_2(\mathbf{R}),$$

which parameterizes affine lattices $\Lambda \subset \mathbf{R}^2$ with $\mathbf{R}^2/\Lambda$ of area 1. This space admits a unique probability measure $\widetilde{\mu}$ that is invariant under the right action of $\mathrm{ASL}_2(\mathbf{R})$ by multiplication.

Now we have, for each $\mathrm{N} \geqslant 1$, a probability measure μ_{N} on M, namely, the law of the random variable $\mathbf{R}/\mathbf{Z} \to \mathrm{M}$ defined by $x \mapsto \Lambda_{\mathrm{N},x}$. What Ratner Theory provides is a very powerful set of tools to prove that certain probability measures on M (or on similar spaces constructed with groups more general than $\mathrm{ASL}_2(\mathbf{R})$ and suitable quotients) are equal to the canonical measure $\widetilde{\mu}$. This is applied, essentially, to all possible limits of subsequences of (μ_{N}), to show that these must coincide with $\widetilde{\mu}$, which leads to the conclusion that the whole sequence converges in law to $\widetilde{\mu}$. It then follows that

$$\mu(\{x \in \mathbf{R}/\mathbf{Z} \mid \mathrm{L}_{\mathrm{N}}(x) < t\}) \to \widetilde{\mu}(\{\Lambda \in \mathrm{M} \mid \mathrm{M} \cap \Delta_t \neq \emptyset\}).$$

This gives, in principle, an explicit form of the gap distribution. To compute it exactly is an "exercise" in euclidean geometry – which is by no means easy!

7.6 And Even More ...

And there are even more interactions between probability theory and number theory than what our point of view considers... Here are some examples, which we order, roughly speaking, in terms of how close they are from the perspective of this book:

- Applications of limit theorems for arithmetic probability measures to other problems of analytic number theory: we have given a few examples in exercises (see Exercises 2.3.5 or 3.3.4), but there are many more of course.

- Using probabilistic ideas to *model* arithmetic objects, and make conjectures or prove theorems concerning those; in contrast with our point of view, it is not always expected in such cases that there should exist actual limit theorems comparing the model with the actual arithmetic phenomena. A typical example is the so-called "Cramér model" for the distribution of primes, which is known to lead to wrong conclusions in some cases, but is often close enough to the truth to be used to suggest how certain problems might behave (see, for instance, the survey of Pintz [94]).
- Using number-theoretic ideas to *derandomize* certain constructions or algorithms. There are indeed a number of very interesting results that use the "randomness" of specific arithmetic objects to give deterministic constructions, or deterministic proofs of existence, for mathematical objects that might have first been shown to exist using probabilistic ideas. Examples include the construction of expander graphs by Margulis (see, e.g., [74, §4.4]), or of Ramanujan graphs by Lubotzky, Phillips and Sarnak [84], or in a different vein, the construction of explicit "ultraflat" trigonometric polynomials (in the sense of Kahane) by Bombieri and Bourgain [13], or the construction of explicit functions modulo a prime with smallest possible Gowers norms by Fouvry, Kowalski and Michel [42].

Appendix A

Analysis

In Chapters 3 and 4, we use a number of facts of analysis, and especially complex analysis, which are not necessarily included in most introductory graduate courses. We review them here and give some details of the proofs (when they are sufficiently elementary and enlightening) or detailed references.

A.1 Summation by Parts

Analytic number theory makes very frequent use of "summation by parts," which is a discrete form of integration by parts. We state the version that we use.

Lemma A.1.1 (Summation by parts) *Let $(a_n)_{n\geqslant 1}$ be a sequence of complex numbers and $f\colon [0,+\infty[\to \mathbf{C}$ a function of class C^1. For all $x\geqslant 0$, define*

$$\mathrm{M}_a(x)=\sum_{1\leqslant n\leqslant x} a_n.$$

For $x\geqslant 0$, we then have

$$\sum_{1\leqslant n\leqslant x} a_n f(n)=\mathrm{M}_a(x)f(x)-\int_1^x \mathrm{M}_a(t)f'(t)dt. \tag{A.1}$$

If $\mathrm{M}_a(x)f(x)$ tends to 0 as $x\to+\infty$, then we have

$$\sum_{n\geqslant 1} a_n f(n)=-\int_1^{+\infty}\mathrm{M}_a(t)f'(t)dt,$$

provided either the series or the integral converges absolutely, in which case both of them do.

Using this formula, one can exploit known information (upper bounds or asymptotic formulas) concerning the summation function M_a, typically when the sequence (a_n) is irregular, in order to understand the summation function for $a_n f(n)$ for many sufficiently regular functions f.

The reader should attempt to write a proof of this lemma, but we give the details for completeness.

Proof Let $\mathrm{N} \geqslant 0$ be the integer such that $\mathrm{N} \leqslant x < \mathrm{N}+1$. We have

$$\sum_{1 \leqslant n \leqslant x} a_n f(n) = \sum_{1 \leqslant n \leqslant \mathrm{N}} a_n f(n).$$

By the usual integration by parts formula, we then note that

$$\mathrm{M}_a(x) f(x) - \int_1^x \mathrm{M}_a(t) f'(t) dt = \mathrm{M}_a(\mathrm{N}) f(\mathrm{N}) - \int_1^{\mathrm{N}} \mathrm{M}_a(t) f'(t) dt$$

(because M_a is constant on the interval $\mathrm{N} \leqslant t \leqslant x$). We therefore reduce to the case $x = \mathrm{N}$. We then have

$$\begin{aligned}
\sum_{n \leqslant \mathrm{N}} a_n f(n) &= \sum_{1 \leqslant n \leqslant \mathrm{N}} (\mathrm{M}_a(n) - \mathrm{M}_a(n-1)) f(n) \\
&= \sum_{1 \leqslant n \leqslant \mathrm{N}} \mathrm{M}_a(n) f(n) - \sum_{0 \leqslant n \leqslant \mathrm{N}-1} \mathrm{M}_a(n) f(n+1) \\
&= \mathrm{M}_a(\mathrm{N}) f(\mathrm{N}) + \sum_{1 \leqslant n \leqslant \mathrm{N}-1} \mathrm{M}_a(n)(f(n) - f(n+1)) \\
&= \mathrm{M}_a(\mathrm{N}) f(\mathrm{N}) - \sum_{1 \leqslant n \leqslant \mathrm{N}-1} \mathrm{M}_a(n) \int_n^{n+1} f'(t) dt \\
&= \mathrm{M}_a(\mathrm{N}) f(\mathrm{N}) - \int_1^{\mathrm{N}} \mathrm{M}_a(t) f'(t) dt,
\end{aligned}$$

which concludes the first part of the lemma. The last assertion follows immediately by letting $x \to +\infty$, in view of the assumption on the limit of $\mathrm{M}_a(x) f(x)$. ☐

A.2 The Logarithm

In Chapters 3 and 4, we sometimes use the logarithm for complex numbers. Since this is not a globally defined function on $\mathbf{C}^\times$, we clarify here what we mean.

Definition A.2.1 Let $z \in \mathbf{C}$ be a complex number with $|z| < 1$. We define

$$\log(1 - z) = -\sum_{k \geqslant 1} \frac{z^k}{k}.$$

Proposition A.2.2 (1) *For any complex number z such that $|z| < 1$, we have*

$$e^{\log(1-z)} = 1 - z.$$

(2) *Let $(z_n)_{n\geqslant 1}$ be a sequence of complex numbers such that $|z_n| < 1$. If*

$$\sum_n |z_n| < +\infty,$$

then

$$\prod_{n\geqslant 1}(1 - z_n) = \exp\Bigl(\sum_{n\geqslant 1}\log(1 - z_n)\Bigr).$$

(3) *For $|z| \leqslant \frac{1}{2}$, we have $|\log(1 - z)| \leqslant 2|z|$.*

Proof Part (1) is standard since the series used in the definition is the Taylor series of the logarithm around 1 (evaluated at $-z$), and this power series has radius of convergence 1.

Part (2) is then simply a consequence of the continuity of the exponential and the fact that the product is convergent under the assumption on $(z_n)_{n\geqslant 1}$.

For (3), we note that for $|z| < 1$, we have

$$\log(1 - z) = -z\left(1 + \frac{z}{2} + \frac{z^2}{3} + \cdots - \frac{z^{k-1}}{k} + \cdots\right)$$

so that if $|z| \leqslant \frac{1}{2}$, we get

$$|\log(1 - z)| \leqslant |z|\left(1 + \frac{1}{4} + \frac{1}{8} + \cdots + \frac{1}{2^{k-1}} + \cdots\right) \leqslant 2|z|.$$ □

A.3 Mellin Transform

The Mellin transform is a multiplicative analogue of the Fourier transform, to which it can indeed in principle be reduced. We consider it only in simple cases. Let

$$\varphi\colon [0, +\infty[\longrightarrow \mathbf{C}$$

be a continuous function that decays faster than any polynomial at infinity (for instance, a function with compact support). Then the Mellin transform $\hat{\varphi}$ of φ is the holomorphic function defined by the integral

$$\hat{\varphi}(s) = \int_0^{+\infty} \varphi(x)x^s\frac{dx}{x},$$

for all those $s \in \mathbf{C}$ for which the integral makes sense, which under our assumption includes all complex numbers with $\mathrm{Re}(s) > 0$.

The basic properties of the Mellin transform that are relevant for us are summarized in the next proposition:

Proposition A.3.1 *Let $\varphi\colon [0, +\infty[\longrightarrow \mathbf{C}$ be smooth and assume that φ and all its derivatives decay faster than any polynomial at infinity.*

(1) *The Mellin transform $\hat{\varphi}$ extends to a meromorphic function on $\mathrm{Re}(s) > -1$, with at most a simple pole at $s = 0$ with residue $\varphi(0)$.*

(2) *For any real numbers $-1 < \mathrm{A} < \mathrm{B}$, the Mellin transform has rapid decay in the strip $\mathrm{A} \leqslant \mathrm{Re}(s) \leqslant \mathrm{B}$, in the sense that for any integer $k \geqslant 1$, there exists a constant $\mathrm{C}_k \geqslant 0$ such that*

$$|\hat{\varphi}(s)| \leqslant \mathrm{C}_k(1 + |t|)^{-k}$$

for all $s = \sigma + it$ with $\mathrm{A} \leqslant \sigma \leqslant \mathrm{B}$ and $|t| \geqslant 1$.

(3) *For any $\sigma > 0$ and any $x \geqslant 0$, we have the Mellin inversion formula*

$$\varphi(x) = \frac{1}{2i\pi} \int_{(\sigma)} \hat{\varphi}(s) x^{-s} ds.$$

In the last formula, the notation $\int_{(\sigma)} (\cdots) ds$ refers to an integral over the vertical line $\mathrm{Re}(s) = \sigma$, oriented upward.

Proof (1) We integrate by parts in the definition of $\hat{\varphi}(s)$ for $\mathrm{Re}(s) > 0$ and obtain

$$\hat{\varphi}(s) = \left[\varphi(x)\frac{x^s}{s}\right]_0^{+\infty} - \frac{1}{s}\int_0^{+\infty} \varphi'(x) x^{s+1}\frac{dx}{x} = -\frac{1}{s}\int_0^{+\infty} \varphi'(x) x^{s+1}\frac{dx}{x}$$

since φ and φ' decay faster than any polynomial at ∞. It follows that $\psi(s) = s\hat{\varphi}(s)$ is holomorphic for $\mathrm{Re}(s) > -1$, and hence that $\hat{\varphi}(s)$ is meromorphic in this region. Since

$$\lim_{s\to 0} s\hat{\varphi}(s) = \psi(0) = -\int_0^{+\infty} \varphi'(x) dx = \varphi(0),$$

it follows that there is at most a simple pole with residue $\varphi(0)$ at $s = 0$.

(2) Iterating the integration by parts $k \geqslant 2$ times, we obtain for $\mathrm{Re}(s) > -1$ the relation

$$\hat{\varphi}(s) = \frac{(-1)^k}{s(s+1)\cdots(s+k)} \int_0^{+\infty} \varphi^{(k)}(x) x^{s+k}\frac{dx}{x}.$$

Hence, for $\mathrm{A} \leqslant \sigma \leqslant \mathrm{B}$ and $|t| \geqslant 1$, we obtain the bound

$$|\hat{\varphi}(s)| \ll \frac{1}{(1+|t|)^k} \int_0^{+\infty} |\varphi^{(k)}(x)| x^{\mathrm{B}+k}\frac{dx}{x} \ll \frac{1}{(1+|t|)^k}.$$

(3) We interpret $\hat{\varphi}(s)$, for $s = \sigma + it$ with $\sigma > 0$ fixed, as a Fourier transform; by means of the change of variable $x = e^y$, we have

$$\hat{\varphi}(s) = \int_0^{+\infty} \varphi(x) x^{\sigma} x^{it} \frac{dx}{x} = \int_{\mathbf{R}} \varphi(e^y) e^{\sigma y} e^{iyt} dy,$$

which shows that $t \mapsto \hat{\varphi}(\sigma + it)$ is the Fourier transform (with the above normalization) of the function $g(y) = \varphi(e^y)e^{\sigma y}$. Note that g is smooth and tends to zero very rapidly at infinity (for $y \to -\infty$, this is because φ is bounded close to 0, but $e^{\sigma y}$ then tends exponentially fast to 0). Therefore the Fourier inversion formula holds, and for any $y \in \mathbf{R}$, we obtain

$$\varphi(e^y)e^{\sigma y} = \frac{1}{2\pi} \int_{\mathbf{R}} \hat{\varphi}(\sigma + it) e^{-ity} dt.$$

Putting $x = e^y$, this translates to

$$\varphi(x) = \frac{1}{2\pi} \int_{\mathbf{R}} \hat{\varphi}(\sigma + it) x^{-\sigma - it} dt = \frac{1}{2i\pi} \int_{(\sigma)} \hat{\varphi}(s) x^{-s} ds.$$

□

One of the most important functions of analysis is classically defined as a Mellin transform. This is the Gamma function of Euler, which is essentially the Mellin transform of the exponential function, or more precisely of $\exp(-x)$. In other words, we have

$$\Gamma(s) = \int_0^{+\infty} e^{-x} x^s \frac{dx}{x}$$

for all complex numbers s such that $\mathrm{Re}(s) > 0$. Proposition A.3.1 shows that Γ extends to a meromorphic function on $\mathrm{Re}(s) > -1$, with a simple pole at $s = 0$ with residue 1. In fact, much more is true:

Proposition A.3.2 *The function $\Gamma(s)$ extends to a meromorphic function on $\mathbf{C}$ with only simple poles at $s = -k$ for all integers $k \geqslant 0$, with residue $(-1)^k/k!$. It satisfies*

$$\Gamma(s + 1) = s\Gamma(s)$$

for all $s \in \mathbf{C}$, with the obvious meaning if s or $s + 1$ is a pole, and in particular we have

$$\Gamma(n) = (n - 1)!$$

for all integers $n \geqslant 0$.

Moreover, the function $1/\Gamma$ is entire.

Proof It suffices to prove that $\Gamma(s+1) = s\Gamma(s)$ for $\mathrm{Re}(s) > 0$. Indeed, this formula proves, by induction on $k \geqslant 1$, that Γ has an analytic continuation to $\mathrm{Re}(s) > -k$, with a simple pole at $-k+1$, where the residue r_{-k+1} satisfies

$$r_{-k+1} = \frac{r_{-k+2}}{-k+1}.$$

This easily gives every statement in the proposition. And the formula we want is just a simple integration by parts away:

$$\Gamma(s+1) = \int_0^{+\infty} e^{-x}x^s dx = \left[-e^{-x}x^s\right]_0^{+\infty} + s\int_0^{+\infty} e^{-x}x^{s-1}dx = s\Gamma(s).$$

Since Γ is meromorphic, its inverse $1/\Gamma$ is also meromorphic; for the proof that $1/\Gamma$ is in fact entire (i.e., that $\Gamma(s) \neq 0$ for $s \in \mathbf{C}$), we refer to [116, p. 149] (it follows, e.g., from the formula

$$\Gamma(s)\Gamma(1-s) = \frac{\pi}{\sin(\pi s)},$$

valid for all $s \in \mathbf{C}$, since the known poles of $\Gamma(1-s)$ are compensated by those of $1/\sin(\pi s)$). □

An important feature of the Gamma function, which is often quite important, is that its asymptotic behavior in very wide ranges of the complex plane is very clearly understood. This is the so-called *Stirling formula.*

Proposition A.3.3 *Let $\alpha > 0$ be a real number and let X_α be either the set of $s \in \mathbf{C}$ such that $\mathrm{Re}(s) > \alpha$ or the set of $s \in \mathbf{C}$ such that $|\mathrm{Im}(s)| > \alpha$. We have*

$$\log\Gamma(s) = s\log s - s - \frac{1}{2}\log s + \frac{1}{2}\log 2\pi + \mathrm{O}(|s|^{-1}),$$

$$\frac{\Gamma'(s)}{\Gamma(s)} = \log s - \frac{1}{2s} + \mathrm{O}(s^{-2})$$

for any $s \in \mathrm{X}_\alpha$.

For a proof, see for instance [16, Ch. VII, Prop. 4].

A.4 Dirichlet Series

We present in this section some of the basic analytic properties of series of the type

$$\sum_{n\geqslant 1} a_n n^{-s},$$

where $a_n \in \mathbf{C}$ for $n \geqslant 1$. These are called *Dirichlet series*, and we refer to Titchmarsh's book [116, Ch. 9] for basic information about these functions.

If $a_n = 0$ for n large enough (so that there are only finitely many terms), the series converges of course for all s, and the resulting function is called a *Dirichlet polynomial.*

Lemma A.4.1 *Let $(a_n)_{n\geqslant 1}$ be a sequence of complex numbers. Let $s_0 \in \mathbf{C}$. If the series*

$$\sum_{n\geqslant 1} a_n n^{-s_0}$$

converges, then the series

$$\sum_{n\geqslant 1} a_n n^{-s}$$

converges uniformly on compact subsets of $\mathrm{U} = \{s \in \mathbf{C} \mid \mathrm{Re}(s) > \mathrm{Re}(s_0)\}$. *In particular the function*

$$f(s) = \sum_{n\geqslant 1} a_n n^{-s}$$

is holomorphic on U.

Sketch of proof We may assume (by considering $a_n n^{-s_0}$ instead of a_n) that $s_0 = 0$. For any integers $\mathrm{N} < \mathrm{M}$, let

$$s_{\mathrm{N},\mathrm{M}} = a_{\mathrm{N}} + \cdots + a_{\mathrm{M}}.$$

By Cauchy's criterion, we have $s_{\mathrm{N},\mathrm{M}} \to 0$ as $\mathrm{N}, \mathrm{M} \to +\infty$. Suppose that $\sigma = \mathrm{Re}(s) > 0$. Let $\mathrm{N} < \mathrm{M}$ be integers. By the elementary summation by parts formula (Lemma A.1.1), we have

$$\sum_{\mathrm{N}\leqslant n\leqslant \mathrm{M}} a_n n^{-s} = a_{\mathrm{M}}\mathrm{M}^{-s} - \sum_{\mathrm{N}\leqslant n<\mathrm{M}} ((n+1)^{-s} - n^{-s}) s_{\mathrm{N},n}.$$

It is however also elementary that

$$|(n+1)^{-s} - n^{-s}| = \left| s\int_n^{n+1} x^{-s-1}dx \right| \leqslant \frac{|s|}{\sigma}(n^{-\sigma} - (n+1)^{-\sigma}). \quad \text{(A.2)}$$

Hence

$$\left| \sum_{\mathrm{N}\leqslant n\leqslant \mathrm{M}} a_n n^{-s} \right| \leqslant \frac{|s|}{\sigma} \max_{\mathrm{N}\leqslant n\leqslant \mathrm{M}} |s_{\mathrm{N},n}| \left(\frac{1}{\mathrm{N}^\sigma} - \frac{1}{\mathrm{M}^\sigma} \right).$$

It therefore follows by Cauchy's criterion that the Dirichlet series $f(s)$ converges uniformly in any region in $\mathbf{C}$ defined by the condition

$$\frac{|s|}{\sigma} \leqslant \mathrm{A}$$

for some $A > 0$. This includes, for a suitable value of A, any compact subset of the half-plane $\{s \in \mathbf{C} \mid \sigma > 0\}$. □

In general, the convergence is *not* absolute. We can see in this lemma a first instance of a fairly general principle concerning Dirichlet series: if some particular property holds for some $s_0 \in \mathbf{C}$ (or for all s_0 with some fixed real part), then it holds – or even a stronger property holds – for any s with $\mathrm{Re}(s) > \mathrm{Re}(s_0)$.

This principle also applies in many cases to the possible analytic continuation of Dirichlet series beyond the region of convergence. The next proposition is another example, concerning the size of the Dirichlet series.

Proposition A.4.2 *Let $\sigma \in \mathbf{R}$ be a real number and let $(a_n)_{n\geqslant 1}$ be a bounded sequence of complex numbers such that the Dirichlet series*

$$f(s) = \sum_{n\geqslant 1} a_n n^{-s}$$

converges for $\mathrm{Re}(s) > \sigma$. *Then, for any $\sigma_1 > \sigma$, we have*

$$|f(s)| \ll 1 + |t|$$

uniformly for $\mathrm{Re}(s) \geqslant \sigma_1$.

Proof We may assume that $\sum a_n$ converges by replacing (a_n) by $(a_n n^{-\tau})$ for some $\tau > \sigma$. The partial sums

$$s_N = a_1 + \cdots + a_N$$

are then bounded. Let $s \in \mathbf{C}$ be such that $\sigma = \mathrm{Re}(s) > 0$. Then we have by partial summation

$$\sum_{n=1}^{N} \frac{a_n}{n^s} = \sum_{n=1}^{M} \frac{a_n}{n^s} + \sum_{n=M+1}^{N} \left(\frac{1}{n^s} - \frac{1}{(n+1)^s}\right) s_n - \frac{s_M}{(M+1)^s} + \frac{s_N}{(N+1)^s}$$

for any integers $M \leqslant n$. Letting $N \to +\infty$, as we may, we get

$$f(s) = \sum_{n=1}^{M} \frac{a_n}{n^s} + \sum_{n>M} \left(\frac{1}{n^s} - \frac{1}{(n+1)^s}\right) s_n - \frac{s_M}{(M+1)^s}.$$

Applying (A.2), this leads to

$$|f(s)| \ll \sum_{n=1}^{M} \frac{1}{n^{\sigma}} + \frac{|s|}{\sigma} \sum_{n>M} \left(\frac{1}{n^{\sigma}} - \frac{1}{(n+1)^{\sigma}} \right) + \frac{1}{(M+1)^{\sigma}}$$
$$\ll M + tM^{-\sigma} + 1,$$

and the desired bounds follow by taking $M=\lceil \mathrm{Im}(s) \rceil$ (see also [116, 9.33]). □

In order to express in a practical manner a Dirichlet series outside of its region of convergence, one can use *smooth partial sums*, which exploit harmonic analysis.

Proposition A.4.3 *Let* $\varphi\colon [0, +\infty[\longrightarrow [0,1]$ *be a smooth function with compact support such that* $\varphi(0) = 1$. *Let* $\hat{\varphi}$ *denote its Mellin transform. Let* $\sigma > 0$ *be given with* $0 < \sigma_0 < 1$, *and let* $(a_n)_{n\geqslant 1}$ *be any sequence of complex numbers with* $|a_n| \leqslant 1$ *such that the Dirichlet series*

$$\sum_{n\geqslant 1} a_n n^{-s}$$

extends to a holomorphic function $f(s)$ *in the region* $\mathrm{Re}(s) > \sigma_0$ *with at most a simple pole at* $s = 1$ *with residue* $c \in \mathbf{C}$.

For $N \geqslant 1$, *define*

$$f_N(s) = \sum_{n\geqslant 1} a_n \varphi\left(\frac{n}{N}\right) n^{-s}.$$

Let σ *be a real number such that* $\sigma_0 < \sigma < 1$. *Then we have*

$$f(s) - f_N(s) = -\frac{1}{2i\pi} \int_{(-\delta)} f(s+w) N^{w} \hat{\varphi}(w) dw - cN^{1-s}\hat{\varphi}(1-s)$$

for any $s = \sigma + it$ *and any* $\delta > 0$ *such that* $-\delta + \sigma > \sigma_0$.

It is of course possible that $c = 0$, which corresponds to a Dirichlet series that is holomorphic for $\mathrm{Re}(s) > \sigma_0$.

This result gives a convergent approximation of $f(s)$, inside the strip $\mathrm{Re}(s) > \sigma_1$, using the *finite* sums $f_N(s)$ – the point is that $|N^{w}| = N^{-\delta}$, so that the polynomial growth of f on vertical lines combined with the fast decay of the Mellin transform $\hat{\varphi}$ shows that the integral on the right tends to 0 as $N \to +\infty$. Moreover, the shape of the formula makes it very accessible to further manipulations, as done in Chapter 3.

Proof Fix $\alpha > 1$ such that the Dirichlet series $f(s)$ converges absolutely for $\mathrm{Re}(s) = \alpha$. By the Mellin inversion formula, followed by exchanging the order of the sum and integral, we have

$$\begin{aligned} f_{\mathrm{N}}(s) &= \sum_{n\geqslant 1} a_n n^{-s} \times \frac{1}{2i\pi}\int_{(\alpha)} \mathrm{N}^w n^{-w}\hat{\varphi}(w)dw \\ &= \frac{1}{2i\pi}\int_{(-\delta)} \Big(\sum_{n\geqslant 1} a_n n^{-s-w}\Big)\mathrm{N}^w\hat{\varphi}(w)dw \\ &= \frac{1}{2i\pi}\int_{(-\delta)} f(s+w)\mathrm{N}^w\hat{\varphi}(w)dw, \end{aligned}$$

where the absolute convergence justifies the exchange of sum and integral.

Now consider some $\mathrm{T} \geqslant 1$ and some δ such that $0 < \delta < 1$. Let $\mathcal{R}_{\mathrm{T}}$ be the rectangle in $\mathbf{C}$ with sides $[\alpha - i\mathrm{T}, \alpha + i\mathrm{T}]$, $[\alpha + i\mathrm{T}, -\delta + i\mathrm{T}]$, $[-\delta + i\mathrm{T}, -\delta - \mathrm{T}]$, $[-\delta - i\mathrm{T}, \alpha - i\mathrm{T}]$, oriented counterclockwise. Inside this rectangle, the function

$$w \mapsto f(s+w)\mathrm{N}^w\hat{\varphi}(w)$$

is meromorphic. It has a simple pole at $w = 0$, by our choice of δ and the properties of the Mellin transform of φ given by Proposition A.3.1, and the residue at $w = 0$ is $\varphi(0)f(s) = f(s)$, again by Proposition A.3.1. If $c \neq 0$, it may also have a simple pole at $w = 1-s$, with residue equal to $c\mathrm{N}^{1-s}\hat{\varphi}(1-s)$.

Cauchy's theorem therefore implies that

$$f_{\mathrm{N}}(s) = f(s) + \frac{1}{2i\pi}\int_{\mathcal{R}_{\mathrm{T}}} f(s+w)\mathrm{N}^w\hat{\varphi}(w)dw + c\mathrm{N}^{1-s}\hat{\varphi}(1-s).$$

Now we let $\mathrm{T} \to +\infty$. Our assumptions imply that $w \mapsto f(s+w)$ has polynomial growth on the strip $-\delta \leqslant \mathrm{Re}(w) \leqslant \alpha$, and therefore the fast decay of $\hat{\varphi}$ (Proposition A.3.1 again) shows that the contribution of the two horizontal segments to the integral along $\mathcal{R}_{\mathrm{T}}$ tends to 0 as $\mathrm{T} \to +\infty$. Taking into account orientation, we get

$$f(s) - f_{\mathrm{N}}(s) = -\frac{1}{2i\pi}\int_{(-\delta)} f(s+w)\mathrm{N}^w\hat{\varphi}(w)dw - c\mathrm{N}^{1-s}\hat{\varphi}(1-s),$$

as claimed. □

We also recall the formula for the product of two Dirichlet series, which involves the so-called *Dirichlet convolution* (see also Section C.1 for more properties and examples of this operation).

Proposition A.4.4 *Let $(a(n))_{n\geqslant 1}$ and $(b(n))_{n\geqslant 1}$ be sequences of complex numbers. For any $s \in \mathbf{C}$ such that the Dirichlet series*

$$\mathrm{A}(s) = \sum_{n \geqslant 1} a(n) n^{-s} \quad \text{and} \quad \mathrm{B}(s) = \sum_{n \geqslant 1} b(n) n^{-s},$$

converge absolutely, we have

$$\mathrm{A}(s)\mathrm{B}(s) = \sum_{n \geqslant 1} c(n) n^{-s},$$

where

$$c(n) = \sum_{\substack{d \mid n \\ d \geqslant 1}} a(d) b\left(\frac{n}{d}\right).$$

We will denote $c(n) = (a \star b)(n)$ and often abbreviate the definition by writing

$$(a \star b)(n) = \sum_{d \mid n} a(d) b\left(\frac{n}{d}\right) \quad \text{or} \quad (a \star b)(n) = \sum_{de = n} a(d) b(e).$$

Proof Formally, this is quite clear:

$$\mathrm{A}(s)\mathrm{B}(s) = \Big(\sum_{n \geqslant 1} a(n) n^{-s}\Big)\Big(\sum_{n \geqslant 1} b(m) m^{-s}\Big)$$
$$= \sum_{m,n \geqslant 1} a(n) b(m) (nm)^{-s} = \sum_{k \geqslant 1} k^{-s} \Big(\sum_{mn = k} a(n) b(m)\Big) = \mathrm{C}(s).$$

The assumptions are sufficient to allow us to rearrange the double series so that these manipulations are valid. □

A.5 Density of Certain Sets of Holomorphic Functions

Let D be a nonempty open disc in **C** and $\bar{\mathrm{D}}$ its closure. We denote by H(D) the Banach space of all continuous functions $f : \bar{\mathrm{D}} \longrightarrow \mathbf{C}$ that are holomorphic in D, with the norm

$$\|f\|_\infty = \sup_{z \in \bar{\mathrm{D}}} |f(z)|.$$

We also denote by C(K) the Banach space of continuous functions on a compact space K, also with the norm

$$\|f\|_\infty = \sup_{x \in \mathrm{K}} |f(x)|$$

(so that there is no risk of confusion if K = D and we apply this to a function that also belongs to H(D)). We denote by C(K)′ the dual of C(K), namely, the space of continuous linear functionals $\mathrm{C(K)} \longrightarrow \mathbf{C}$. An element $\mu \in \mathrm{C(K)}'$ can also be interpreted as a *complex measure* on K (by the Riesz–Markov Theorem; see, e.g., [40, Th. 7.17]), and in this interpretation, one would write

$$\mu(f) = \int_{\mathrm{K}} f(x) d\mu(x)$$

for $f \in \mathrm{C(K)}$.

Theorem A.5.1 *Let* D *be as above. Let* $(f_n)_{n\geqslant 1}$ *be a sequence of elements of* H(D) *with*

$$\sum_{n\geqslant 1} \| f_n \|_\infty^2 < +\infty.$$

Let X *be the set of sequences* (α_n) *of complex numbers with* $|\alpha_n| = 1$ *such that the series*

$$\sum_{n\geqslant 1} \alpha_n f_n$$

converges in H(D).

Assume that X *is not empty and that, for any continuous linear functional* $\mu \in \mathrm{C}(\bar{\mathrm{D}})'$ *such that*

$$\sum_{n\geqslant 1} |\mu(f_n)| < +\infty, \tag{A.3}$$

the Laplace transform of μ *is identically* 0. *Then, for any* $\mathrm{N} \geqslant 1$, *the set of series*

$$\sum_{n\geqslant \mathrm{N}} \alpha_n f_n$$

for (α_n) *in* X *is* dense *in* H(D).

Here, the Laplace transform of μ is defined by

$$g(z) = \mu(w \mapsto e^{wz})$$

for $z \in \mathbf{C}$. In the interpretation of μ as a complex measure, which can be viewed as a complex measure on $\mathbf{C}$ that is supported on $\bar{\mathrm{D}}$, one would write

$$g(z) = \int_{\mathbf{C}} e^{wz} d\mu(w).$$

Proof This result is proved, for instance, in [4, Lemma 5.2.9], except that only the case $\mathrm{N} = 1$ is considered. However, if the assumptions hold for $(f_n)_{n\geqslant 1}$, they hold equally for $(f_n)_{n>\mathrm{N}}$, hence the general case follows. ☐

We will use the last part of the following lemma as a criterion to establish that the Laplace transform is zero in certain circumstances.

Lemma A.5.2 *Let* K *be a complex subset of* **C** *and* $\mu \in \mathrm{C}(\mathrm{K})'$ *a continuous linear functional. Let*

$$g(z) = \int e^{wz} d\mu(z) = \mu(w \mapsto e^{wz})$$

be its Laplace transform.

(1) *The function* g *is an entire function on* **C***, that is, it is holomorphic on* **C**.

(2) *We have*

$$\limsup_{|z|\to+\infty} \frac{\log|g(z)|}{|z|} < +\infty.$$

(3) *If* $g \neq 0$*, then*

$$\limsup_{r\to+\infty} \frac{\log|g(r)|}{r} \geqslant \inf_{z\in\mathrm{K}} \mathrm{Re}(z).$$

Proof (1) Let $z \in \mathbf{C}$ be fixed. For $h \neq 0$, we have

$$\frac{g(z+h)-g(z)}{h} = \mu(f_h),$$

where $f_h(w) = (e^{w(z+h)} - e^{wz})/h$. We have

$$f_h(w) \to we^{wz}$$

as $h \to 0$, and the convergence is uniform on K. Hence we get

$$\frac{g(z+h)-g(z)}{h} \longrightarrow \mu(w \mapsto we^{wz}),$$

which shows that g is holomorphic at z with derivative $\mu(w \mapsto we^{wz})$. Since z is arbitrary, this means that g is entire.

(2) We have

$$|g(z)| \leqslant \|\mu\| \, \|w \mapsto e^{wz}\|_\infty \leqslant \|\mu\| e^{|z|\mathrm{M}}$$

where $\mathrm{M} = \sup_{w\in\mathrm{K}} |w|$, and therefore

$$\limsup_{|z|\to+\infty} \frac{\log|g(z)|}{|z|} \leqslant \mathrm{M} < +\infty.$$

(3) This is proved, for instance, in [4, Lemma 5.2.2], using relatively elementary properties of entire functions satisfying growth conditions such as those in (2). □

Finally, we will use the following theorem of Bernstein, extending a result of Pólya.

Theorem A.5.3 *Let* $g\colon \mathbf{C} \longrightarrow \mathbf{C}$ *be an entire function such that*

$$\limsup_{|z|\to+\infty} \frac{\log|g(z)|}{|z|} < +\infty.$$

Let (r_k) *be a sequence of positive real numbers, and let* α, β *be real numbers such that*

(1) *we have* $\alpha\beta < \pi$;

(2) *we have*

$$\limsup_{\substack{y\in\mathbf{R}\\ |y|\to+\infty}} \frac{\log|g(iy)|}{|y|} \leqslant \alpha;$$

(3) *we have* $|r_k - r_l| \gg |k - l|$ *for all* $k, l \geqslant 1$, *and* $r_k/k \to \beta$.

Then it follows that

$$\limsup_{k\to+\infty} \frac{\log|g(r_k)|}{r_k} = \limsup_{r\to+\infty} \frac{\log|g(r)|}{r}.$$

This is explained in Lemma [4, 5.2.3].

Example A.5.4 Taking $g(z) = \sin(\pi z)$, with $\alpha = 1$, $r_n = n\pi$ so that $\beta = \pi$, we see that the first condition is best possible.

We also use a relatively elementary lemma due to Hurwitz on zeros of holomorphic functions.

Lemma A.5.5 *Let* D *be a nonempty open disc in* **C**. *Let* (f_n) *be a sequence of holomorphic functions in* H(D). *Assume* f_n *converges to* f *in* H(D). *If* $f_n(z) \neq 0$ *all* $n \geqslant 1$ *and* $z \in \mathrm{D}$, *then either* $f = 0$ *or* f *does not vanish on* D.

Proof We assume that f is not zero, and show that it has no zero in D. Let $z_0 \in \mathrm{D}$ be fixed, and let C be a circle of radius $r > 0$ centered at z_0 such that $\mathrm{C} \subset \mathrm{D}$ and such that f has no zero, except possibly z_0, in the disc with boundary C. We have $\delta = \inf_{z\in\mathrm{C}} |f(z)| > 0$. For n large enough, we get

$$\sup_{z\in\mathrm{C}} |f(z) - f_n(z)| < \delta,$$

and then the relation $f = f - f_n + f_n$ combined with Rouché's Theorem (see, e.g., [116, 3.42]) shows that f has the same number of zeros as f_n in the disc bounded by C. This means that f has no zeros there and, in particular, that $f(z_0) \neq 0$. ☐

Appendix B

Probability

This appendix summarizes the probabilistic notions that are most important in the book. Although many readers will not need to be reminded of the basic definitions, they might still refer to it to check some easy probabilistic statements whose proof we have included here to avoid disrupting the arguments in the main part of the book. For convergence in law, we will refer mostly to the book of Billingsley [10] and, for random series and similar topics, to that of Li and Queffélec [83].

B.1 The Riesz Representation Theorem

Let X be a *locally compact* topological space (such as $\mathbf{R}^d$ for $d \geqslant 1$). We recall that *Radon measures* on X are certain measures for which compact subsets of X have finite measure, and which satisfy some regularity property (the latter requirement being unnecessary if any open set in X is a countable union of compact sets, as is the case of **R** for instance).

The Riesz representation theorem interprets Radon measures in terms of the corresponding integration functional. It can be taken as a definition (and it is indeed the definition in Bourbaki's theory of integration [17]); for a proof in the usual context where measures are defined "set-theoretically," see, for example, [106, Th 2.14].

Theorem B.1.1 *Let* X *be a locally compact topological space and* $\mathrm{C}_c(\mathrm{X})$ *the space of compactly supported continuous functions on* X. *For any linear form* $\lambda\colon \mathrm{C}_c(\mathrm{X}) \to \mathbf{C}$ *such that* $\lambda(f) \geqslant 0$ *if* $f \geqslant 0$, *there exists a unique Radon measure* μ *on* X *such that*

$$\lambda(f) = \int_{\mathrm{X}} f \, d\mu$$

for all $f \in \mathrm{C}_c(\mathrm{X})$.

Let $k \geqslant 1$ be an integer. If $X = \mathbf{R}^k$ (or an open set in $\mathbf{R}^k$), then Radon measures can be identified by the integration of much more regular functions. For instance, we have the following (see, e.g., [28, Th. 3.18]):

Proposition B.1.2 *Let $C_c^\infty(\mathbf{R}^k)$ be the space of smooth compactly-supported functions on $\mathbf{R}$. For any linear form $\lambda\colon C_c^\infty(\mathbf{R}^k) \to \mathbf{C}$ such that $\lambda(f) \geqslant 0$ if $f \geqslant 0$, there exists a unique Radon measure μ on $\mathbf{R}^k$ such that*

$$\lambda(f) = \int_{\mathbf{R}^k} f \, d\mu$$

for all $f \in C_c^\infty(\mathbf{R}^k)$.

Remark B.1.3 When applying either form of the Riesz representation theorem, we may wish to identify whether the measure μ obtained from the linear form λ is a probability measure on X or not. This is the case if and only

$$\sup_{\substack{f \in C_c(X) \\ 0 \leqslant f \leqslant 1}} \lambda(f) = 1,$$

(see, e.g., [17, Ch. 4, § 1, n°8]) where, in the setting of Proposition B.1.2, we may also restrict f to be smooth.

Moreover, if a positive linear form $\lambda\colon C_c(X) \to \mathbf{C}$ admits an extension to a linear form $\lambda\colon C_b(X) \to \mathbf{C}$, where $C_b(X)$ is the space of continuous and bounded functions on X, which is still positive (so $\lambda(f) \geqslant 0$ if $f \in C_b(X)$ is non-negative), then the measure μ associated to λ is a probability measure if and only if $\lambda(1) = 1$, where 1 on the left is the constant function. (This is natural enough, but it is not entirely obvious; the underlying reason is that the positivity implies that

$$|\lambda(f)| \leqslant \|f\|_\infty \lambda(1)$$

where $\|f\|_\infty$ is the supremum norm for a bounded continuous function, so that λ is a continuous linear form on the Banach space $C_b(X)$.)

B.2 Support of a Measure

Let M be a topological space. If M is either second countable (i.e., there is basis of open sets that is countable) or compact, then any Borel measure μ on M has a well-defined closed *support*, denoted $\mathrm{supp}(\mu)$, which is characterized by either of the following properties: (1) it is the complement of the largest open set U, with respect to inclusion, such that $\mu(U) = 0$; or (2) it is the set of those $x \in M$ such that, for any open neighborhood U of x, we have $\mu(U) > 0$.

If X is a random variable with values in M, we will say that the *support of* X is the support of the law of X, which is a probability measure on M.

We need the following elementary property of the support of a measure:

Lemma B.2.1 *Let* M *and* N *be topological spaces that are each either second countable or compact. Let* μ *be a probability measure on* M, *and let* $f : \mathrm{M} \longrightarrow \mathrm{N}$ *be a continuous map. The support of* $f_*(\mu)$ *is the closure of* $f(\mathrm{supp}(\mu))$.

We recall that given a probability measure μ on M and a continuous map $f : \mathrm{M} \to \mathrm{N}$, the *image measure* $f_*(\mu)$ is defined by

$$f_*(\mu)(\mathrm{A}) = \mu(f^{-1}(\mathrm{A}))$$

for a measurable set $\mathrm{A} \subset \mathrm{N}$, and it satisfies

$$\int_{\mathrm{N}} \varphi(x) d(f_*\mu)(x) = \int_{\mathrm{M}} \varphi(f(y)) d\mu(y)$$

for $\varphi \geqslant 0$ and measurable, or $\varphi \circ f$ integrable with respect to μ.

Proof First, if $y = f(x)$ for some $x \in \mathrm{supp}(\mu)$, and if U is an open neighborhood of y, then we can find an open neighborhood $\mathrm{V} \subset \mathrm{M}$ of x such that $f(\mathrm{V}) \subset \mathrm{U}$. Then $(f_*\mu)(\mathrm{U}) \geqslant \mu(\mathrm{V}) > 0$. This shows that y belongs to the support of $f_*\mu$. Since the support is closed, we deduce that $\overline{f(\mathrm{supp}(\mu))} \subset \mathrm{supp}(f_*\mu)$.

For the converse, let $y \in \mathrm{N}$ be in the support of $f_*\mu$. For any open neighborhood U of y, we have $\mu(f^{-1}(\mathrm{U})) = (f_*\mu)(\mathrm{U}) > 0$. This implies that $f^{-1}(\mathrm{U}) \cap \mathrm{supp}(\mu)$ is not empty, and since U is arbitrary, that y belongs to the closure of $f(\mathrm{supp}(\mu))$. □

Recall that a family $(\mathrm{X}_i)_{i \in \mathrm{I}}$ of random variables, each taking possibly values in a different metric space M_i, is *independent* if, for any finite subset $\mathrm{J} \subset \mathrm{I}$, the joint distribution of $(\mathrm{X}_j)_{j \in \mathrm{J}}$ is the measure on $\prod \mathrm{M}_j$ which is the product measure of the laws of the X_j.

Lemma B.2.2 *Let* $\mathrm{X} = (\mathrm{X}_i)_{i \in \mathrm{I}}$ *be a finite family of random variables with values in a topological space* M *that is compact or second countable. Viewed as a random variable taking values in* M^{I}, *we have*

$$\mathrm{supp}(\mathrm{X}) = \prod_{i \in \mathrm{I}} \mathrm{supp}(\mathrm{X}_i).$$

Proof If $x = (x_i) \in \mathrm{M}^{\mathrm{I}}$, then an open neighborhood U of x contains a product set $\prod \mathrm{U}_i$, where U_i is an open neighborhood of x_i in M. Then we have

$$\mathbf{P}(\mathrm{X} \in \mathrm{U}) \geqslant \mathbf{P}(\mathrm{X} \in \prod_i \mathrm{U}_i) = \prod_i \mathbf{P}(\mathrm{X}_i \in \mathrm{U}_i)$$

by independence. If $x_i \in \mathrm{supp}(X_i)$ for each i, then this is > 0, and hence $x \in \mathrm{supp}(X)$.

Conversely, if $x \in \mathrm{supp}(X)$, then for any $j \in I$, and any open neighborhood U of x_j, the set

$$V = \{y = (y_i)_{i \in I} \in M^I \mid y_j \in U\} \subset M^I$$

is an open neighborhood of x. Hence we have $\mathbf{P}(X \in V) > 0$, and since $\mathbf{P}(X \in V) = \mathbf{P}(X_i \in U)$, it follows that x_j is in the support of X_j. □

B.3 Convergence in Law

Let M be a metric space. We view it as given with the Borel σ-algebra generated by open sets, and we denote by $C_b(M)$ the Banach space of bounded complex-valued continuous functions on M, with the norm

$$\|f\|_\infty = \sup_{x \in M} |f(x)|.$$

Given a sequence (μ_n) of probability measures on M, and a probability measure μ on M, one says that μ_n converges weakly to μ if and only if, for any bounded and continuous function $f : M \longrightarrow \mathbf{R}$, we have

$$\int_M f(x) d\mu_n(x) \longrightarrow \int_M f(x) d\mu(x). \tag{B.1}$$

If $(\Omega, \Sigma, \mathbf{P})$ is a probability space and $(X_n)_{n \geqslant 1}$ is a sequence of M-valued random variables, and if X is an M-valued random variable, then one says that (X_n) converges in law to X if and only if the measures $X_n(\mathbf{P})$ converge weakly to $X(\mathbf{P})$. If μ is a probability measure on M, then we will also say that X_n converges to μ if the measures $X_n(\mathbf{P})$ converge weakly to μ.

The probabilistic versions of (B.1) in those cases is that

$$\mathbf{E}(f(X_n)) \longrightarrow \mathbf{E}(f(X)) = \int_M f d\mu \tag{B.2}$$

for all functions $f \in C_b(M)$.

Remark B.3.1 If $M = \mathbf{R}$, convergence in law is often introduced in terms of the distribution function $F_X(x) = \mathbf{P}(X \leqslant x)$ of a real-valued random variable X. Precisely, it is classical (see, e.g., [9, Th. 25.8]) that a a sequence of real-valued random variables (X_N) converges in law to a random variable X if and only if $F_{X_m}(x) \to F_X(x)$ for all $x \in \mathbf{R}$ such that F_X is continuous at x (which is true for all but at most countably many x, namely, all x such that $\mathbf{P}(X = x) = 0$).

The definition immediately implies the following very useful fact, which we state in probabilistic language (we will refer to it as the composition principle).

Proposition B.3.2 *Let* M *be a metric space. Let* (X_n) *be a sequence of* M*-valued random variables such that* X_n *converges in law to a random variable* X*. For any metric space* N *and any continuous function* $\varphi\colon M \to N$*, the* N*-valued random variables* $\varphi \circ X_n$ *converge in law to* $\varphi \circ X$.

Proof For any continuous and bounded function $f : N \longrightarrow \mathbf{C}$, the composite $f \circ \varphi$ is bounded and continuous on M, and therefore convergence in law implies that

$$\mathbf{E}(f(\varphi(X_n))) \longrightarrow \mathbf{E}(f(\varphi(X))).$$

By definition, this formula, valid for all f, means that $\varphi(X_n)$ converges in law to $\varphi(X)$. □

Checking the condition (B.2) for all $f \in C_b(M)$ may be difficult. A number of convenient criteria, and properties, of convergence in law are related to weakening this requirement to only *certain* "test functions" f, which may be more regular, or have special properties. We will discuss some of these in the next sections.

One often important consequence of convergence in law is a simple relation with the support of the limit of a sequence of random variables.

Lemma B.3.3 *Let* M *be a second countable or compact topological space. Let* (X_n) *be a sequence of* M*-valued random variables, defined on some probability spaces* Ω_n*. Assume that* (X_n) *converges in law to some random variable* X*, and let* $N \subset M$ *be the support of the law of* X.

(1) *For any* $x \in N$ *and for any open neighborhood* U *of* x*, we have*

$$\liminf_{n\to+\infty} \mathbf{P}(X_n \in U) > 0,$$

and in particular there exists some $n \geqslant 1$ *and some* $\omega \in \Omega_n$ *such that* $X_n(\omega) \in U$.

(2) *For any* $x \in M$ *not belonging to* N*, there exists an open neighborhood* U *of* x *such that*

$$\limsup_{n\to+\infty} \mathbf{P}(X_n \in U) = 0.$$

Proof For (1), a standard equivalent form of convergence in law is that, for any open set $U \subset M$, we

$$\liminf_{n\to+\infty} \mathbf{P}(X_n \in U) \geqslant \mathbf{P}(X \in U)$$

(see [10, Th. 2.1, (i) and (iv)]). If $x \in N$ and U is an open neighborhood of x, then by definition we have $\mathbf{P}(X \in U) > 0$, and therefore

$$\liminf_{n\to+\infty} \mathbf{P}(X_n \in U) > 0.$$

For (2), if $x \in M$ is not in N, there exists an open neighborhood V of x such that $\mathbf{P}(X \in V) = 0$. For some $\delta > 0$, this neighborhood contains the closed ball C of radius δ around f, and by [10, Th. 2.1., (i) and (iii)], we have

$$0 \leqslant \limsup_{p\to+\infty} \mathbf{P}(X_n \in C) \leqslant \mathbf{P}(X \in C) = 0,$$

hence the second assertion with U the open ball of radius δ. □

Another useful relation between support and convergence in law is the following:

Corollary B.3.4 *Let* M *be a second countable or compact topological space. Let* (X_n) *be a sequence of* M*-valued random variables, defined on some probability spaces* Ω_n *such that* (X_n) *converges in law to a random variable* X*. Let* g *be a continuous function on* M *such that* $g(X_n)$ *converges in probability to zero; that is, we have*

$$\lim_{n\to+\infty} \mathbf{P}_n(|g(X_n)| > \delta) = 0$$

for all $\delta > 0$*. The support of* X *is then contained in the zero set of* g*.*

Proof Let N be the support of X. Suppose that there exists $x \in N$ such that $|g(x)| = \delta > 0$. Since the set of all $y \in M$ such that $|g(y)| > \delta$ is an open neighborhood of x, we have

$$\liminf_{n\to+\infty} \mathbf{P}(|g(X_n)| > \delta) > 0$$

by the previous lemma; this contradicts the assumption, which implies that

$$\lim_{n\to+\infty} \mathbf{P}(|g(X_n)| > \delta) = 0.$$

□

Remark B.3.5 Another proof is that it is well known (and elementary) that convergence in probability implies convergence in law, so in the situation of the corollary, the sequence $(g(X_n))$ converges to 0 in law. Since it also converges to $g(\mathrm{X})$ by composition, we have $\mathbf{P}(g(\mathrm{X}) \neq 0) = 0$, which precisely means that the support of X is contained in the zero set of g.

We also recall an important definition that is a property of weak-compactness for a family of probability measures (or random variables).

Definition B.3.6 (Tightness) Let M be a complete separable metric space. Let $(\mu_i)_{i\in \mathrm{I}}$ be a family of probability measures on M. One says that (μ_i) is *tight* if for any $\varepsilon > 0$, there exists a compact subset $\mathrm{K} \subset \mathrm{M}$ such that $\mu_i(\mathrm{K}) \geqslant 1 - \varepsilon$ for all $i \in \mathrm{I}$.

It is a non-obvious fact that a single probability measure on a complete separable metric space is tight (see [10, Th. 1.3]).

B.4 Perturbation and Convergence in Law

As we suggested in Section 1.2, we will often prove convergence in law of the sequences of random variables that interest us by showing that they are obtained by "perturbation" of other sequences that are more accessible. In this section, we explain how to handle some of these perturbations.

A very useful tool for this purpose is the following property, which is a first example of reducing the proof of convergence in law to more regular test functions than all bounded continuous functions.

Let M be a metric space, with distance d. Recall that a continuous function $f\colon \mathrm{M} \to \mathbf{C}$ is said to be a *Lipschitz function* if there exists a real number $\mathrm{C} \geqslant 0$ such that

$$|f(x) - f(y)| \leqslant \mathrm{C}d(x,y)$$

for all $(x,y) \in \mathrm{M} \times \mathrm{M}$. We then say that C is a *Lipschitz constant* for f (it is, of course, not unique).

Proposition B.4.1 *Let* M *be a complete separable metric space. Let* (X_n) *be a sequence of* M*-valued random variables, and* μ *a probability measure on* M*. Then* X_n *converges in law to* μ *if and only if we have*

$$\mathbf{E}(f(\mathrm{X}_n)) \to \int_{\mathrm{M}} f(x)\,d\mu(x)$$

for all bounded Lipschitz functions $f\colon \mathrm{M} \longrightarrow \mathbf{C}$.

In other words, it is enough to prove the convergence property (B.2) for Lipschitz test functions.

Proof A classical argument shows that convergence in law of (X_n) to μ is equivalent to

$$\mu(F) \geqslant \limsup_{n\to+\infty} \mathbf{P}(X_n \in F) \tag{B.3}$$

for all closed subsets F of M (see, e.g., [10, Th. 2.1, (iii)]).

However, the proof that convergence in law *implies* this property uses only Lipschitz test functions f (see, e.g., [10, (ii)⇒(iii), p. 16, and (1.1), p. 8], where it is only stated that the relevant functions f are uniformly continuous, but this is shown by checking that they are Lipschitz). Hence the assumption that (B.2) holds for Lipschitz functions implies (B.3) for all closed subsets F, and consequently it implies convergence in law. □

We can now deduce various corollaries concerning perturbation of sequences that converge in law.

The first result along these lines is quite standard, and the second is a bit more ad hoc but will be convenient in Chapter 6.

Corollary B.4.2 *Let* M *be a separable Banach space. Let* (X_n) *and* (Y_n) *be sequences of* M*-valued random variables. Assume that the sequence* (X_n) *converges in law to a random variable* X.

If the sequence (Y_n) *converges in probability to* 0*, or if* (Y_n) *converges to* 0 *in* L^p *for some fixed* $p \geqslant 1$*, with the possibility that* $p = +\infty$*, then the sequence* $(X_n + Y_n)_n$ *converges in law to* X *in* M.

Proof Let $f\colon M \longrightarrow \mathbf{C}$ be a bounded Lipschitz continuous function, and C a Lipschitz constant of f. For any n, we have

$$\begin{aligned} |\,\mathbf{E}(f(X_n+Y_n)) - \mathbf{E}(f(X))| &\leqslant |\,\mathbf{E}(f(X_n+Y_n)) - \mathbf{E}(f(X_n))| \\ &\quad + |\,\mathbf{E}(f(X_n)) - \mathbf{E}(f(X))|. \end{aligned} \tag{B.4}$$

First assume that (Y_n) converges to 0 in L^p, and that $p < +\infty$. Then we obtain

$$\begin{aligned} |\,\mathbf{E}(f(X_n+Y_n)) - \mathbf{E}(f(X))| &\leqslant C\,\mathbf{E}(|Y_n|) + |\,\mathbf{E}(f(X_n)) - \mathbf{E}(f(X))| \\ &\leqslant C\,\mathbf{E}(|Y_n|^p)^{1/p} + |\,\mathbf{E}(f(X_n)) - \mathbf{E}(f(X))| \end{aligned}$$

which converges to 0, hence $(X_n + Y_n)$ converges in law to X. If $p = +\infty$, a similar argument, left to the reader, applies.

Suppose now that (Y_n) converges to 0 in probability. Let $\varepsilon > 0$. For n large enough, the second term in (B.4) is $\leqslant \varepsilon$ since X_n converges in law to X. For the first, we fix another parameter $\delta > 0$ and write

$$|\mathbf{E}(f(X_n + Y_n)) - \mathbf{E}(f(X_n))| \leqslant C\delta + 2\|f\|_\infty \mathbf{P}(|Y_n| > \delta)$$

by separating the integral depending on whether $|Y_n| \leqslant \delta$ or not. Take $\delta = C^{-1}\varepsilon$, so the first term here is $\leqslant \varepsilon$. Then since (Y_n) converges in probability to 0, we have

$$2\|f\|_\infty \mathbf{P}(|Y_n| > \delta) \leqslant \varepsilon$$

for all n large enough, and therefore

$$|\mathbf{E}(f(X_n + Y_n)) - \mathbf{E}(f(X))| \leqslant 3\varepsilon$$

for all n large enough. The result now follows from Proposition B.4.1. □

Here is the second variant, where we do not attempt to optimize the assumptions.

Corollary B.4.3 *Let $m \geqslant 1$ be an integer. Let (X_n) and (Y_n) be sequences of $\mathbf{R}^m$-valued random variables, let (α_n) be a sequence in $\mathbf{R}^m$ and (β_n) a sequence of real numbers. Assume*

(1) *the sequence (X_n) converges in law to a random variable X, and $\|X_n\|$ is bounded by a constant $N \geqslant 0$, independent of n;*

(2) *for all n, we have $\|Y_n\| \leqslant \beta_n$;*

(3) *we have $\alpha_n \to (1, \ldots, 1)$ and $\beta_n \to 0$ as $n \to +\infty$.*

Then the sequence $(\alpha_n \cdot X_n + Y_n)_n$ converges in law to X in $\mathbf{R}^m$, where here $\cdot$ denotes the componentwise product of vectors.[1]

Proof We begin as in the previous corollary. Let $f : \mathbf{R}^m \longrightarrow \mathbf{C}$ be a bounded Lipschitz continuous function, and C its Lipschitz constant. For any n, we now have

$$|\mathbf{E}(f(\alpha_n \cdot X_n + Y_n)) - \mathbf{E}(f(X))| \leqslant |\mathbf{E}(f(\alpha_n \cdot X_n + Y_n)) - \mathbf{E}(f(\alpha_n \cdot X_n))|$$
$$+ |\mathbf{E}(f(\alpha_n \cdot X_n)) - \mathbf{E}(f(X_n))| + |\mathbf{E}(f(X_n)) - \mathbf{E}(f(X))|.$$

The last term tends to 0 since X_n converges in law to X. The second is at most

$$C|\alpha_n - (1, \ldots, 1)\| \mathbf{E}(\|X_n\|) \leqslant CN\|\alpha_n - (1, \ldots, 1)\| \to 0,$$

and the first is at most

$$C\,\mathbf{E}(\|Y_n\|) \leqslant C\beta_n \to 0.$$

The result now follows from Proposition B.4.1. □

[1] I.e., we have $(a_1, \ldots, a_m) \cdot (b_1, \ldots, b_m) = (a_1 b_1, \ldots, a_m b_m)$.

The last instance of perturbations is slightly different. It amounts, in practice, to using some "auxiliary parameter m" to approximate a sequence of random variables; when the error in such an approximation is suitably small, and the approximations converge in law for each fixed m, we obtain convergence in law.

Proposition B.4.4 *Let* M *be a finite-dimensional Banach space. Let* $(\mathrm{X}_n)_{n\geqslant 1}$ *and* $(\mathrm{X}_{n,m})_{n\geqslant m\geqslant 1}$ *be* M*-valued random variables. Define* $\mathrm{E}_{n,m} = \mathrm{X}_n - \mathrm{X}_{n,m}$. *Assume that*

(1) *for each* $m \geqslant 1$, *the random variables* $(\mathrm{X}_{n,m})_{n\geqslant m}$ *converge in law to a random variable* Y_m*;*

(2) *we have*

$$\lim_{m\to+\infty} \limsup_{n\to+\infty} \mathbf{E}(\|\mathrm{E}_{n,m}\|) = 0.$$

Then the sequences (X_n) *and* (Y_m) *converge in law as* $n \to +\infty$ *and have the same limit distribution.*

The second assumption means in practice that

$$\mathbf{E}(\|\mathrm{E}_{n,m}\|) \leqslant f(n,m)$$

where $f(n,m) \to 0$ as m tends to $+\infty$, uniformly for $n \geqslant m$.

A statement of this kind can be found also, for instance, in [10, Th. 3.2], but the latter assumes that it is already known that (Y_m) converges in law.

Proof We begin by proving that (X_n) converges in law. Let $f : \mathrm{M} \longrightarrow \mathbf{R}$ be a bounded Lipschitz continuous function, and C a Lipschitz constant for f. For any $n \geqslant 1$ and any $m \leqslant n$, we have

$$|\,\mathbf{E}(f(\mathrm{X}_n)) - \mathbf{E}(f(\mathrm{X}_{n,m}))| \leqslant \mathrm{C}\,\mathbf{E}(\|\mathrm{E}_{n,m}\|),$$

hence

$$\mathbf{E}(f(\mathrm{X}_{n,m})) - \mathrm{C}\,\mathbf{E}(\|\mathrm{E}_{n,m}\|) \leqslant \mathbf{E}(f(\mathrm{X}_n)) \leqslant \mathbf{E}(f(\mathrm{X}_{n,m})) + \mathrm{C}\,\mathbf{E}(\|\mathrm{E}_{n,m}\|).$$

Fix first $m \geqslant 1$. By the first assumption, the expectations $\mathbf{E}(f(\mathrm{X}_{n,m}))$ converge to $\mathbf{E}(f(\mathrm{Y}_m))$ as $n \to +\infty$. Then these inequalities imply that we have

$$\limsup_{n\to+\infty} \mathbf{E}(f(\mathrm{X}_n)) - \liminf_{n\to+\infty} \mathbf{E}(f(\mathrm{X}_n)) \leqslant 2\mathrm{C} \limsup_{n\to+\infty} \mathbf{E}(\|\mathrm{E}_{n,m}\|)$$

(because any limit of a convergent subsequence of $\mathbf{E}(f(\mathrm{X}_n))$ will lie in an interval of length at most the right-hand side). Letting m go to infinity, the second assumption allows us to conclude that

$$\limsup_{n\to+\infty} \mathbf{E}(f(\mathrm{X}_n)) - \liminf_{n\to+\infty} \mathbf{E}(f(\mathrm{X}_n)) = 0,$$

so that the sequence $(\mathbf{E}(f(\mathrm{X}_n)))_{n\geqslant 1}$ converges.

Now consider the map μ defined on bounded Lipschitz functions on M by

$$\mu(f) = \lim_{n\to+\infty} \mathbf{E}(f(\mathrm{X}_n)).$$

It is elementary that μ is linear and that it is positive (in the sense that if $f \geqslant 0$, we have $\mu(f) \geqslant 0$) and satisfies $\mu(1) = 1$. By the Riesz representation theorem (see Proposition B.1.2 and Remark B.1.3, noting that a finite-dimensional Banach space is locally compact), it follows that μ "is" a probability measure on M. It is then tautological that (X_n) converges in law to a random vector X with probability law μ by Proposition B.4.1.

It remains to prove that the sequence (Y_m) also converges in law with limit X. We again consider the Lipschitz function f, with Lipschitz constant C, and write

$$|\,\mathbf{E}(f(\mathrm{X}_n)) - \mathbf{E}(f(\mathrm{X}_{n,m}))| \leqslant \mathrm{C}\,\mathbf{E}(\|\mathrm{E}_{n,m}\|).$$

For a fixed m, we let $n \to +\infty$. Since we have proved that (X_n) converges to X, we deduce by the first assumption that

$$|\,\mathbf{E}(f(\mathrm{X})) - \mathbf{E}(f(\mathrm{Y}_m))| \leqslant \mathrm{C} \limsup_{n\to+\infty} \mathbf{E}(\|\mathrm{E}_{n,m}\|).$$

Since the right-hand side converges to 0 by the second assumption, we conclude that

$$\mathbf{E}(f(\mathrm{Y}_m)) \to \mathbf{E}(f(\mathrm{X}))$$

and, finally, that (Y_m) converges to X. □

Remark B.4.5 If one knows that Y_n converges in law, one also obtains convergence in law (by a straightforward adaptation of the previous argument) if Assumption (2) of the proposition is replaced by

$$\lim_{m\to+\infty} \limsup_{n\to+\infty} \mathbf{P}(\|\mathrm{E}_{n,m}\| > \delta) = 0 \tag{B.5}$$

for any $\delta > 0$; see again [10, Th. 3.2].

Remark B.4.6 Although we have stated all results in the case where the random variables are defined on the same probability space, the proofs do not rely on this fact, and the statements apply also if they are defined on spaces depending on n, with obvious adaptations of the assumptions. For instance, in the last statement, we can take X_n and $X_{n,m}$ to be defined on a space Ω_n (independent of m) and the second assumption means that

$$\lim_{m\to+\infty} \limsup_{n\to+\infty} \mathbf{E}_n(|E_{n,m}|) = 0.$$

B.5 Convergence in Law in a Finite-Dimensional Vector Space

We will use two important criteria for convergence in law for random variables with values in a finite-dimensional real vector space V, which both amount to testing (B.1) for a restricted set of functions. Another important criterion applies to variables with values in a compact topological group and is reviewed in Section B.6.

The first result is valid in all cases and is based on the Fourier transform. Given an integer $m \geqslant 1$ and a probability measure μ on $\mathbf{R}^m$, recall that the *characteristic function* (or *Fourier transform*) of μ is the function

$$\varphi_\mu \colon \mathbf{R}^m \longrightarrow \mathbf{C}$$

defined by

$$\varphi_\mu(t) = \int_{\mathbf{R}^m} e^{it\cdot x} d\mu(x),$$

where $t \cdot x = t_1x_1 + \cdots + t_mx_m$ is the standard inner product. This is a continuous bounded function on $\mathbf{R}^m$. For a random vector X with values in $\mathbf{R}^m$, we denote by φ_X the characteristic function of $X(\mathbf{P})$, namely,

$$\varphi_X(t) = \mathbf{E}(e^{it\cdot X}).$$

We state two (obviously equivalent) versions of P. Lévy's theorem for convenience:

Theorem B.5.1 (Lévy Criterion) *Let $m \geqslant 1$ be an integer.*

(1) *Let (μ_n) be a sequence of probability measures on $\mathbf{R}^m$, and let μ be a probability measure on $\mathbf{R}^m$. Then (μ_n) converges weakly to μ if and only if, for any $t \in \mathbf{R}^m$, we have*

$$\varphi_{\mu_n}(t) \longrightarrow \varphi_\mu(t)$$

as $n \to +\infty$.

(2) *Let $(\Omega, \Sigma, \mathbf{P})$ be a probability space. Let $(X_n)_{n\geqslant 1}$ be $\mathbf{R}^m$-valued random vectors on Ω, and let X be an $\mathbf{R}^m$-valued random vector. Then (X_n) converges in law to X if and only if, for all $t \in \mathbf{R}^m$, we have*

$$\mathbf{E}(e^{it\cdot X_n}) \longrightarrow \mathbf{E}(e^{it\cdot X}).$$

For a proof, see, for example, [9, Th. 26.3] in the case $m = 1$.

Remark B.5.2 In fact, the precise version of Lévy's Theorem does not require to know in advance the limit of the sequence: if a sequence (μ_n) of probability measures is such that, for all $t \in \mathbf{R}^m$, we have

$$\varphi_{\mu_n}(t) \longrightarrow \varphi(t)$$

for *some* function φ, *and* if φ is continuous at 0, then one can show that φ is the characteristic function of a probability measure μ (and hence that μ_n converges weakly to μ); see for instance [9, p. 350, cor. 1]. So, for instance, it is not necessary to know beforehand that $\varphi(t) = e^{-t^2/2}$ is the characteristic function of a probability measure in order to prove the Central Limit Theorem using Lévy's Criterion.

Lemma B.5.3 *Let $m \geqslant 1$ be an integer. Let $(X_n)_{n\geqslant 1}$ be a sequence of random variables with values in $\mathbf{R}^m$ on some probability space. Let (β_n) be a sequence of positive real numbers such that $\beta_n \to 0$ as $n \to +\infty$. If (X_n) converges in law to an $\mathbf{R}^m$-valued random variable X, then for any sequence (Y_n) of $\mathbf{R}^m$-valued random variables such that $\|X_n - Y_n\|_\infty \leqslant \beta_n$ for all $n \geqslant 1$, the random variables Y_n converge to X.*

Proof We use Lévy's criterion. We fix $t \in \mathbf{R}^m$ and write

$$\mathbf{E}(e^{it\cdot Y_n}) - \mathbf{E}(e^{it\cdot X}) = \mathbf{E}(e^{it\cdot Y_n} - e^{it\cdot X_n}) + \mathbf{E}(e^{it\cdot X_n} - e^{it\cdot X}).$$

By Lévy's Theorem and our assumption on the convergence of the sequence (X_n), the second term on the right converges to 0 as $n \to +\infty$. For the first, we can simply apply the dominated convergence theorem to derive the same conclusion: we have

$$\|X_n - Y_n\|_\infty \leqslant \beta_n \to 0,$$

hence

$$e^{it\cdot Y_n} - e^{it\cdot X_n} = e^{it\cdot Y_n}\big(1 - e^{it\cdot (X_n - Y_n)}\big) \to 0$$

(pointwise) as $n \to +\infty$. Moreover, we have

$$\left|e^{it\cdot Y_n} - e^{it\cdot X_n}\right| \leqslant 2$$

for all $n \geqslant 1$. Hence the dominated convergence theorem implies that the expectation $\mathbf{E}(e^{it\cdot Y_n} - e^{it\cdot X_n})$ converges to 0.

Lévy's Theorem applied once more allows us to conclude that (Y_n) converges in law to X, as claimed. □

The second convergence criterion is known as the *method of moments*. It is more restrictive than Lévy's criterion, but is sometimes analytically more flexible, especially because it is often more manageable when there is no independence assumptions.

Definition B.5.4 (Mild measure) Let μ be a probability measure on $\mathbf{R}^m$. We will say that μ is *mild*[2] if the *absolute moments*

$$\mathrm{M}_{\boldsymbol{k}}^{a}(\mu) = \int_{\mathbf{R}^m} |x_1|^{k_1} \cdots |x_m|^{k_m} d\mu(x_1, \dots, x_m)$$

exist for all tuples of nonnegative integers $\boldsymbol{k} = (k_1, \dots, k_m)$ and if there exists $\delta > 0$ such that the power series

$$\sum \sum_{k_i \geqslant 0} \mathrm{M}_{\boldsymbol{k}}^{a}(\mu) \frac{z_1^{k_1} \cdots z_m^{k_m}}{k_1! \cdots k_m!}$$

converges in the region

$$\{(z_1, \dots, z_m) \in \mathbf{C}^m \mid |z_i| \leqslant \delta\}.$$

If a measure μ is mild, then it follows in particular that the moments

$$\mathrm{M}_{\boldsymbol{k}}(\mu) = \int_{\mathbf{R}^m} x_1^{k_1} \cdots x_m^{k_m} d\mu(x_1, \dots, x_m)$$

exist for all $\boldsymbol{k} = (k_1, \dots, k_m)$ with k_i nonnegative integers.

If X is a random variable, we will say as usual that a random vector $\mathrm{X} = (\mathrm{X}_1, \dots, \mathrm{X}_m)$ is *mild* if its law $\mathrm{X}(\mathbf{P})$ is mild. The moments and absolute moments are then

$$\mathrm{M}_{\boldsymbol{k}}(\mathrm{X}) = \mathbf{E}(\mathrm{X}_1^{k_1} \cdots \mathrm{X}_m^{k_m}), \quad \text{and} \quad \mathrm{M}_{\boldsymbol{k}}^{a}(\mathrm{X}) = \mathbf{E}(|\mathrm{X}_1|^{k_1} \cdots |\mathrm{X}_m|^{k_m}).$$

We again give two versions of the method of moments for weak convergence when the limit is mild:

Theorem B.5.5 (Method of moments) *Let $m \geqslant 1$ be an integer.*

(1) *Let (μ_n) be a sequence of probability measures on $\mathbf{R}^m$ such that all moments $\mathrm{M}_{\boldsymbol{k}}(\mu_n)$ exist, and let μ be a probability measure on $\mathbf{R}^m$.* Assume that μ is mild. *Then (μ_n) converges weakly to μ if for any m-tuple $\boldsymbol{k}$ of nonnegative integers, we have*

$$\mathrm{M}_{\boldsymbol{k}}(\mu_n) \longrightarrow \mathrm{M}_{\boldsymbol{k}}(\mu)$$

as $n \to +\infty$.

[2] There doesn't seem to be an especially standard name for this notion.

(2) *Let* $(\Omega, \Sigma, \mathbf{P})$ *be a probability space. Let* $(X_n)_{n\geqslant 1}$ *be* $\mathbf{R}^m$*-valued random vectors on* Ω *such that all moments* $M_{\boldsymbol{k}}(X_n)$ *exist, and let* Y *be an* $\mathbf{R}^m$*-valued random vector.* Assume that Y is mild. *Then* (X_n) *converges in law to* Y *if for any m-tuple* $\boldsymbol{k}$ *of non-negative integers, we have*

$$\mathbf{E}(X_{n,1}^{k_1} \cdots X_{n,m}^{k_m}) \longrightarrow \mathbf{E}(Y_1^{k_1} \cdots Y_n^{k_m}).$$

For a proof (in the case $m = 1$), see, for instance, [9, Th. 30.2 and Th. 30.1].

This only gives one implication in comparison with the Lévy Criterion. It is often useful to have a converse, and here is one such statement:

Theorem B.5.6 (Converse of the method of moments) *Let* $(\Omega, \Sigma, \mathbf{P})$ *be a probability space.*

Let $m \geqslant 1$ *be an integer and let* $(X_n)_{n\geqslant 1}$ *be a sequence of* $\mathbf{R}^m$*-valued random vectors on* Ω *such that all moments* $M_{\boldsymbol{k}}(X_n)$ *exist and such that there exist constants* $c_{\boldsymbol{k}} \geqslant 0$ *with*

$$\mathbf{E}(|X_{n,1}|^{k_1} \cdots |X_{n,m}|^{k_m}) \leqslant c_{\boldsymbol{k}} \tag{B.6}$$

for all $n \geqslant 1$. *Assume that* X_n *converges in law to a random vector* Y. *Then* Y *is mild and for any m-tuple* $\boldsymbol{k}$ *of nonnegative integers, we have*

$$\mathbf{E}(X_{n,1}^{k_1} \cdots X_{n,m}^{k_m}) \longrightarrow \mathbf{E}(Y_1^{k_1} \cdots Y_n^{k_m}).$$

Proof See [9, Th 25.12 and Cor.] for a proof (again for $m = 1$). □

Example B.5.7 This converse applies, in particular, if (X_n) is a sequence of real-valued random variables given by

$$X_n = \frac{B_1 + \cdots + B_n}{\sigma_n}$$

where the variables $(B_i)_{i\geqslant 1}$ are independent and satisfy

$$\mathbf{E}(B_i) = 0, \qquad |B_i| \leqslant 1, \qquad \sigma_n^2 = \sum_{i=1}^{n} \mathbf{V}(B_i) \to +\infty \text{ as } n \to +\infty.$$

Then the Central Limit Theorem (see Theorem B.7.2) implies that the sequence (X_n) converges in law to a standard Gaussian random variable Y. Moreover, X_n is bounded (by n/σ_n), so all its moments exist. We will check that this sequence satisfies the uniform integrability condition (B.6), from which we deduce the convergence of moments

$$\lim_{n\to+\infty} \mathbf{E}(X_n^k) = \mathbf{E}(Y^k)$$

for all integers $k \geqslant 0$ (the moments of Y are described explicitly in Proposition B.7.3).

For any $k \geqslant 0$, there exists a constant $C_k \geqslant 0$ such that

$$|x|^k \leqslant C_k(e^x + e^{-x})$$

for all $x \in \mathbf{R}$. It follows that if we can show that there exists $D \geqslant 0$ such that

$$\mathbf{E}(e^{X_n}) \leqslant D \quad \text{and} \quad \mathbf{E}(e^{-X_n}) \leqslant D \tag{B.7}$$

for all $n \geqslant 1$, then we obtain $\mathbf{E}(|X_n|^k) \leqslant 2C_kD$ for all n, which gives the desired conclusion. Note that, since X_n is bounded, these expectations make sense, and moreover, we may assume that we only consider n large enough so that $\sigma_n \geqslant 1$.

To prove (B.7), fix more generally $t \in [-1, 1]$. Since the (B_i) are independent random variables, we have

$$\mathbf{E}(e^{tX_n}) = \prod_{i=1}^{n} \mathbf{E}\left(\exp\left(\frac{tB_i}{\sigma_n}\right)\right).$$

Since we assumed that $\sigma_n \geqslant 1$ and $|B_i| \leqslant 1$, we have $|tB_i/\sigma_n| \leqslant 1$, hence

$$\exp\left(\frac{tB_i}{\sigma_n}\right) \leqslant 1 + \frac{tB_i}{\sigma_n} + \frac{t^2B_i^2}{\sigma_n^2}$$

(because $e^x \leqslant 1 + x + x^2$ for $|x| \leqslant 1$, as can be checked using basic calculus). We then obtain further

$$\mathbf{E}(e^{tX_n}) \leqslant \prod_{i=1}^{n} \left(1 + \frac{t^2}{\sigma_n^2}\mathbf{E}(B_i^2)\right)$$

since $\mathbf{E}(B_i) = 0$. Using $1 + x \leqslant e^x$, this leads to

$$\mathbf{E}(e^{tX_n}) \leqslant \exp\left(\frac{t^2}{\sigma_n^2}\sum_{i=1}^{n}\mathbf{E}(B_i^2)\right) = \exp(t^2).$$

Applying this with $t = 1$ and $t = -1$, we get (B.7) with $D = e$, hence also (B.6), for all n large enough.

Remark B.5.8 In the case $m = 2$, one often deals with random variables that are naturally seen as complex-valued, instead of $\mathbf{R}^2$-valued. In that case, it is sometimes quite useful to use the complex moments

$$\tilde{M}_{k_1,k_2}(X) = \mathbf{E}(X^{k_1}\bar{X}^{k_2})$$

of a $\mathbf{C}$-valued random variable instead of $M_{k_1,k_2}(X)$. The corresponding statements are that X is mild if and only if the power series

$$\sum\sum_{k_1,k_2 \geqslant 0} \tilde{M}_{k_1,k_2}(X)\frac{z_1^{k_1}z_2^{k_2}}{k_1!\,k_2!}$$

converges in a region

$$\{(z_1, z_2) \in \mathbf{C} \mid |z_1| \leqslant \delta, \quad |z_2| \leqslant \delta\}$$

for some $\delta > 0$, and that if X is mild, then (X_n) converges weakly to X if

$$\tilde{M}_{k_1,k_2}(X_n) \longrightarrow \tilde{M}_{k_1,k_2}(X)$$

for all $k_1, k_2 \geqslant 0$. Example B.5.7 extends to the complex-valued case.

Example B.5.9 (1) Any bounded random vector is mild. Indeed, if $\|X\|_\infty \leqslant B$, say, then we get

$$M_{\boldsymbol{k}}^a(X) \leqslant B^{k_1+\cdots+k_m},$$

and therefore

$$\sum_{k_i \geqslant 0} \sum M_{\boldsymbol{k}}^a(\mu) \frac{|z_1|^{k_1} \cdots |z_m|^{k_m}}{k_1! \cdots k_m!} \leqslant e^{B|z_1|+\cdots+B|z_m|},$$

so that the power series converges, in that case, for all $z \in \mathbf{C}^m$.

(2) Any Gaussian random vector is mild (see Section B.7).

(3) If X is mild, and Y is another random vector with $|Y_i| \leqslant |X_i|$ (almost surely) for all i, then Y is also mild.

B.6 The Weyl Criterion

One important special case of convergence in law is known as *equidistribution* in the context of topological groups in particular. We only consider compact groups here for simplicity. Let G be such a group. Then there exists on G a unique Borel probability measure μ_G which is invariant under left (and right) translations: for any integrable function $f : G \longrightarrow \mathbf{C}$ and for any fixed $g \in G$, we have

$$\int_G f(gx) d\mu_G(x) = \int_G f(xg) d\mu_G(x) = \int_G f(x) d\mu_G(x).$$

This measure is called the (probability) Haar measure on G (see, e.g., [17, VII, §1, n. 2, th. 1 and prop. 2]).

If a G-valued random variable X is distributed according to μ_G, one says that X is *uniformly distributed* on G.

Example B.6.1 (1) Let $G = \mathbf{S}^1$ be the multiplicative group of complex numbers of modulus 1. This group is isomorphic to $\mathbf{R}/\mathbf{Z}$ by the isomorphism $\theta \mapsto e(\theta)$. The measure μ_G is then identified with the Lebesgue measure $d\theta$ on $\mathbf{R}/\mathbf{Z}$. In other words, for any integrable function $f\colon \mathbf{S}^1 \to \mathbf{C}$, we have

$$\int_{\mathbf{S}^1} f(z)d\mu_{\mathrm{G}}(z) = \int_{\mathbf{R}/\mathbf{Z}} f(e(\theta))d\theta = \int_0^1 f(e(\theta))d\theta.$$

(2) If $(\mathrm{G}_i)_{i\in\mathrm{I}}$ is any family of compact groups, each with a probability Haar measure μ_i, then the (possibly infinite) tensor product

$$\bigotimes_{i\in\mathrm{I}} \mu_i$$

is the probability Haar measure μ on the product G of the groups G_i. Probabilistically, one would interpret this as saying that μ is the law of a family (X_i) of *independent* random variables, where each X_i is uniformly distributed on G_i.

(3) Let G be the nonabelian compact group $\mathrm{SU}_2(\mathbf{C})$, that is,

$$\mathrm{G} = \left\{ \begin{pmatrix} \alpha & \bar{\beta} \\ -\beta & \bar{\alpha} \end{pmatrix} \mid \alpha,\, \beta \in \mathbf{C},\, |\alpha|^2 + |\beta|^2 = 1 \right\}.$$

Writing $\alpha = a + ib$, $\beta = c + id$, we can identify G, as a topological space, with the unit 3-sphere

$$\{(a,b,c,d) \in \mathbf{R}^4 \mid a^2 + b^2 + c^2 + d^2 = 1\}$$

in $\mathbf{R}^4$. Then the left multiplication by some element on G is the restriction of a rotation of $\mathbf{R}^4$. Hence the surface (Lebesgue) measure μ_0 on the 3-sphere is a Borel-invariant measure on G. By uniqueness, we see that the probability Haar measure on G is

$$\mu = \frac{1}{2\pi^2}\mu_0$$

(since the surface area of the 3-sphere is $2\pi^2$).

Consider now the trace $\mathrm{Tr}\colon \mathrm{G} \longrightarrow \mathbf{R}$, which is given by $(a,b,c,d) \mapsto 2a$ in the sphere coordinates. One can show that the direct image $\mathrm{Tr}_*(\mu)$ is the so-called *Sato–Tate* measure

$$\mu_{\mathrm{ST}} = \frac{1}{\pi}\sqrt{1 - \frac{x^2}{4}}dx,$$

supported on $[-2,2]$ (for probabilists, this is also a semicircle law); equivalently, if we write the trace as

$$\mathrm{Tr}(g) = 2\cos(\theta)$$

for a unique $\theta \in [0, \pi]$, then this measure is identified with the measure

$$\frac{2}{\pi} \sin^2 \theta d\theta$$

on $[0, \pi]$ (for a proof, see, e.g., [18, Ch. 9, p. 58, example]). One obtains from either description of μ_{ST} the expectation and variance

$$\int_{\mathbf{R}} t d\mu_{\mathrm{ST}} = 0 \quad \text{and} \quad \int_{\mathbf{R}} t^2 d\mu_{\mathrm{ST}} = 1. \tag{B.8}$$

(4) If G is a finite group, then the probability Haar measure is just the normalized counting measure: for any function f on G, the integral of f is

$$\frac{1}{|\mathrm{G}|} \sum_{x \in \mathrm{G}} f(x).$$

For a topological group G, a *unitary character* χ of G is a continuous homomorphism

$$\chi : \mathrm{G} \longrightarrow \mathbf{S}^1.$$

The *trivial character* is the character $g \mapsto 1$ of G. The set of all characters of G is denoted $\widehat{\mathrm{G}}$. It has a structure of abelian group by multiplication of functions. If G is locally compact, then $\widehat{\mathrm{G}}$ is a locally compact topological group with the topology of uniform convergence on compact sets (for this theory, see, e.g., [19]).

In general, $\widehat{\mathrm{G}}$ may be reduced to the trivial character (this is the case if $\mathrm{G} = \mathrm{SL}_2(\mathbf{R})$, for instance). Assume now that G is locally compact and abelian. Then it is a fundamental fact (known as *Pontryagin duality*; see, e.g., [70, §7.3] for a survey or [19, II, §1, n. 5, th. 2] for the details) that there are "many" characters, in a suitable sense. If G is compact, then a simple version of this assertion is that $\widehat{\mathrm{G}}$ is an orthonormal basis of the space $\mathrm{L}^2(\mathrm{G}, \mu)$, where μ is the probability Haar measure on G.

For an integrable function $f \in \mathrm{L}^1(\mathrm{G}, \mu)$, its *Fourier transform* is the function $\widehat{f} : \widehat{\mathrm{G}} \longrightarrow \mathbf{C}$ defined by

$$\widehat{f}(\chi) = \int_{\mathrm{G}} f(x) \overline{\chi(x)} d\mu(x)$$

for all $\chi \in \widehat{\mathrm{G}}$. For a compact commutative group G, and $f \in \mathrm{L}^2(\mathrm{G}, \mu)$, we have

$$f = \sum_{\chi \in \widehat{\mathrm{G}}} \widehat{f}(\chi) \chi,$$

as a series converging in $L^2(G,\mu)$. It follows easily that a function $f \in L^1(G,\mu)$ is almost everywhere constant if and only if $\widehat{f}(\chi) = 0$ for all $\chi \neq 1$.

The following relation is immediate from the invariance of Haar measure: for f integrable and any fixed $y \in G$, if we let $g(x) = f(xy)$, then g is well defined as an integrable function, and

$$\widehat{g}(\chi) = \int_G f(xy)\overline{\chi(x)}d\mu(x) = \chi(y)\int_G f(x)\overline{\chi(x)}d\mu(x) = \chi(y)\widehat{f}(y).$$

Example B.6.2 (1) The characters of $\mathbf{S}^1$ are given by

$$z \mapsto z^m$$

for $m \in \mathbf{Z}$. Equivalently, the characters of $\mathbf{R}/\mathbf{Z}$ are given by $x \mapsto e(hx)$, where $e(z) = \exp(2i\pi z)$. More generally, the characters of $(\mathbf{R}/\mathbf{Z})^n$ are of the form

$$x = (x_1, \ldots, x_n) \mapsto e(h_1x_1 + \cdots + h_nx_n) = e(h \cdot x)$$

for some (unique) $h \in \mathbf{Z}^n$ (see, e.g., [19, p. 236, cor. 3]).

(2) If $(G_i)_{i\in I}$ is any family of compact groups, each with the probability Haar measure μ_i, then the characters of the product G of the G_i are given in a unique way as follows: take a finite subset S of I, and for any $i \in I$, pick a nontrivial character χ_i of G_i, then define

$$\chi(x) = \prod_{i\in S} \chi_i(x_i)$$

for any $x = (x_i)_{i\in I}$ in G. Here the trivial character corresponds to $S = \emptyset$. See, for example, [70, Example 5.6.10] for a proof.

In particular, if I is a finite set, this computation shows that the group of characters of G is isomorphic to the product of the groups of characters of the G_i and that the isomorphism is such that a family (χ_i) of characters of the groups G_i is mapped to the character

$$(x_i) \mapsto \prod_{i\in I} \chi_i(x_i).$$

(3) If G is a finite abelian group, then the group $\widehat{G}$ of characters of G is also finite, and it is isomorphic to G. This can be seen from the structure theorem for finite abelian groups, which shows that any finite abelian group is a direct product of some finite cyclic groups (see, e.g., [103, Th. B-3.13]) combined with the previous example and the explicit computation of the dual group of a finite cyclic group $\mathbf{Z}/q\mathbf{Z}$ for $q \geqslant 1$: an isomorphism from $\mathbf{Z}/q\mathbf{Z}$ to $\widehat{\mathbf{Z}/q\mathbf{Z}}$ is given by sending $a \pmod q$ to the character

$$x \mapsto e\left(\frac{ax}{q}\right),$$

which is well defined because replacing a and x by other integers congruent modulo q does not change the value of $e(ax/q)$.

In this case, one can also prove elementarily that the characters form an orthonormal basis of the finite-dimensional vector space C(G) of complex-valued functions on G, which in this case has the inner product

$$\langle f,g\rangle = \frac{1}{|\mathrm{G}|}\sum_{x\in\mathrm{G}} f(x)\overline{g(x)}.$$

Indeed, one can also reduce to the case of cyclic groups by checking that there is a unique isomorphism

$$\mathrm{C}(\mathrm{G}_1)\otimes\mathrm{C}(\mathrm{G}_2)\to\mathrm{C}(\mathrm{G}_1\times\mathrm{G}_2)$$

such that a pure tensor $f_1\otimes f_2$ is mapped to the function $(x_1,x_2)\mapsto f_1(x_1)f_2(x_2)$. In particular, the characters of $\mathrm{G}_1\times\mathrm{G}_2$ (which belong to $\mathrm{C}(\mathrm{G}_1\times\mathrm{G}_2)$) correspond under this isomorphism to the pure tensors $\chi_1\otimes\chi_2$.

In addition, under this isomorphism, we have

$$\langle f_1\otimes f_2, g_1\otimes g_2\rangle = \langle f_1,g_1\rangle\,\langle f_2,g_2\rangle$$

for any functions f_i and g_i on G_i. This implies that if the characters of G_1 and those of G_2 form orthornomal bases of their respective spaces of functions, then so do the characters of $\mathrm{G}_1\times\mathrm{G}_2$.

And in the case of $\mathrm{G}=\mathbf{Z}/q\mathbf{Z}$, we can simply compute using the explicit description of the characters $\chi_a\colon x\mapsto e(ax/q)$ for $a\in\mathbf{Z}/q\mathbf{Z}$ that

$$\langle\chi_a,\chi_b\rangle = \frac{1}{q}\sum_{0\leqslant x\leqslant q-1} e\Big(\frac{ax}{q}\Big)e\Big(-\frac{bx}{q}\Big),$$

which is equal to 1 if $a=b$, and otherwise is

$$\frac{1-e(q(a-b)/q)}{1-e((a-b)/q)}$$

by summing a finite geometric sum, and is therefore zero, as we wanted.

The Weyl Criterion is a criterion for a sequence of G-valued random variables to converge in law to a uniformly distributed random variable on G. We state it for compact abelian groups only:

Theorem B.6.3 (Weyl's Criterion) *Let* G *be a compact abelian group. A sequence* (X_n) *of* G*-valued random variables converges in law to a uniformly distributed random variable on* G *if and only if, for any nontrivial character* χ *of* G*, we have*

$$\lim_{n\to+\infty}\mathbf{E}(\chi(\mathrm{X}_n))\longrightarrow 0.$$

Remark B.6.4 (1) Note that the orthogonality of characters implies that

$$\int_{G} \chi(x) d\mu_{G}(x) = \langle \chi, 1 \rangle = 0$$

for any nontrivial character χ of G. Hence the Weyl Criterion has the same flavor of Lévy's Criterion (note that, for any $t \in \mathbf{R}^m$, the function $x \mapsto e^{ix \cdot t}$ is a character of $\mathbf{R}^m$).

(2) If G is compact, but not necessarily abelian, there is a version of the Weyl Criterion using as "test functions" the traces of irreducible finite-dimensional representations of G (see [70, §5.5] for an account).

The best-known example of application of the Weyl Criterion is to prove the following equidistribution theorem of Kronecker:

Theorem B.6.5 (Kronecker) *Let $d \geqslant 1$ be an integer. Let z be an element of $\mathbf{R}^d$ and let* T (resp. $\widetilde{\mathrm{T}}$) *be the closure of the subgroup of $(\mathbf{R}/\mathbf{Z})^d$ generated by the class of z* (resp. generated by the classes of the elements yz for $y \in \mathbf{R}$).

(1) *As $\mathrm{N} \to +\infty$, the probability measures on $(\mathbf{R}/\mathbf{Z})^d$ defined by*

$$\frac{1}{\mathrm{N}} \sum_{0 \leqslant n < \mathrm{N}} \delta_{nz}$$

converge in law to the probability Haar measure on T.

(2) *Let λ denote the Lebesgue measure on* **R**. *As $\mathrm{X} \to +\infty$, the probability measures μ_{X} on $(\mathbf{R}/\mathbf{Z})^d$ defined by*

$$\mu_{\mathrm{X}}(\mathrm{A}) = \frac{1}{\mathrm{X}} \lambda(\{x \in [0, \mathrm{X}] \mid xz \in \mathrm{A}\}),$$

for a measurable subset A *of $(\mathbf{R}/\mathbf{Z})^d$, converge in law to the probability Haar measure on* $\widetilde{\mathrm{T}}$.

Proof We only prove the "continuous" version in (2), since the first one is easier (and better known). First note that each probability measure μ_{X} has support contained in $\widetilde{\mathrm{T}}$ by definition, so it can be viewed as a measure on $\widetilde{\mathrm{T}}$.

From the theory of compact abelian groups, we know that any character χ of $\widetilde{\mathrm{T}}$ can be extended to a character of $(\mathbf{R}/\mathbf{Z})^d$ (see, e.g., [19, p. 226, th. 4], applied to the exact sequence $1 \to \widetilde{\mathrm{T}} \to (\mathbf{R}/\mathbf{Z})^d$), which is therefore of the form

$$v \mapsto e(n \cdot v)$$

for some $n \in \mathbf{Z}^d$ (by Example B.6.2, (1)). We then have

$$\int_{(\mathbf{R}/\mathbf{Z})^d} \chi(v) d\mu_{\mathrm{X}}(v) = \frac{1}{\mathrm{X}} \int_0^{\mathrm{X}} e((x\,n) \cdot z) dx.$$

Suppose that χ is a nontrivial character of $\widetilde{\mathrm{T}}$; since the classes of yz for $y \in \mathbf{R}$ generate a dense subgroup of $\widetilde{\mathrm{T}}$, we have then $n \cdot z \neq 0$. Hence

$$\int_{(\mathbf{R}/\mathbf{Z})^d} \chi(v) d\mu_{\mathrm{X}}(v) = \frac{1}{\mathrm{X}} \frac{e((\mathrm{X}\,n) \cdot z) - 1}{2i\pi n \cdot z} \to 0$$

as $\mathrm{X} \to +\infty$. We conclude by an application of the Weyl Criterion. □

Example B.6.6 In order to apply Theorem B.6.5 in practice, we need to identify the subgroup T (or $\widetilde{\mathrm{T}}$). The following special cases are quite often sufficient (writing $z = (z_1, \ldots, z_d) \in \mathbf{R}^d$):

1. we have $\mathrm{T} = (\mathbf{R}/\mathbf{Z})^d$ if and only if $(1, z_1, \ldots, z_d)$ are **Q**-linearly independent;
2. we have $\widetilde{\mathrm{T}} = (\mathbf{R}/\mathbf{Z})^d$ if and only if $(z_1, \ldots, z_d)$ are **Q**-linearly independent.

For instance, if $d = 1$, then the first condition means that $z = z_1$ is irrational, and the second means that z is nonzero.

We check (1), leaving (2) as an exercise. If $(1, z_1, \ldots, z_d)$ are *not* **Q**-linearly independent, then multiplying a nontrivial linear dependency relation with a suitable nonzero integer, we obtain a relation

$$m_0 + \sum_{i=1}^{d} m_i z_i = 0,$$

where $m_i \in \mathbf{Z}$ and not all m_i are zero, in fact, not all m_i with $i \geqslant 1$ are zero (since this would also imply that $m_0 = 0$). Then the class of nz modulo $\mathbf{Z}^d$ is, for all $n \in \mathbf{Z}$, an element of the proper closed subgroup

$$\{x = (x_1, \ldots, x_d) \in (\mathbf{R}/\mathbf{Z})^d \mid m_1 x_1 + \cdots + m_d x_d = 0\},$$

which implies that T is also contained in that subgroup, hence is not all of $(\mathbf{R}/\mathbf{Z})^d$.

Conversely, a simple argument is to check that if $(1, z_1, \ldots, z_d)$ are **Q**-linearly independent, then a direct application of the Weyl Criterion proves that the probability measures

$$\frac{1}{\mathrm{N}} \sum_{0 \leqslant n < \mathrm{N}} \delta_{nz}$$

converge in law to the probability Haar measure on $(\mathbf{R}/\mathbf{Z})^d$ (because nontrivial characters of this group correspond to $(m_i) \in \mathbf{Z}^d$, and the integral against the measure above is

$$\frac{1}{N}\sum_{n=1}^{N} e((m_1 z_1 + \cdots + m_d z_d)n),$$

where the real number $m_1 z_1 + \cdots + m_d z_d$ is not an integer by the linear independence, so that the sum tends to 0 by summing a finite geometric series).

B.7 Gaussian Random Variables

By definition, a random vector X with values in $\mathbf{R}^m$ is called a *(centered) Gaussian vector* if there exists a nonnegative quadratic form Q on $\mathbf{R}^m$ such that the characteristic function φ_X of X is of the form

$$\varphi_X(t) = e^{-Q(t)/2}$$

for $t \in \mathbf{R}^m$. The quadratic form can be recovered from X by the relation

$$Q(t_1, \ldots, t_m) = \sum_{1 \leqslant i,j \leqslant m} a_{i,j} t_i t_j,$$

with $a_{i,j} = \mathbf{E}(X_i X_j)$, and the (symmetric) matrix $(a_{i,j})_{1\leqslant i,j\leqslant m}$ is called the *correlation matrix* of X. The components X_i of X are independent if and only if $a_{i,j} = 0$ if $i \neq j$, that is, if and only if the components of X are orthogonal.

If X is a Gaussian random vector, then X is mild, and in fact

$$\sum_{k} M_m(X)\frac{t_1^{k_1}\cdots t_m^{k_m}}{k_1!\cdots k_m!} = \mathbf{E}(e^{t\cdot X}) = e^{Q(t)/2}$$

for $t \in \mathbf{R}^m$, so that the power series converges on all of $\mathbf{C}^m$. The Laplace transform $\psi_X(z) = \mathbf{E}(e^{z\cdot X})$ is also defined for all $z \in \mathbf{C}^m$, and in fact

$$\mathbf{E}(e^{z\cdot X}) = e^{Q(z)/2}. \tag{B.9}$$

For $m = 1$, this means that a random variable is a centered Gaussian if and only if there exists $\sigma \geqslant 0$ such that

$$\varphi_X(t) = e^{-\sigma^2 t/2}, \tag{B.10}$$

and in fact we have

$$\mathbf{E}(X^2) = \mathbf{V}(X) = \sigma^2.$$

If $\sigma = 1$, then we say that X is a *standard Gaussian random variable* (also sometimes called a standard normal random variable). We then have

$$\mathbf{P}(a < \mathrm{X} < b) = \frac{1}{\sqrt{2\pi}} \int_a^b e^{-x^2/2} dx$$

for all real numbers $a < b$.

Exercise B.7.1 We recall a standard proof of the fact that the measure on $\mathbf{R}$ given by

$$\mu = \frac{1}{\sqrt{2\pi}} e^{-x^2/2} dx$$

is indeed a Gaussian probability measure with variance 1.

(1) Define

$$\varphi(t) = \varphi_\mu(t) = \frac{1}{\sqrt{2\pi}} \int_{\mathbf{R}} e^{itx - x^2/2} dx$$

for $t \in \mathbf{R}$. Prove that φ is of class C^1 on $\mathbf{R}$ and satisfies $\varphi'(t) = -t\varphi(t)$ for all $t \in \mathbf{R}$ and $\varphi(0) = 1$.

(2) Deduce that $\varphi(t) = e^{-t^2/2}$ for all $t \in \mathbf{R}$. [**Hint**: This is an elementary argument with ordinary differential equations, but because the order is 1, one can define $g(t) = e^{t^2/2}\varphi(t)$ and check by differentiation that $g'(t) = 0$ for all $t \in \mathbf{R}$.]

We will use the following simple version of the Central Limit Theorem:

Theorem B.7.2 *Let* $\mathrm{B} \geqslant 0$ *be a fixed real number. Let* (X_n) *be a sequence of independent real-valued random variables with* $|\mathrm{X}_n| \leqslant \mathrm{B}$ *for all* n. *Let*

$$\alpha_n = \mathbf{E}(\mathrm{X}_n) \quad \text{and} \quad \beta_n = \mathbf{V}(\mathrm{X}_n^2).$$

Let $\sigma_\mathrm{N} \geqslant 0$ *be defined by*

$$\sigma_\mathrm{N}^2 = \beta_1 + \cdots + \beta_\mathrm{N}$$

for $\mathrm{N} \geqslant 1$. *If* $\sigma_\mathrm{N} \to +\infty$ *as* $n \to +\infty$, *then the random variables*

$$\mathrm{Y}_\mathrm{N} = \frac{(\mathrm{X}_1 - \alpha_1) + \cdots + (\mathrm{X}_\mathrm{N} - \alpha_\mathrm{N})}{\sigma_\mathrm{N}}$$

converge in law to a standard Gaussian random variable.

Proof Although this is a very simple case of the general Central Limit Theorem for sums of independent random variables (indeed, even of Lyapunov's well-known version), we give a proof using Lévy's criterion for convenience.

First of all, we may assume that $\alpha_n = 0$ for all n by replacing X_n by $X_n - \alpha_n$ (up to replacing B by 2B, since $|\alpha_n| \leqslant B$).

By independence of the variables (X_n), the characteristic function φ_N of Y_N is given by

$$\varphi_N(t) = \mathbf{E}(e^{itY_N}) = \prod_{1 \leqslant n \leqslant N} \mathbf{E}(e^{itX_n/\sigma_N}).$$

Fix $t \in \mathbf{R}$. Since tX_n/σ_N is bounded (because t is fixed), we have a Taylor expansion around 0 of the form

$$e^{itX_n/\sigma_N} = 1 + \frac{itX_n}{\sigma_N} - \frac{t^2X_n^2}{2\sigma_N^2} + O\left(\frac{|t|^3|X_n|^3}{\sigma_N^3}\right),$$

for $1 \leqslant n \leqslant N$.

Consequently, we obtain

$$\varphi_{X_n}\left(\frac{t}{\sigma_N}\right) = \mathbf{E}(e^{itX_n/\sigma_N}) = 1 - \frac{1}{2}\left(\frac{t}{\sigma_N}\right)^2 \mathbf{E}(X_n^2) + O\left(\left(\frac{|t|}{\sigma_N}\right)^3 \mathbf{E}(|X_n|^3)\right).$$

Observe that with our assumption, we have

$$\mathbf{E}(|X_n|^3) \leqslant B\,\mathbf{E}(X_n^2) = B\beta_n.$$

Moreover, for N large enough (depending on t, but t is fixed), the modulus of

$$-\frac{1}{2}\left(\frac{t}{\sigma_N}\right)^2 \mathbf{E}(X_n^2) + O\left(\left(\frac{|t|}{\sigma_N}\right)^3 \mathbf{E}(|X_n|^3)\right)$$

is less than 1, so that we can use Proposition A.2.2 and deduce that

$$\begin{aligned}
\varphi_N(t) &= \exp\left(\sum_{n=1}^{N} \log \mathbf{E}(e^{itX_n/\sigma_N})\right) \\
&= \exp\left(-\frac{t^2}{2\sigma_N}\sum_{n=1}^{N} \beta_n + O\left(\frac{B|t|^3}{\sigma_N^3}\sum_{n=1}^{N} \beta_n\right)\right) \\
&= \exp\left(-\frac{t^2}{2} + O\left(\frac{B|t|^3}{\sigma_N}\right)\right) \longrightarrow \exp(-t^2/2)
\end{aligned}$$

as $N \to +\infty$; we conclude, then, by Lévy's Criterion and (B.10). □

If one uses directly the method of moments to get convergence in law to a Gaussian random variable, it is useful to know the values of their moments. We only state the one-dimensional and the simplest complex case:

Proposition B.7.3 (1) *Let* X *be a real-valued Gaussian random variable with expectation* 0 *and variance* σ^2. *For* $k \geqslant 0$, *we have*

$$\mathbf{E}(\mathrm{X}^k) = \begin{cases} 0 & \text{if } k \text{ is odd,} \\ \sigma^k \frac{k!}{2^{k/2}(k/2)!} = \sigma^k \cdot (1 \cdot 3 \cdot \dots \cdot (k-1)) & \text{if } k \text{ is even.} \end{cases}$$

(1) *Let* X *be a complex-valued Gaussian random variable with covariance matrix*

$$\begin{pmatrix} \sigma & 0 \\ 0 & \sigma \end{pmatrix}$$

for some $\sigma > 0$. *For* $k \geqslant 0$ *and* $l \geqslant 0$, *we have*

$$\mathbf{E}(\mathrm{X}^k \bar{\mathrm{X}}^l) = \begin{cases} 0 & \text{if } k \neq l, \\ \sigma^k 2^k k! & \text{if } k = l. \end{cases}$$

Exercise B.7.4 Prove this proposition.

B.8 Sub-Gaussian Random Variables

Gaussian random variables have many remarkable properties. It is a striking fact that a number of these, especially with respect to integrability properties, are shared by a much more general class of random variables.

Definition B.8.1 (Sub-Gaussian random variable) Let $\sigma > 0$ be a real number. A real-valued random variable X is σ^2-sub-Gaussian if we have

$$\mathbf{E}(e^{t\mathrm{X}}) \leqslant e^{\sigma^2 t^2/2}$$

for all $t \in \mathbf{R}$. A complex-valued random variable X is σ^2-sub-Gaussian if $\mathrm{X} = \mathrm{Y} + i\mathrm{Z}$ with Y and Z real-valued σ^2-sub-Gaussian random variables.

If X is a real σ^2-sub-Gaussian random variable, then we obtain immediately good Gaussian-type upper bounds for the tail of the distribution: for any $b > 0$, using first a general auxiliary parameter $t > 0$, we have

$$\mathbf{P}(\mathrm{X} > b) \leqslant \mathbf{P}(e^{t\mathrm{X}} > e^{bt}) \leqslant \frac{\mathbf{E}(e^{t\mathrm{X}})}{e^{bt}} \leqslant e^{\sigma^2 t^2/2 - bt},$$

and selecting $t = -\frac{1}{2}b^2/\sigma^2$, we get

$$\mathbf{P}(\mathrm{X} > b) \leqslant e^{-\frac{1}{2}b^2/\sigma^2}.$$

The right-hand side is a standard upper bound for the probability $\mathbf{P}(\mathrm{N} > b)$ for a centered Gaussian random variable N with variance σ^2, so this inequality justifies the name "sub-Gaussian."

A Gaussian random variable is sub-Gaussian by (B.9). But there are many more examples, in particular the random variables described in the next proposition.

Proposition B.8.2 (1) *Let* X *be a complex-valued random variable and* $m > 0$ *a real number such that* $\mathbf{E}(\mathrm{X}) = 0$ *and* $|\mathrm{X}| \leqslant m$. *Then* X *is* m^2*-sub-Gaussian.*

(2) *Let* X_1 *and* X_2 *be independent random variables such that* X_i *is* σ_i^2*-sub-Gaussian. Then* $\mathrm{X}_1 + \mathrm{X}_2$ *is* $(\sigma_1^2 + \sigma_2^2)$*-sub-Gaussian.*

Proof (1) We may assume that X is real-valued, and by considering $m^{-1}\mathrm{X}$ instead of X, we may assume that $|\mathrm{X}| \leqslant 1$, and of course that X is not almost surely 0. In particular, the Laplace transform $\varphi(t) = \mathbf{E}(e^{t\mathrm{X}})$ is well defined, and $\varphi(t) > 0$ for all $t \in \mathbf{R}$. Moreover, it is easy to check that φ is smooth on $\mathbf{R}$ with

$$\varphi'(t) = \mathbf{E}(\mathrm{X}e^{t\mathrm{X}}) \quad \text{and} \quad \varphi''(t) = \mathbf{E}(\mathrm{X}^2 e^{t\mathrm{X}})$$

(by differentiating under the integral sign) and in particular

$$\varphi(0) = 1 \quad \text{and} \quad \varphi'(0) = \mathbf{E}(\mathrm{X}) = 0.$$

We now define $f(t) = \log(\varphi(t)) - \frac{1}{2}t^2$. The function f is also smooth and satisfies $f(0) = f'(0) = 0$. Moreover, we compute that

$$f''(t) = \frac{\varphi''(t)\varphi(t) - \varphi'(t)^2 - \varphi(t)^2}{\varphi(t)^2}.$$

The formula for φ'' and the condition $|\mathrm{X}| \leqslant 1$ imply that $0 \leqslant \varphi''(t) \leqslant \varphi(t)$ for all $t \in \mathbf{R}$. Therefore

$$\varphi''(t)\varphi(t) - \varphi'(t)^2 - \varphi(t)^2 \leqslant -\varphi'(t)^2 \leqslant 0,$$

and hence $f''(t) \leqslant 0$ for all $t \in \mathbf{R}$. This means that the derivative of f is decreasing, so that $f'(t) \leqslant 0$ for $t \geqslant 0$, and $f'(t) \geqslant 0$ for $t \leqslant 0$. Thus f is nondecreasing when $t \leqslant 0$ and nonincreasing when $t \geqslant 0$. In particular, we have $f(t) \leqslant f(0) = 0$ for all $t \in \mathbf{R}$, which translates exactly to the condition $\mathbf{E}(e^{t\mathrm{X}}) \leqslant e^{t^2/2}$ defining a sub-Gaussian random variable.

(2) Since X_1 and X_2 are independent and sub-Gaussian, we have

$$\mathbf{E}(e^{t(\mathrm{X}_1+\mathrm{X}_2)}) = \mathbf{E}(e^{t\mathrm{X}_1})\,\mathbf{E}(e^{t\mathrm{X}_2}) \leqslant \exp\left(\tfrac{1}{2}(\sigma_1^2 + \sigma_2^2)t^2\right)$$

for any $t \in \mathbf{R}$. ◻

Proposition B.8.3 *Let $\sigma > 0$ be a real number, and let X be a σ^2-sub-Gaussian random variable, either real or complex-valued. For any integer $k \geqslant 0$, there exists $c_k \geqslant 0$ such that*

$$\mathbf{E}(|X|^k) \leqslant c_k \sigma^k.$$

Proof The random variable $Y = \sigma^{-1}X$ is 1-sub-Gaussian. As in the proof of Theorem B.5.6 (2), we observe that there exists $c_k \geqslant 0$ such that

$$|Y|^k \leqslant c_k(e^{X_k} + e^{-X_k}),$$

and therefore

$$\sigma^{-k}\,\mathbf{E}(|X|^k) = \mathbf{E}(|Y|^k) \leqslant c_k(e^{1/2} + e^{-1/2}),$$

which gives the result. □

Remark B.8.4 A more precise argument leads to specific values of c_k. For instance, if X is real-valued, one can show that the inequality holds with $c_k = k2^{k/2}\Gamma(k/2)$.

B.9 Poisson Random Variables

Let $\lambda > 0$ be a real number. A random variable X is said to have a Poisson distribution with parameter $\lambda \in [0, +\infty[$ if and only if it is integral-valued and if, for any integer $k \geqslant 0$, we have

$$\mathbf{P}(X = k) = e^{-\lambda}\frac{\lambda^k}{k!}.$$

One checks immediately that

$$\mathbf{E}(X) = \lambda \quad \text{and} \quad \mathbf{V}(X) = \lambda$$

and that the characteristic function of X is

$$\varphi_X(t) = e^{-\lambda}\sum_{k\geqslant 0} e^{ikt}\frac{\lambda^k}{k!} = \exp(\lambda(e^{it} - 1)). \tag{B.11}$$

Proposition B.9.1 *Let (λ_n) be a sequence of real numbers such that $\lambda_n \to +\infty$ as $n \to +\infty$. Then*

$$\frac{X_n - \lambda_n}{\sqrt{\lambda_n}}$$

converges in law to a standard Gaussian random variable.

Proof Use the Lévy Criterion: the characteristic function φ_n of X_n is given by

$$\varphi_n(t) = \mathbf{E}(e^{it(X_n-\lambda_n)/\sqrt{\lambda_n}}) = \exp\Big(-it\sqrt{\lambda_n} + \lambda_n(e^{it/\sqrt{\lambda_n}} - 1)\Big)$$

for $t \in \mathbf{R}$, by (B.11). Since

$$-\frac{it}{\sqrt{\lambda_n}} + \lambda_n(e^{it/\sqrt{\lambda_n}} - 1) = it\sqrt{\lambda_n} + \lambda_n\left(\frac{it}{\sqrt{\lambda_n}} - \frac{t^2}{2\lambda_n} + \mathrm{O}\left(\frac{|t|^3}{\lambda_n^{3/2}}\right)\right)$$

$$= -\frac{t^2}{2} + \mathrm{O}\left(\frac{|t|^3}{\lambda_n^{1/2}}\right),$$

we obtain $\varphi_n(t) \to \exp(-t^2/2)$, which is the characteristic function of a standard Gaussian random variable. □

B.10 Random Series

We will need some fairly elementary results on certain random series, especially concerning almost sure convergence. We first have a well-known sufficient criterion of Kolmogorov for convergence in the case of independent summands:

Theorem B.10.1 (Kolmogorov) *Let $(X_n)_{n\geqslant 1}$ be a sequence of independent complex-valued random variables such that both series*

$$\sum_{n\geqslant 1} \mathbf{E}(X_n), \tag{B.12}$$

$$\sum_{n\geqslant 1} \mathbf{V}(X_n) \tag{B.13}$$

converge. Then the series

$$\sum_{n\geqslant 1} X_n$$

converges almost surely, and hence also in law. Moreover, its sum X is square integrable and has expectation $\sum \mathbf{E}(X_n)$.

Proof By replacing X_n with $X_n - \mathbf{E}(X_n)$, we reduce to the case where $\mathbf{E}(X_n) = 0$ for all n. Assuming that this is the case, we will prove that the series converges almost surely by checking that the sequence of partial sums

$$S_N = \sum_{1\leqslant n\leqslant N} X_n$$

is almost surely a Cauchy sequence. For this purpose, denote

$$Y_{N,M} = \sup_{1\leqslant k\leqslant M} |S_{N+k} - S_N|$$

for N, M $\geqslant$ 1. For fixed N, $Y_{N,M}$ is an increasing sequence of random variables; we denote by $Y_N = \sup_{k\geqslant 1} |S_{N+k} - S_N|$ its limit. Because of the estimate

$$|S_{N+k} - S_{N+l}| \leqslant |S_{N+k} - S_N| + |S_{N+l} - S_N| \leqslant 2Y_N$$

for $N \geqslant 1$ and $k, l \geqslant 1$, we have

$$\begin{aligned}\{(S_N)_{N\geqslant 1} \text{ is not Cauchy}\} &= \bigcup_{k\geqslant 1}\bigcap_{N\geqslant 1}\bigcup_{k\geqslant 1}\bigcup_{l\geqslant 1}\{|S_{N+k} - S_{N+l}| > 2^{-k}\}\\ &\subset \bigcup_{k\geqslant 1}\bigcap_{N\geqslant 1}\{Y_N > 2^{-k-1}\}.\end{aligned}$$

It is therefore sufficient to prove that

$$\mathbf{P}\Big(\bigcap_{N\geqslant 1}\{Y_N > 2^{-k-1}\}\Big) = 0$$

for each $k \geqslant 1$, or what amounts to the same thing, to prove that for any $\varepsilon > 0$, we have

$$\lim_{N\to+\infty} \mathbf{P}(Y_N > \varepsilon) = 0$$

(which means that Y_N converges to 0 in probability).

We begin by estimating $\mathbf{P}(Y_{N,M} > \varepsilon)$. *If* $Y_{N,M}$ was defined as $S_{N+M} - S_N$ (without the sup over $k \leqslant M$), this would be easy using the Markov inequality. To handle it, we use Kolmogorov's Maximal Inequality (see Lemma B.10.3): since the $(X_n)_{N+1\leqslant n\leqslant N+M}$ are independent, this shows that for any $\varepsilon > 0$, we have

$$\mathbf{P}(Y_{N,M} > \varepsilon) = \mathbf{P}\Big(\sup_{k\leqslant M}\Big|\sum_{1\leqslant n\leqslant k} X_{N+n}\Big| > \varepsilon\Big) \leqslant \frac{1}{\varepsilon^2}\sum_{n=N+1}^{N+M} \mathbf{V}(X_n).$$

Letting $M \to +\infty$, we obtain

$$\mathbf{P}(Y_N > \varepsilon) \leqslant \frac{1}{\varepsilon^2}\sum_{n\geqslant N+1} \mathbf{V}(X_n).$$

From the assumption on the convergence of the series of variance, this tends to 0 as $N \to +\infty$, which shows that the partial sums converge almost surely as claimed.

Now let $X = \sum X_n$ be the sum of the series, defined almost surely. For $N \geqslant 1$ and $M \geqslant 1$, we have

$$\|S_{M+N} - S_N\|_{L^2}^2 = \mathbf{E}\left(\left|\sum_{n=N+1}^{N+M} X_n\right|^2\right) = \sum_{n=N+1}^{N+M} \mathbf{E}(|X_n|^2) = \sum_{n=N+1}^{N+M} \mathbf{V}(X_n).$$

The assumption that (B.13) converges therefore implies that $(S_N)_{N\geqslant 1}$ is a Cauchy sequence in L^2, hence converges. Its limit necessarily coincides (almost surely) with the sum X, which shows that X is square-integrable. It follows that it is integrable and that its expectation can be computed as the sum of $\mathbf{E}(X_n)$. □

Remark B.10.2 (1) This result is a special case of Kolmogorov's Three Series Theorem, which gives a necessary and sufficient condition for almost sure convergence of a series of independent complex random variables (X_n); namely, it is enough that for some $c > 0$, and necessary that for all $c > 0$, the series

$$\sum_n \mathbf{P}(|X_n| > c), \qquad \sum_n \mathbf{E}(X_n^c), \qquad \sum_n \mathbf{V}(X_n^c)$$

converge, where $X_n^c = X_n$ if $|X_n| \leqslant c$ and $X_n^c = 0$ otherwise (see, e.g., [9, Th. 22.8] for the full proof, or try to reduce it to the previous case).

(2) It is worth mentioning two further results for context: (1) the event "the series converges" is an asymptotic event, in the sense that it doesn't depend on any finite number of the random variables – Kolmogorov's Zero–One Law then shows that this event can only have probability 0 or 1 – and (2) a theorem of P. Lévy shows that, again for independent summands, the almost sure convergence is equivalent to convergence in law or to convergence in probability. For proofs and discussion of these facts, see, for instance, [83, §0.III].

Here is Kolmogorov's maximal inequality:

Lemma B.10.3 *Let* $M \geqslant 1$ *be an integer,* $Y_1, \ldots, Y_M$ *independent complex random variables in* L^2 *with* $\mathbf{E}(Y_n) = 0$ *for all* n*. Then for any* $\varepsilon > 0$*, we have*

$$\mathbf{P}\left(\sup_{1\leqslant k\leqslant M} |Y_1 + \cdots + Y_k| > \varepsilon\right) \leqslant \frac{1}{\varepsilon^2}\sum_{n=1}^{M} \mathbf{V}(Y_n).$$

Proof Let

$$S_n = Y_1 + \cdots + Y_n$$

for $1 \leqslant n \leqslant \mathrm{M}$. We define a random variable T with values in $[0, +\infty]$ by $\mathrm{T} = \infty$ if $|\mathrm{S}_n| \leqslant \varepsilon$ for all $n \leqslant \mathrm{M}$, and otherwise

$$\mathrm{T} = \inf\{n \leqslant \mathrm{M} \mid |\mathrm{S}_n| > \varepsilon\}.$$

We then have

$$\Big\{ \sup_{1 \leqslant k \leqslant \mathrm{M}} |\mathrm{Y}_1 + \cdots + \mathrm{Y}_k| > \varepsilon \Big\} = \bigcup_{1 \leqslant n \leqslant \mathrm{M}} \{\mathrm{T} = n\},$$

and the union is disjoint. In particular, we get

$$\mathbf{P}\Big(\sup_{1 \leqslant k \leqslant \mathrm{M}} |\mathrm{S}_k| > \varepsilon \Big) = \sum_{n=1}^{\mathrm{M}} \mathbf{P}(\mathrm{T} = n).$$

We now note that $|\mathrm{S}_n|^2 \geqslant \varepsilon^2$ on the event $\{\mathrm{T} = n\}$, so that we can also write

$$\mathbf{P}\Big(\sup_{1 \leqslant k \leqslant \mathrm{M}} |\mathrm{S}_k| > \varepsilon \Big) \leqslant \frac{1}{\varepsilon^2} \sum_{n=1}^{\mathrm{M}} \mathbf{E}(|\mathrm{S}_n|^2 \mathbf{1}_{\{\mathrm{T}=n\}}). \tag{B.14}$$

We claim next that

$$\mathbf{E}(|\mathrm{S}_n|^2 \mathbf{1}_{\{\mathrm{T}=n\}}) \leqslant \mathbf{E}(|\mathrm{S}_\mathrm{M}|^2 \mathbf{1}_{\{\mathrm{T}=n\}}) \tag{B.15}$$

for all $n \leqslant \mathrm{M}$.

Indeed, if we write $\mathrm{S}_\mathrm{M} = \mathrm{S}_n + \mathrm{R}_n$, the independence assumption shows that R_n is independent of $(\mathrm{X}_1, \ldots, \mathrm{X}_n)$ and in particular is independent of the indicator function of the event $\{\mathrm{T} = n\}$, which only depends on $\mathrm{X}_1, \ldots, \mathrm{X}_n$. Moreover, we have $\mathbf{E}(\mathrm{R}_n) = 0$. Now, taking the modulus square in the definition and multiplying by this indicator function, we get

$$|\mathrm{S}_\mathrm{M}|^2 \mathbf{1}_{\{\mathrm{T}=n\}} = |\mathrm{S}_n|^2 \mathbf{1}_{\{\mathrm{T}=n\}} + \mathrm{S}_n \overline{\mathrm{R}}_n \mathbf{1}_{\{\mathrm{T}=n\}} + \overline{\mathrm{S}}_n \mathrm{R}_n \mathbf{1}_{\{\mathrm{T}=n\}} + |\mathrm{R}_n|^2 \mathbf{1}_{\{\mathrm{T}=n\}}.$$

Taking then the expectation, and using the positivity of the last term, this gives

$$\mathbf{E}(|\mathrm{S}_\mathrm{M}|^2 \mathbf{1}_{\{\mathrm{T}=n\}}) \geqslant \mathbf{E}(|\mathrm{S}_n|^2 \mathbf{1}_{\{\mathrm{T}=n\}}) + \mathbf{E}(\mathrm{S}_n \overline{\mathrm{R}}_n \mathbf{1}_{\{\mathrm{T}=n\}}) + \mathbf{E}(\overline{\mathrm{S}}_n \mathrm{R}_n \mathbf{1}_{\{\mathrm{T}=n\}}).$$

But, by independence, we have

$$\mathbf{E}(\mathrm{S}_n \overline{\mathrm{R}_n} \mathbf{1}_{\{\mathrm{T}=n\}}) = \mathbf{E}(\mathrm{S}_n \mathbf{1}_{\{\mathrm{T}=n\}})\, \mathbf{E}(\overline{\mathrm{R}_n}) = 0,$$

and similarly, $\mathbf{E}(\overline{\mathrm{S}_n} \mathrm{R}_n \mathbf{1}_{\{\mathrm{T}=n\}}) = 0$. Thus we get the bound (B.15).

Using this in (B.14), this gives

$$\mathbf{P}\Big(\sup_{1 \leqslant k \leqslant \mathrm{M}} |\mathrm{S}_k| > \varepsilon \Big) \leqslant \frac{1}{\varepsilon^2} \sum_{n=1}^{\mathrm{M}} \mathbf{E}(|\mathrm{S}_\mathrm{M}|^2 \mathbf{1}_{\{\mathrm{T}=n\}}) \leqslant \frac{1}{\varepsilon^2} \mathbf{E}(|\mathrm{S}_\mathrm{M}|^2)$$

by positivity once again. ☐

Exercise B.10.4 Deduce from Kolmogorov's Theorem the nontrivial direction of the Borel–Cantelli Lemma: if $(A_n)_{n\geqslant 1}$ is a sequence of independent events such that

$$\sum_{n\geqslant 1} \mathbf{P}(A_n) = +\infty,$$

then an element of the underlying probability space belongs almost surely to infinitely many of the sets A_n.

The second result we need is more subtle. It concerns similar series, but *without* the independence assumption, which is replaced by an orthogonality condition.

Theorem B.10.5 (Menshov–Rademacher) *Let* (X_n) *be a sequence of complex-valued random variables such that* $\mathbf{E}(X_n) = 0$ *and*

$$\mathbf{E}(X_n\overline{X_m}) = \begin{cases} 0 & \text{if } n \neq m, \\ 1 & \text{if } n = m. \end{cases}$$

Let (a_n) *be any sequence of complex numbers such that*

$$\sum_{n\geqslant 1} |a_n|^2(\log n)^2 < +\infty.$$

Then the series

$$\sum_{n\geqslant 1} a_n X_n$$

converges almost surely, and hence also in law.

Remark B.10.6 Consider the probability space $\Omega = \mathbf{R}/\mathbf{Z}$ with the Lebesgue measure and the random variables $X_n(t) = e(nt)$ for $n \in \mathbf{Z}$. One easily sees (adapting to double-sided sequences and symmetric partial sums) that Theorem B.10.5 implies that the series

$$\sum_{n\in\mathbf{Z}} a_n e(nt)$$

converges almost everywhere (with respect to Lebesgue measure), provided

$$\sum_{n\in\mathbf{Z}} |a_n|^2(\log |n|)^2 < +\infty.$$

This may be proved more directly (see, e.g., [121, III, th. 4.4]) using properties of Fourier series, but it is far from obvious. Note that, in this case, a very famous theorem of Carleson shows that the condition may be replaced with

$\sum |a_n|^2 < +\infty$. On the other hand, Menshov proved that Theorem B.10.5 *cannot* in general be relaxed in this way: for general orthonormal sequences, the term $(\log n)^2$ cannot be replaced by any positive function $f(n)$ such that $f(n) = o((\log n)^2)$, even for $\mathbf{R}/\mathbf{Z}$.

We begin with a lemma that will play an auxiliary role similar to Kolmogorov's inequality.

Lemma B.10.7 *Let* $(X_1, \ldots, X_N)$ *be orthonormal random variables,* $(a_1, \ldots, a_N)$ *be complex numbers, and* $S_k = a_1X_1 + \cdots + a_kX_k$ *for* $1 \leqslant k \leqslant N$. *We have*

$$\mathbf{E}\Big(\max_{1 \leqslant k \leqslant N} |S_k|^2\Big) \ll (\log N)^2 \sum_{n=1}^{N} |a_n|^2,$$

where the implied constant is absolute.

Proof The basic ingredient is a simple combinatorial property, which we present a bit abstractly. We claim that there exists a family $\mathcal{J}$ of discrete intervals

$$I = \{n_I, \ldots, m_I - 1\}, \qquad m_I - n_I \geqslant 1,$$

for $I \in \mathcal{J}$, with the following two properties:

(1) any interval $1 \leqslant n \leqslant M$ with $M \leqslant N$ is the disjoint union of $\ll \log N$ intervals $I \in \mathcal{J}$;

(2) an integer n with $1 \leqslant n \leqslant N$ belongs to $\ll \log N$ intervals in $\mathcal{J}$;

and in both cases the implied constant is independent of N.

To see this, let $n \geqslant 1$ be such that $2^{n-1} \leqslant N \leqslant 2^n$ (so that $n \ll \log N$), and consider for instance the family $\mathcal{J}$ of dyadic intervals

$$I_{i,j} = \{n \mid 1 \leqslant n \leqslant N \text{ and } i2^j \leqslant n < (i+1)2^j\}$$

for $0 \leqslant j \leqslant n$ and $1 \leqslant i \leqslant 2^{n-j}$. (The proofs of both properties in this case are left to the reader.)

Now, having fixed such a collection of intervals, we denote by T the smallest integer between 1 and N such that

$$\max_{1 \leqslant k \leqslant N} |S_k| = |S_T|.$$

By our first property of the intervals $\mathcal{J}$, we can write

$$S_T = \sum_{I} \tilde{S}_I,$$

where I runs over a set of $\ll \log N$ disjoint intervals in $\mathcal{J}$ and

$$\tilde{S}_I = \sum_{n\in I} a_n X_n$$

is the corresponding partial sum. By the Cauchy–Schwarz inequality, and the first property again, we get

$$|S_T|^2 \ll (\log N) \sum_{I} |\tilde{S}_I|^2 \ll (\log N) \sum_{I\in\mathcal{J}} |\tilde{S}_I|^2.$$

Taking the expectation and using orthonormality, we derive

$$\begin{aligned} \mathbf{E}\Big(\max_{1\leqslant k\leqslant N} |S_k|^2\Big) &= \mathbf{E}(|S_T|^2) \ll (\log N) \sum_{I\in\mathcal{J}} \mathbf{E}(|\tilde{S}_I|^2) \\ &= (\log N) \sum_{I\in\mathcal{J}} \sum_{n\in I} |a_n|^2 \ll (\log N)^2 \sum_{1\leqslant n\leqslant N} |a_n|^2 \end{aligned}$$

by the second property of the intervals $\mathcal{J}$. □

Proof of Theorem B.10.5 If the factor $(\log N)^2$ in Lemma B.10.7 was replaced by $(\log n)^2$ inside the sum, we would proceed just like the deduction of Theorem B.10.1 from Lemma B.10.3. Since this is not the case, a slightly different argument is needed.

We define

$$S_n = a_1X_1 + \cdots + a_nX_n$$

for $n \geqslant 1$. For $j \geqslant 0$, we also define the dyadic sum

$$\tilde{S}_j = \sum_{2^j\leqslant n<2^{j+1}} a_nX_n = S_{2^{j+1}-1} - S_{2^j}.$$

We first note that the series

$$T = \sum_{j\geqslant 0} (j+1)^2 |\tilde{S}_j|^2$$

converges almost surely. Indeed, since it is a series of nonnegative terms, it suffices to show that $\mathbf{E}(T) < +\infty$. But we have

$$\begin{aligned} \mathbf{E}(T) &= \sum_{j\geqslant 0} (j+1)^2\, \mathbf{E}(|\tilde{S}_j|^2) \\ &= \sum_{j\geqslant 0} (j+1)^2 \sum_{2^j\leqslant n<2^{j+1}} |a_n|^2 \ll \sum_{n\geqslant 1} |a_n|^2 (\log 2n)^2 < +\infty \end{aligned}$$

by orthonormality and by the assumption of the theorem.

Next, we observe that for $j \geqslant 0$ and $k \geqslant 0$, we have

$$|S_{2^{j+k}} - S_{2^j}| \leqslant \sum_{i=j}^{j+k-1} |\tilde{S}_i| \leqslant \Big(\sum_{j \leqslant i < j+k} \frac{1}{(i+1)^2} \Big)^{1/2} |T|^{1/2} \ll \Big(\frac{|T|}{j+1} \Big)^{1/2}$$

by the Cauchy–Schwarz inequality. We conclude that the sequence (S_{2^j}) is almost surely a Cauchy sequence and hence converges almost surely to a random variable S.

Finally, to prove that (S_n) converges almost surely to S, we observe that for any $n \geqslant 1$, and $j \geqslant 0$ such that $2^j \leqslant n < 2^{j+1}$, we have

$$|S_n - S_{2^j}| \leqslant M_j = \max_{2^j < k \leqslant 2^{j+1}} \Big| \sum_{m=2^j}^{k} a_n X_n \Big|. \qquad \text{(B.16)}$$

Lemma B.10.7 implies that

$$\mathbf{E}\Big(\sum_{j \geqslant 0} M_j^2 \Big) = \sum_{j \geqslant 0} \mathbf{E}(M_j^2) \ll \sum_{n \geqslant 1} (\log 2n)^2 |a_n|^2 < +\infty,$$

which means in particular that M_j tends to 0 as $j \to +\infty$ almost surely. From (B.16) and the convergence of $(S_{2^j})_j$ to S, we deduce that (S_n) converges almost surely to S. This finishes the proof. □

We will also use information on the support of the distribution of a random series with independent summands.

Proposition B.10.8 *Let* B *be a separable Banach space. Let* $(X_n)_{n \geqslant 1}$ *be a sequence of* independent B*-valued random variables such that the series* $X = \sum X_n$ *converges almost surely.*[3] *The support of the law of* X *is the closure of the set of all convergent series of the form* $\sum x_n$*, where* x_n *belongs to the support of the law of* X_n *for all* $n \geqslant 1$.

Proof For $N \geqslant 1$, we write

$$S_N = \sum_{n=1}^{N} X_n, \qquad R_N = X - S_N.$$

The variables S_N and R_N are independent.

First, we observe that Lemmas B.2.1 and B.2.2 imply that the support of S_N is the closure of the set of elements $x_1 + \cdots + x_N$ with $x_n \in \operatorname{supp}(X_n)$ for $1 \leqslant n \leqslant N$ (apply Lemma B.2.1 to the law of $(X_1, \ldots, X_N)$ on B^N, which has

[3] Recall that by the result of P. Lévy mentioned in Remark B.10.2, this is in fact equivalent to convergence in law.

support the product of the $\mathrm{supp}(X_n)$ by Lemma B.2.2, and to the addition map $B^N \to B$).

We first prove that all convergent series $\sum x_n$ with $x_n \in \mathrm{supp}(X_n)$ belong to the support of X, hence the closure of this set is contained in the support of X, as claimed. Thus let $x = \sum x_n$ be of this type. Let $\varepsilon > 0$ be fixed.

For all N large enough, we have

$$\Big\| \sum_{n>N} x_n \Big\| < \varepsilon,$$

and it follows that $x_1 + \cdots + x_N$, which belongs to the support of S_N as first remarked also belongs to the open ball U_ε of radius ε around x. Hence

$$\mathbf{P}(S_N \in U_\varepsilon) > 0$$

for all N large enough (U_ε is an open neighborhood of some element in the support of S_N).

Now the almost sure convergence implies (by the dominated convergence theorem, for instance) that $\mathbf{P}(\|R_N\| > \varepsilon) \to 0$ as $N \to +\infty$. Therefore, taking N suitably large, we get

$$\begin{aligned}\mathbf{P}(\|X - x\| < 2\varepsilon) &\geqslant \mathbf{P}(\|S_N - x\| < \varepsilon \text{ and } \|R_N\| < \varepsilon)\\ &= \mathbf{P}(\|S_N - x\| < \varepsilon)\,\mathbf{P}(\|R_N\| < \varepsilon) > 0\end{aligned}$$

(by independence). Since ε is arbitrary, this shows that $x \in \mathrm{supp}(X)$, as claimed.

Conversely, let $x \in \mathrm{supp}(X)$. For any $\varepsilon > 0$, we have

$$\mathbf{P}\Big(\Big\| \sum_{n\geqslant 1} X_n - x \Big\| < \varepsilon\Big) > 0.$$

Since, for any $n_0 \geqslant 1$, we have

$$\mathbf{P}\Big(\Big\| \sum_{n\geqslant 1} X_n - x \Big\| < \varepsilon \text{ and } X_{n_0} \notin \mathrm{supp}(X_{n_0})\Big) = 0,$$

this means in fact that

$$\mathbf{P}\Big(\Big\| \sum_{n\geqslant 1} X_n - x \Big\| < \varepsilon \text{ and } X_n \in \mathrm{supp}(X_n) \text{ for all } n\Big) > 0.$$

In particular, we can find $x_n \in \mathrm{supp}(X_n)$ such that the series $\sum x_n$ converges and

$$\Big\| \sum_{n\geqslant 1} x_n - x \Big\| < \varepsilon,$$

and hence x belongs to the closure of the set of convergent series $\sum x_n$ with x_n in the support of X_n for all n. □

B.11 Some Probability in Banach Spaces

We consider in this section some facts about probability in a (complex) Banach space V. Most are relatively elementary. For simplicity, we will always assume that V is separable (so that, in particular, Borel measures on V have a well-defined support).

The first result concerns series

$$\sum_n X_n$$

where (X_n) is a sequence of *symmetric* random variables, which means that for any $N \geqslant 1$, and for any choice $(\varepsilon_1, \dots, \varepsilon_N)$ of signs $\varepsilon_n \in \{-1, 1\}$ for $1 \leqslant n \leqslant N$, the random vectors

$$(X_1, \dots, X_N) \text{ and } (\varepsilon_1 X_1, \dots, \varepsilon_N X_N)$$

have the same distribution.

Symmetric random variables have remarkable properties. For instance, we have:

Proposition B.11.1 (Lévy) *Let* V *be a separable Banach space with norm* $\|\cdot\|$, *and* (X_n) *a sequence of* V*-valued random variables. Assume that the sequence* (X_n) *is symmetric. Let*

$$S_N = X_1 + \cdots + X_N$$

for $N \geqslant 1$. *For* $N \geqslant 1$ *and* $\varepsilon > 0$, *we have*

$$\mathbf{P}(\max_{1 \leqslant n \leqslant N} \|S_N\| > \varepsilon) \leqslant 2\,\mathbf{P}(\|S_N\| > \varepsilon).$$

This result is known as *Lévy's reflection principle* and can be compared with Kolmogorov's maximal inequality (Lemma B.10.3).

Proof (1) Similarly to the proof of Lemma B.10.3, we define a random variable T by $T = \infty$ if $\|S_n\| \leqslant \varepsilon$ for all $n \leqslant N$, and otherwise

$$T = \inf\{n \leqslant N \mid \|S_n\| > \varepsilon\}.$$

Assume $T = k$ and consider the random variables

$$X'_n = X_n \text{ for } 1 \leqslant n \leqslant k, \quad X'_n = -X_n \text{ for } k+1 \leqslant n \leqslant N.$$

By assumption, the sequence $(X'_n)_{1 \leqslant n \leqslant N}$ has the same distribution as $(X_n)_{1 \leqslant n \leqslant N}$. Let S'_n denote the partial sums of the sequence (X'_n), and T' the analogue of T for the sequence (X'_n). The event $\{T' = k\}$ is the same as $\{T = k\}$ since $X'_n = X_n$ for $n \leqslant k$. On the other hand, we have

$$S'_N = X_1 + \cdots + X_k - X_{k+1} - \cdots - X_N = 2S_k - S_N.$$

Therefore

$$\mathbf{P}(\|S_N\| > \varepsilon \text{ and } T = k) = \mathbf{P}(\|S'_N\| > \varepsilon \text{ and } T' = k) = \mathbf{P}(\|2S_k - S_N\| > \varepsilon \text{ and } T = k).$$

By the triangle inequality we have

$$\{T = k\} \subset \{\|S_N\| > \varepsilon \text{ and } T = k\} \cup \{\|2S_k - S_N\| > \varepsilon \text{ and } T = k\}.$$

We deduce

$$\begin{aligned}
\mathbf{P}(\max_{1 \leqslant n \leqslant N} \|S_n\| > \varepsilon) &= \sum_{k=1}^{N} \mathbf{P}(T = k) \\
&\leqslant \sum_{k=1}^{N} \mathbf{P}(\|S_N\| > \varepsilon \text{ and } T = k) \\
&\quad + \sum_{k=1}^{N} \mathbf{P}(\|2S_n - S_K\| > \varepsilon \text{ and } T = k) \\
&= 2\,\mathbf{P}(\|S_N\| > \varepsilon).
\end{aligned}$$

□

We now consider the special case where the Banach space V is $C([0,1])$, the space of complex-valued continuous functions on $[0,1]$ with the norm

$$\|f\|_\infty = \sup_{t \in [0,1]} |f(t)|.$$

For a $C([0,1])$-valued random variable X and any fixed $t \in [0,1]$, we will denote by $X(t)$ the complex-valued random variable that is the value of the random function X at t, that is, $X(t) = e_t \circ X$, where $e_t : C([0,1]) \longrightarrow \mathbf{C}$ is the evaluation at t.

Definition B.11.2 (Convergence of finite distributions) Let (X_n) be a sequence of $C([0,1])$-valued random variables, and let X be a $C([0,1])$-valued random variable. One says that (X_n) converges to X *in the sense of finite distributions* if and only if, for all integers $k \geqslant 1$, and for all

$$0 \leqslant t_1 < \cdots < t_k \leqslant 1,$$

the random vectors $(X_n(t_1), \ldots, X_n(t_k))$ converge in law to $(X(t_1), \ldots, X(t_k))$, in the sense of convergence in law in $\mathbf{C}^k$.

One sufficient condition for convergence of finite distributions is the following:

Lemma B.11.3 *Let* (X_n) *be a sequence of* $\mathrm{C}([0,1])$*-valued random variables, and let* X *be a* $\mathrm{C}([0,1])$*-valued random variable, all defined on the same probability space. Assume that, for any* $t \in [0,1]$*, the random variables* $(\mathrm{X}_n(t))$ *converge in* L^1 *to* $\mathrm{X}(t)$*. Then* (X_n) *converges to* X *in the sense of finite distributions.*

Proof Fix $k \geqslant 1$ and

$$0 \leqslant t_1 < \cdots < t_k \leqslant 1.$$

Let φ be a Lipschitz function on $\mathbf{C}^k$ (given the distance associated to the norm

$$\|(z_1, \ldots, z_k)\| = \sum_i |z_i|,$$

for instance) with Lipschitz constant $\mathrm{C} \geqslant 0$. Then we have

$$\left| \mathbf{E}(\varphi(\mathrm{X}_n(t_1), \ldots, \mathrm{X}_n(t_k))) - \mathbf{E}(\varphi(\mathrm{X}(t_1), \ldots, \mathrm{X}(t_k))) \right| \leqslant \mathrm{C} \sum_{i=1}^{k} \mathbf{E}(|\mathrm{X}_n(t_i) - \mathrm{X}(t_i)|),$$

which tends to 0 as $n \to +\infty$ by our assumption. Hence Proposition B.4.1 shows that $(\mathrm{X}_n(t_1), \ldots, \mathrm{X}_n(t_k))$ converges in law to $(\mathrm{X}(t_1), \ldots, \mathrm{X}(t_k))$. This proves the lemma. □

Convergence in finite distributions is a necessary condition for convergence in law of (X_n) to X, but it is not sufficient: a simple example (see [10, Example 2.5]) consists in taking the random variable X_n to be the *constant* random variable equal to the function f_n that is piecewise linear on $[0, 1/n]$, $[1/n, 1/(2n)]$ and $[1/(2n), 1]$, and such that $0 \mapsto 0$, $1/n \mapsto 1$, $1/(2n) \mapsto 0$ and $1 \mapsto 0$. Then it is elementary that X_n converges to the constant zero random variable in the sense of finite distributions, but that X_n does not converge in law to 0 (because f_n does not converge uniformly to 0).

Nevertheless, under the additional condition of *tightness* of the sequence of random variables (see Definition B.3.6), the convergence of finite distributions implies convergence in law.

Theorem B.11.4 *Let* (X_n) *be a sequence of* $\mathrm{C}([0,1])$*-valued random variables, and let* X *be a* $\mathrm{C}([0,1])$*-valued random variable. Suppose that* (X_n) *converges to* X *in the sense of finite distributions. Then* (X_n) *converges in law* *to* X *in the sense of* C([0, 1])*-valued random variables if and only if* (X_n) *is tight.*

For a proof, see, for example, [10, Th. 7.1]. The key ingredient is Prokhorov's Theorem (see [10, Th. 5.1]), which states that a tight family of random variables is relatively compact in the space $\mathcal{P}$ of probability measures on C([0, 1]), given the topology of convergence in law. To see how this implies the result, we note that convergence in the sense of finite distributions of a sequence implies at least that it has *at most* one limit in $\mathcal{P}$ (because probability measures on C([0, 1]) are uniquely determined by the family of their finite distributions; see [10, Ex. 1.3]). Suppose now that there exists a continuous bounded function f on C([0, 1]) such that

$$\mathbf{E}(f(X_n))$$

does not converge to $\mathbf{E}(f(X))$. Then there exists $\delta > 0$ and some subsequence (X_{n_k}) that satisfies $|\mathbf{E}(f(X_{n_k}) - f(X))| \geqslant \delta$ for all k. This subsequence also converges to X in the sense of finite distributions and by relative compactness admits a further subsequence that converges in law; but the limit of that further subsequence must then be X, which contradicts the inequality above.

Remark B.11.5 For certain purposes, it is important to observe that this proof of convergence in law is indirect and does not give quantitative estimates.

We will also use a variant of this result involving Fourier series. A minor issue is that we wish to consider functions f on [0, 1] that are not necessarily periodic, in the sense that $f(0)$ might differ from $f(1)$. However, we will have $f(0) = 0$. To account for this, we use the identity function in addition to the periodic exponentials to represent continuous functions with $f(0) = 0$.

We denote by $C_0([0, 1])$ the subspace of C([0, 1]) of functions vanishing at 0. We denote by e_0 the function $e_0(t) = t$, and for $h \neq 0$, we put $e_h(t) = e(ht)$. We denote further by $C_0(\mathbf{Z})$ the Banach space of complex-valued functions on $\mathbf{Z}$ converging to 0 at infinity with the sup norm. We define a continuous linear map FT: $C([0,1]) \to C_0(\mathbf{Z})$ by mapping f to the sequence $(\widetilde{f}(h))_{h\in\mathbf{Z}}$ of its Fourier coefficients, where $\widetilde{f}(0) = f(1)$ and for $h \neq 0$ we have

$$\widetilde{f}(h) = \int_0^1 (f(t) - tf(1))e(-ht)dt = \int_0^1 (f - f(1)e_0)e_{-h}.$$

We want to relate convergence in law in $C_0([0, 1])$ with convergence, in law or in the sense of finite distributions, of these "Fourier coefficients" in $C_0(\mathbf{Z})$. Here convergence of finite distributions of a sequence (X_n) of $C_0(\mathbf{Z})$-valued random variables to X means that for any $H \geqslant 1$, the vectors $(X_{n,h})_{|h|\leqslant H}$ converge in law to $(X_h)_{|h|\leqslant H}$, in the sense of convergence in law in $\mathbf{C}^{2H+1}$.

First, since FT is continuous, Proposition B.3.2 gives immediately

Lemma B.11.6 *If* $(X_n)_n$ *is a sequence of* $C_0([0,1])$*-valued random variables that converges in law to a random variable* X*, then* $FT(X_n)$ *converges in law to* FT(X).

Next, we check that the Fourier coefficients determine the law of a $C_0([0,1])$-valued random variable (this is the analogue of [10, Ex. 1.3]).

Lemma B.11.7 *If* X *and* Y *are* $C_0([0,1])$*-valued random variables and if* FT(X) *and* FT(Y) *have the same finite distributions, then* X *and* Y *have the same law.*

Proof For $f \in C_0([0,1])$, the function $g(t) = f(t) - tf(1)$ extends to a 1-periodic continuous function on $\mathbf{R}$. By Féjer's Theorem on the uniform convergence of Cesàro means of Fourier series of continuous periodic functions (see, e.g, [121, III, Th. 3.4]), we have

$$f(t) - tf(1) = \lim_{H\to+\infty} \sum_{|h|\leqslant H} \Big(1 - \frac{|h|}{H}\Big) \widetilde{f}(h)e(ht)$$

uniformly for $t \in [0,1]$. Evaluating at $t = 0$, where the left-hand side vanishes, we deduce that

$$f = \lim_{H\to+\infty} C_H(f),$$

where

$$C_H(f) = f(1)e_0 + \sum_{\substack{|h|\leqslant H\\ h\neq 0}} \Big(1 - \frac{|h|}{H}\Big) \widetilde{f}(h)(e_h - 1).$$

Note that $C_H(f) \in C_0([0,1])$.

We now claim that $C_H(X)$ converges to X as $C_0([0,1])$-valued random variables. Indeed, let φ be a bounded continuous function on $C_0([0,1])$, say, with $|\varphi| \leqslant M$. By the above, we have $\varphi(C_H(X)) \to \varphi(X)$ as $H \to +\infty$ pointwise on $C_0([0,1])$. Since $|\varphi(C_H(X))| \leqslant M$, which is integrable on the underlying probability space, Lebesgue's dominated convergence theorem implies that $\mathbf{E}(\varphi(C_H(X))) \to \mathbf{E}(\varphi(X))$. This proves the claim.

In view of the definition of $C_H(f)$, which only involves finitely many Fourier coefficients, the equality of finite distributions of FT(X) and FT(Y) implies by composition that for any $H \geqslant 1$, the $C_0([0,1])$-valued random variables $C_H(X)$ and $C_H(Y)$ have the same law. Since we have seen that $C_H(X)$

converges in law to X and that $C_H(Y)$ converges in law to Y, it follows that X and Y have the same law. □

Now comes the convergence criterion:

Proposition B.11.8 *Let* (X_n) *be a sequence of* $C_0([0,1])$*-valued random variables, and let* X *be a* $C_0([0,1])$*-valued random variable. Suppose that* $FT(X_n)$ *converges to* FT(X) *in the sense of finite distributions. Then* (X_n) *converges in law to* X *in the sense of* $C_0([0,1])$*-valued random variables if and only if* (X_n) *is tight.*

Proof It is an elementary general fact that if (X_n) converges in law to X, then the family (X_n) is tight. We prove the converse assertion. It suffices to prove that any subsequence of (X_n) has a further subsequence that converges in law to X (see [10, Th. 2.6]). Because (X_n) is tight, so are any of its subsequences. By Prokhorov's Theorem ([10, Th. 5.1]), such a subsequence therefore contains a further subsequence, say, $(X_{n_k})_{k\geqslant 1}$, that converges in law to some probability measure Y. By Lemma B.11.6, the sequence of Fourier coefficients $FT(X_{n_k})$ converges in law to FT(Y). On the other hand, this sequence converges to FT(X) in the sense of finite distributions, by assumption. Hence FT(X) and FT(Y) have the same finite distributions, which implies that X and Y have the same law by Lemma B.11.7. □

Remark B.11.9 The example that was already mentioned before Theorem B.11.4 (namely, [10, Ex. 2.5]) also shows that the convergence of $FT(X_n)$ to FT(X) in the sense of finite distributions is not sufficient to conclude that (X_n) converges in law to X. Indeed, the sequence (X_n) in that example does not converge in law in $C([0,1])$, but for $n \geqslant 1$, the (constant) random variable X_n satisfies $X_n(1) = 0$, and by direct computation, the Fourier coefficients (are constant and) satisfy also $|\widetilde{X}_n(h)| \leqslant n^{-1}$ for all $h \neq 0$, which implies that $FT(X_n)$ converges in law to the constant random variable equal to $0 \in C_0(\mathbf{Z})$.

In applications, we need some criteria to detect tightness. One such criterion is due to Kolmogorov:

Proposition B.11.10 (Kolmogorov's tightness criterion) *Let* (X_n) *be a sequence of* $C([0,1])$*-valued random variables. If there exist real numbers* $\alpha > 0$, $\delta > 0$, *and* $C \geqslant 0$ *such that, for any real numbers* $0 \leqslant s < t \leqslant 1$ *and any* $n \geqslant 1$, *we have*

$$\mathbf{E}(|X_n(t) - X_n(s)|^{\alpha}) \leqslant C|t-s|^{1+\delta}, \tag{B.17}$$

then (X_n) *is tight.*

See, for instance, [81, Th. I.7] for a proof. The statement does not hold if the exponent $1+\delta$ is replaced by 1.

In fact, for some applications (as in [79]), one needs a variant where the single bound (B.17) is replaced by different ones depending on the size of $|t-s|$ relative to n. Such a result does not seem to follow formally from Proposition B.11.10, because the left-hand side in the inequality is not monotonic in terms of α (in contrast with the right-hand side, which is monotonic since $|t-s| \leqslant 1$). We state a result of this type and sketch the proof for completeness.

Proposition B.11.11 (Kolmogorov's tightness criterion, 2) *Let* (X_n) *be a sequence of* $\mathrm{C}([0,1])$*-valued random variables. Suppose that there exist positive real numbers*

$$\alpha_1, \alpha_2, \alpha_3, \quad \beta_2 < \beta_1, \quad \delta, \quad \mathrm{C}$$

such that for any real numbers $0 \leqslant s < t \leqslant 1$ *and any* $n \geqslant 1$, *we have*

$$\mathbf{E}(|\mathrm{X}_n(t) - \mathrm{X}_n(s)|^{\alpha_1}) \leqslant \mathrm{C}|t-s|^{1+\delta} \quad \text{if } 0 \leqslant |t-s| \leqslant n^{-\beta_1}, \tag{B.18}$$

$$\mathbf{E}(|\mathrm{X}_n(t) - \mathrm{X}_n(s)|^{\alpha_2}) \leqslant \mathrm{C}|t-s|^{1+\delta} \quad \text{if } n^{-\beta_1} \leqslant |t-s| \leqslant n^{-\beta_2}, \tag{B.19}$$

$$\mathbf{E}(|\mathrm{X}_n(t) - \mathrm{X}_n(s)|^{\alpha_3}) \leqslant \mathrm{C}|t-s|^{1+\delta} \quad \text{if } n^{-\beta_2} \leqslant |t-s| \leqslant 1. \tag{B.20}$$

Then (X_n) *is tight.*

Sketch of proof For $n \geqslant 1$, let $\mathrm{D}_n \subset [0,1]$ be the set of dyadic rational numbers with denominator 2^n. For $\delta > 0$, let

$$\omega(\mathrm{X}_n, \delta) = \sup\{|\mathrm{X}_n(t) - \mathrm{X}_n(s)| \mid s,t \in [0,1],\ |t-s| \leqslant \delta\}$$

denote the modulus of continuity of X_n, and for $n \geqslant 1$ and $k \geqslant 0$, let

$$\xi_{n,k} = \sup\{|\mathrm{X}_n(t) - \mathrm{X}_n(s)| \mid s,t \in \mathrm{D}_k,\ |s-t| = 2^{-k}\}.$$

We observe that for any $\alpha > 0$, we have

$$\mathbf{E}(\xi_{n,k}^{\alpha}) \leqslant \sum_{\substack{s,t \in \mathrm{D}_k \\ |t-s| = 2^{-k}}} \mathbf{E}(|\mathrm{X}_n(t) - \mathrm{X}_n(s)|^{\alpha}).$$

As in [60, p. 269], the key step is to prove that

$$\lim_{m \to +\infty} \limsup_{n \to +\infty} \mathbf{P}(\omega(\mathrm{X}_n, 2^{-m}) > \eta) = 0$$

for any $\eta > 0$ (the conclusion is then derived from this fact combined with the Ascoli–Arzela Theorem characterizing compact subsets of $\mathrm{C}([0,1])$). It is convenient here to use the notation $\min(a,b) = a \wedge b$. For fixed m and n, we then write

$$\begin{aligned}\mathbf{P}(\omega(\mathrm{X}_n, 2^{-m}) > \eta) \leqslant \mathbf{P}(&\sup_{|t-s|\leqslant 2^{-m}\wedge n^{-\beta_1}} |\mathrm{X}_n(t) - \mathrm{X}_n(s)| > \eta)\\ &+ \mathbf{P}(\sup_{2^{-m}\wedge n^{-\beta_1}<|t-s|\leqslant 2^{-m}\wedge n^{-\beta_2}} |\mathrm{X}_n(t) - \mathrm{X}_n(s)| > \eta)\\ &+ \mathbf{P}(\sup_{2^{-m}\wedge n^{-\beta_2}<|t-s|\leqslant 2^{-m}} |\mathrm{X}_n(t) - \mathrm{X}_n(s)| > \eta)\end{aligned}$$

(because, for a continuous function f on $[0,1]$, if $|f(t) - f(s)| > \eta$ for some (s,t) such that $|t - s| \leqslant 2^{-m}$, then there exist some dyadic rational numbers (s',t'), necessarily with denominator 2^n with $n \geqslant m$, such that $|f(t') - f(s')| > \eta$). Using (B.18), the first term is

$$\leqslant \sum_{\substack{k\geqslant m\\ 2^{-k}\leqslant n^{-\beta_1}}} \frac{\mathbf{E}(\xi_{n,k}^{\alpha_1})}{\eta^{\alpha_1}} \leqslant \frac{\mathrm{C}}{\eta^{\alpha_1}} \sum_{k\geqslant m} 2^k 2^{-k(1+\delta)} \leqslant \frac{\mathrm{C}}{\eta^{\alpha_1}} \frac{1}{1-2^{-\delta}},$$

and similarly, using (B.19), the second (resp. using (B.20), the third) is

$$\leqslant \sum_{\substack{k\geqslant m\\ \beta_2\log_2(n)\leqslant k\leqslant\beta_1\log_2(n)}} \frac{\mathbf{E}(\xi_{n,k}^{\alpha_2})}{\eta^{\alpha_2}} \leqslant \frac{\mathrm{C}}{\eta^{\alpha_2}} \sum_{k\geqslant m} 2^k 2^{-k(1+\delta)} \leqslant \frac{\mathrm{C}}{\eta^{\alpha_2}} \frac{1}{1-2^{-\delta}}$$

(resp. $\leqslant \mathrm{C}\eta^{-\alpha_3}(1-2^{-\delta})^{-1}$). The result follows. □

We will also use the following inequality of Talagrand, which gives a type of sub-Gaussian behavior of sums of random variable in Banach spaces, extending standard properties of real or complex-valued random variables.

Theorem B.11.12 (Talagrand) *Let* V *be a separable real Banach space and* V′ *its dual. Let* $(\mathrm{X}_n)_{n\geqslant 1}$ *be a sequence of independent real-valued random variables with* $|\mathrm{X}_n| \leqslant 1$ *almost surely, and let* $(v_n)_{n\geqslant 1}$ *be a sequence of elements of* V. *Assume that the series* $\sum \mathrm{X}_n v_n$ *converges almost surely in* V. *Let* $m \geqslant 0$ *be a median of*

$$\Big\| \sum_{n\geqslant 1} \mathrm{X}_n v_n \Big\|.$$

Let $\sigma \geqslant 0$ *be the real number such that*

$$\sigma^2 = \sup_{\substack{\lambda\in\mathrm{V}'\\ \|\lambda\|\leqslant 1}} \sum_{n\geqslant 1} |\lambda(v_n)|^2.$$

For any real number $t > 0$, *we have*

$$\mathbf{P}\Big(\Big\|\sum_{n\geqslant 1} \mathrm{X}_n v_n\Big\| \geqslant t\sigma + m\Big) \leqslant 4\exp\Big(-\frac{t^2}{16}\Big).$$

We recall that a median m of a real-valued random variable X is any real number such that

$$\mathbf{P}(\mathrm{X} \geqslant m) \geqslant \frac{1}{2} \quad \text{and} \quad \mathbf{P}(\mathrm{X} \leqslant m) \geqslant \frac{1}{2}.$$

A median always exists. If X is integrable, then Chebychev's inequality

$$\mathbf{P}(\mathrm{X} \geqslant t) \leqslant \frac{\mathbf{E}(|\mathrm{X}|)}{t} \tag{B.21}$$

shows that $m \leqslant 2\,\mathbf{E}(|\mathrm{X}|)$.

Proof This follows easily from [114, Th. 13.2], which concerns finite sums, by passing to the limit. □

The application of this inequality will be the following, which is (in the case $\mathrm{V} = \mathbf{R}$) partly a simple variant of a result of Montgomery-Smith [88].

Proposition B.11.13 *Let* V *be a separable real or complex Banach space and* V' *its dual. Let* $(\mathrm{X}_n)_{n\geqslant 1}$ *be a sequence of independent random variables with* $|\mathrm{X}_n| \leqslant 1$ *almost surely, which are either real- or complex-valued depending on the base field. Let* $(v_n)_{n\geqslant 1}$ *be a sequence of elements of* V. *Assume that the series* $\sum \mathrm{X}_n v_n$ *converges almost surely in* V, *and let* X *be its sum.*

(1) *Assume that*

$$\sum_{n\leqslant \mathrm{N}} \|v_n\| \ll \log(\mathrm{N}), \qquad \sum_{n>\mathrm{N}} \|v_n\|^2 \ll \frac{1}{\mathrm{N}} \tag{B.22}$$

for all $\mathrm{N} \geqslant 1$. *There exists a constant* $c > 0$ *such that for any* $\mathrm{A} > 0$, *we have*

$$\mathbf{P}(\|\mathrm{X}\| > \mathrm{A}) \leqslant c\exp(-\exp(c^{-1}\mathrm{A})).$$

(2) *Assume that* V *is a real Banach space, that* (X_n) *is symmetric, identically distributed, and real-valued, and that there exists* $\lambda \in \mathrm{V}'$ *of norm* 1 *such that*

$$\sum_{n\leqslant \mathrm{N}} |\lambda(v_n)| \gg \log(\mathrm{N}) \tag{B.23}$$

for $\mathrm{N} \geqslant 1$. *Then there exists a constant* $c' > 0$ *such that for any* $\mathrm{A} > 0$, *we have*

$$c^{-1}\exp(-\exp(c\mathrm{A})) \leqslant \mathbf{P}(|\lambda(\mathrm{X})| > \mathrm{A}) \leqslant \mathbf{P}(\|\mathrm{X}\| > \mathrm{A}).$$

Proof We begin with (1), and we first check that we may assume that V is a real Banach space and that the random variables X_n are real-valued. To see this, if V is a complex Banach space, we view it as a real Banach space $V_{\mathbf{R}}$ (by restricting scalar multiplication), and we write $X_n = Y_n + iZ_n$ where Y_n and Z_n are real-valued random variables. Then $X = Y + iZ$ where

$$Y = \sum_{n \geqslant 1} Y_n v_n \quad \text{and} \quad Z = \sum_{n \geqslant 1} Z_n v_n$$

are both almost surely convergent series in $V_{\mathbf{R}}$ with independent real coefficients of absolute value $\leqslant 1$. We then have

$$\mathbf{P}(\|X\| > A) \leqslant \mathbf{P}\left(\|Y\| > \tfrac{1}{2}A \text{ or } \|Z\| > \tfrac{1}{2}A\right) \leqslant \mathbf{P}\left(\|Y\| > \tfrac{1}{2}A\right) + \mathbf{P}\left(\|Z\| > \tfrac{1}{2}A\right)$$

for any $A > 0$, by the triangle inequality. Since the assumptions (B.22) hold independently of whether V is viewed as a real or complex Banach space, we deduce that if (1) holds in the real case, then it also does for complex coefficients.

We now assume that V is a real Banach space. The idea is that if V was simply equal to $\mathbf{R}$, then the series X would be a sub-Gaussian random variable, and standard estimates would give a sub-Gaussian upper bound for $\mathbf{P}(|X| > A)$, of the type $\exp(-cA^2)$. Such a bound would be essentially sharp for a *Gaussian* series. But although this is already quite strong, it is far from the truth here; intuitively, this is because, in the Gaussian case, the lower bound for the probability arises from the small but non-zero probability that a single summand (distributed like a Gaussian) might be very large. This cannot happen for the series X, because each X_n is absolutely bounded.

The actual proof "interpolates" between the sub-Gaussian behavior (given by Talagrand's inequality, when the Banach space is infinite-dimensional) and the boundedness of the coefficients (X_n) of the first few steps. This principle goes back (at least) to Montgomery-Smith [88] and has relations with the theory of interpolation of Banach spaces.

Fix an auxiliary parameter $s \geqslant 1$. We write $X = X^\sharp + X^\flat$, where

$$X^\sharp = \sum_{1 \leqslant n \leqslant s^2} X_n v_n \quad \text{and} \quad X^\flat = \sum_{n > s^2} X_n v_n.$$

Let m be a median of the real random variable $\|X^\flat\|$. Then, for any $\alpha > 0$ and $\beta > 0$, we have

$$\mathbf{P}(\|X\| \geqslant \alpha + \beta + m) \leqslant \mathbf{P}(\|X^\sharp\| \geqslant \alpha) + \mathbf{P}(\|X^\flat\| \geqslant m + \beta),$$

by the triangle inequality. We pick

$$\alpha = 8 \sum_{1 \leqslant n \leqslant s^2} \|v_n\|$$

so that by the assumption $|X_n| \leqslant 1$, we have

$$\mathbf{P}(\|X^\sharp\| \geqslant \alpha) = 0.$$

Then we take $\beta = s\sigma$, where $\sigma \geqslant 0$ is such that

$$\sigma^2 = \sup_{\|\lambda\| \leqslant 1} \sum_{n > s^2} |\lambda(v_n)|^2,$$

where λ runs over the elements of norm $\leqslant 1$ of the dual space V'. By Talagrand's Inequality (Theorem B.11.12), we have

$$\mathbf{P}(\|X^\flat\| \geqslant m + \beta) \leqslant 4 \exp\left(-\frac{s^2}{8}\right).$$

Hence, for all $s \geqslant 1$, we have

$$\mathbf{P}(\|X\| \geqslant \alpha + \beta + m) \leqslant 4 \exp\left(-\frac{s^2}{8}\right).$$

We now select s as large as possible so that $m + \alpha + \beta \leqslant A$. By Chebychev's inequality (B.21), we have

$$m \leqslant 2\,\mathbf{E}(\|X^\flat\|) \leqslant 2 \sum_{1 \leqslant h \leqslant s^2} \|v_n\|$$

so that

$$m + \alpha \ll \sum_{1 \leqslant n \leqslant s^2} \|v_n\| \ll \log(2s) \tag{B.24}$$

for any $s \geqslant 1$ by (B.22). Moreover, for any linear form λ with $\|\lambda\| \leqslant 1$, we have

$$\sum_{n > s^2} |\lambda(v_n)|^2 \ll \sum_{n > s^2} \|v_n\|^2 \ll \frac{1}{s^2}$$

so that $\sigma \ll s^{-1}$ and $\beta = s\sigma \ll 1$. It follows that

$$m + \alpha + \beta \leqslant c \log(cs) \tag{B.25}$$

for some constant $c \geqslant 1$ and all $s \geqslant 1$. We finally select s so that $c \log(cs) = A$, that is,

$$s = \frac{1}{c} \exp\left(\frac{A}{c}\right)$$

(assuming, as we may, that A is large enough so that $s \geqslant 1$) and deduce that

$$\mathbf{P}(\|X\| \geqslant A) \leqslant 4\exp\Big(-\frac{s^2}{8}\Big) = 4\exp\Big(-\frac{1}{8c^2}\exp\Big(\frac{A}{c}\Big)\Big).$$

This gives the desired upper bound.

We now prove (2). Replacing the vectors v_n by the real numbers $\lambda(v_n)$ (recall that (2) is a statement for real Banach spaces and random variables), we may assume that $V = \mathbf{R}$. Let $\alpha > 0$ be such that

$$\sum_{n \leqslant N} |\lambda(v_n)| \geqslant \alpha \log(N)$$

for $N \geqslant 1$, and let β_n be a median of $|X_n|$. We then derive

$$\mathbf{P}(|X| > A) \geqslant \mathbf{P}\Big(X_n \geqslant \beta_n \text{ for } 1 \leqslant n \leqslant e^{(\alpha\beta)^{-1}A} \text{ and } \sum_{n > e^{A/(\alpha\beta)}} v_n X_n \geqslant 0\Big).$$

Since the random variables (X_n) are independent, this leads to

$$\mathbf{P}(|X| > A) \geqslant \Big(\frac{1}{4}\Big)^{\lfloor \exp(A/(\alpha\beta))\rfloor} \mathbf{P}\Big(\sum_{n > e^{A/(\alpha\beta)}} v_n X_n \geqslant 0\Big).$$

Furthermore, since each X_n is symmetric, so is the sum

$$\sum_{n > e^{A/(\alpha\beta)}} v_n X_n,$$

which means that it has probability $\geqslant 1/2$ to be $\geqslant 0$. Therefore we have

$$\mathbf{P}(|X| > A) \geqslant \frac{1}{8} e^{-(\log 4)\exp(A/(\alpha\beta))}.$$

This is of the right form asymptotically, and thus the proof is completed. □

Remark B.11.14 (1) The typical example where the proposition applies is when $\|v_n\|$ is comparable to $1/n$.

(2) Many variations along these lines are possible. For instance, in Chapter 3, we encounter the situation where the vector v_n is zero unless $n = p$ is a prime p, in which case

$$\|v_p\| = \frac{1}{p^\sigma}$$

for some real number σ such that $1/2 < \sigma < 1$. In that case, we have

$$\sum_{n \leqslant N} \|v_n\| \gg \frac{N^{1-\sigma}}{\log N}, \qquad \sum_{n > N} \|v_n\|^2 \ll \frac{1}{N^{2\sigma-1}} \frac{1}{\log N}$$

for $N \geqslant 2$ (by the Prime Number Theorem) instead of (B.22), and the adaptation of the arguments in the proof of the proposition lead to

$$\mathbf{P}(\|X\| > A) \leqslant c \exp\left(-cA^{1/(1-\sigma)}(\log A)^{1/(2(1-\sigma))}\right).$$

(Indeed, check that (B.25) gives here

$$m + \alpha + \beta \ll \frac{s^{2(1-\sigma)}}{\sqrt{\log s}},$$

and we take

$$s = A^{1/(2(1-\sigma))}(\log A)^{1/(4(1-\sigma))}$$

in the final application of Talagrand's inequality.)

On the other hand, in Chapter 5, we have a case where (up to re-indexing), the assumptions (B.22) and (B.23) are replaced by

$$\sum_{n \leqslant N} \|v_n\| \gg (\log N)^2 \quad \text{and} \quad \sum_{n > N} \|v_n\|^2 \ll \frac{\log N}{N}.$$

Then we obtain by the same argument the estimates

$$\mathbf{P}(\|X\| > A) \leqslant c \exp(-\exp(c^{-1}A^{1/2})),$$
$$c^{-1}\exp(-\exp(cA^{1/2})) \leqslant \mathbf{P}(|\lambda(X)| > A) \leqslant \mathbf{P}(\|X\| > A)$$

for some real number $c > 0$.

Appendix C

Number Theory

We review here the facts of number theory that we use and give references for their proofs.

C.1 Multiplicative Functions and Euler Products

Analytic number theory frequently deals with functions f defined for integers $n \geqslant 1$ such that $f(mn) = f(m)f(n)$ whenever m and n are coprime. Any such function that is not identically zero is called a *multiplicative* function.[1] A multiplicative function is uniquely determined by the values $f(p^k)$ for primes p and integers $k \geqslant 1$ and satisfies $f(1) = 1$.

We recall that if f and g are functions defined for positive integers, the Dirichlet convolution $f \star g$ is defined by

$$(f \star g)(n) = \sum_{d \mid n} f(d) g\left(\frac{n}{d}\right).$$

Its key property is that the generating Dirichlet series

$$\sum_{n \geqslant 1} (f \star g)(n) n^{-s}$$

for $f \star g$ is the product of the generating Dirichlet series for f and g (see Proposition A.4.4). In particular, one deduces that the convolution is associative and commutative, and that the function δ such that $\delta(1) = 1$ and $\delta(n) = 0$ for all $n \geqslant 2$ is a neutral element. In other words, for any arithmetic functions f, g, and h, we have

$$f \star g = g \star f, \quad f \star (g \star h) = (f \star g) \star h, \quad f \star \delta = f.$$

[1] We emphasize that it is not required that $f(mn) = f(m)f(n)$ for *all* pairs of positive integers.

Lemma C.1.1 *Let f and g be multiplicative functions. Then the Dirichlet convolution $f \star g$ of f and g is multiplicative. Moreover, the function $f \circledcirc g$ defined by*

$$(f \circledcirc g)(n) = \sum_{[a,b]=n} f(a)g(b)$$

is also multiplicative.

Proof Both statements follow simply from the fact that if n and m are coprime integers, then any divisor d of nm can be uniquely written $d = d'd''$, where d' divides n and d'' divides m. □

Example C.1.2 To get an idea of the behavior of a multiplicative function, it is always useful to write down the values at powers of primes. In the situation of the lemma, the Dirichlet convolution satisfies

$$(f \star g)(p^k) = \sum_{j=0}^{k} f(p^j)g(p^{k-j}),$$

whereas

$$(f \circledcirc g)(p^k) = \sum_{j=0}^{k-1}(f(p^j)g(p^k) + f(p^k)g(p^j)) + f(p^k)g(p^k).$$

In particular, suppose that f and g are supported on squarefree integers, so that $f(p^k) = g(p^k) = 0$ for any prime if $k \geqslant 2$. Then $f \circledcirc g$ is also supported on squarefree integers (this is not necessarily the case for $f \star g$) and satisfies

$$(f \circledcirc g)(p) = f(p) + g(p) + f(p)g(p)$$

for all primes p.

A very important multiplicative function is the Möbius function.

Definition C.1.3 The Möbius function $\mu(n)$ is the multiplicative function supported on squarefree integers such that $\mu(p) = -1$ for all primes p.

In other words, if we factor

$$n = p_1 \cdots p_j,$$

where each p_i is prime, then we have $\mu(n) = 0$ if there exists $i \neq j$ such that $p_i = p_j$, and otherwise $\mu(n) = (-1)^j$.

A key property of multiplicative functions is their *Euler product* expansion, as a product over primes.

Lemma C.1.4 *Let f be a multiplicative function such that*

$$\sum_{n\geqslant 1} |f(n)| < +\infty.$$

Then we have

$$\sum_{n\geqslant 1} f(n) = \prod_{p} \Big(\sum_{j\geqslant 0} f(p^j) \Big),$$

where the product on the right is absolutely convergent. In particular, for all $s \in \mathbf{C}$ such that

$$\sum_{n\geqslant 1} \frac{f(n)}{n^s}$$

converges absolutely, we have

$$\sum_{n\geqslant 1} \frac{f(n)}{n^s} = \prod_{p} (1 + f(p)p^{-s} + \cdots + f(p^k)p^{-ks} + \cdots),$$

where the right-hand side converges absolutely.

Proof For any prime p, the series

$$1 + f(p) + \cdots + f(p^k) + \cdots$$

is a subseries of $\sum f(n)$, so that the absolute convergence of the latter (which holds by assumption) implies that all of these partial series are also absolutely convergent.

We now first assume that $f(n) \geqslant 0$ for all n. Then, for any $\mathrm{N} \geqslant 1$, we have

$$\prod_{p\leqslant \mathrm{N}} \sum_{k\geqslant 0} f(p^k) = \sum_{\substack{n\geqslant 1 \\ p\mid n \Rightarrow p\leqslant \mathrm{N}}} f(n)$$

by expanding the product and using the absolute convergence and the uniqueness of factorization of integers. It follows that

$$\Big| \prod_{p\leqslant \mathrm{N}} \sum_{k\geqslant 0} f(p^k) - \sum_{n\leqslant \mathrm{N}} f(n) \Big| \leqslant \sum_{n>\mathrm{N}} f(n)$$

(since we assume $f(n) \geqslant 0$). This converges to 0 as $\mathrm{N} \to +\infty$, because the series $\sum f(n)$ is absolutely convergent. Thus this case is done.

In the general case, replacing f by $|f|$, the previous argument shows that the product converges absolutely. Then we get in the same manner

$$\Big| \prod_{p\leqslant \mathrm{N}} \sum_{k\geqslant 0} f(p^k) - \sum_{n\leqslant \mathrm{N}} f(n) \Big| \leqslant \sum_{n>\mathrm{N}} |f(n)| \longrightarrow 0$$

as $\mathrm{N} \to +\infty$. ☐

Corollary C.1.5 *For any $s \in \mathbf{C}$ such that* $\mathrm{Re}(s) > 1$, *we have*

$$\sum_{n \geqslant 1} n^{-s} = \prod_p \frac{1}{1 - p^{-s}}, \tag{C.1}$$

$$\sum_{n \geqslant 1} \mu(n) n^{-s} = \prod_p (1 - p^{-s}) = \frac{1}{\zeta(s)}. \tag{C.2}$$

Example C.1.6 The fact that the Dirichlet series for the Möbius function is the inverse of the Riemann zeta function, combined with the link between multiplication and Dirichlet convolution, leads to the so-called *Möbius inversion formula*: for arithmetic functions f and g, the relations

$$g(n) = \sum_{d \mid n} f(d) \quad \text{and} \quad f(n) = \sum_{d \mid n} \mu(d) g\left(\frac{n}{d}\right)$$

(for all $n \geqslant 1$) are equivalent. (Indeed, the first means that $g = f \star 1$, where 1 is the constant function, and the second that $f = g \star \mu$; since $\mu \star 1 = \delta$, which is the multiplicative function version of the identity $\zeta(s)^{-1} \cdot \zeta(s) = 1$, the equivalence of the two follows from the associativity of the convolution.)

Example C.1.7 Let f and g be multiplicative functions supported on squarefree integers defining absolutely convergent series. Then for $\sigma > 0$, we have

$$\sum_{n \geqslant 1} \frac{f(m)g(n)}{[m,n]^{\sigma}} = \sum_{d \geqslant 1} \frac{(f \circledcirc g)(d)}{d^{\sigma}} = \prod_p \left(1 + (f(p) + g(p) + f(p)g(p))p^{-\sigma}\right).$$

For instance, consider the case where f and g are both the Möbius function μ. Then $\mu \circledcirc \mu$ is supported on squarefree numbers and takes value $-1-1+1 = -1$ at primes and so is in fact equal to μ. We obtain the nice formula

$$\sum_{n \geqslant 1} \frac{\mu(m)\mu(n)}{[m,n]^s} = \sum_{d \geqslant 1} \frac{(f \circledcirc g)(d)}{d^s} = \prod_p \left(1 - \frac{1}{p^s}\right) = \sum_{n \geqslant 1} \mu(n) n^{-s} = \frac{1}{\zeta(s)}$$

for $\mathrm{Re}(s) > 1$.

Example C.1.8 Another very important multiplicative arithmetic function is the Euler function φ defined by $\varphi(q) = |(\mathbf{Z}/q\mathbf{Z})^{\times}|$ for $q \geqslant 1$. This function is multiplicative, by the Chinese Remainder Theorem, which implies that there exists an isomorphism of groups

$$(\mathbf{Z}/q_1 q_2 \mathbf{Z})^{\times} \simeq (\mathbf{Z}/q_1 \mathbf{Z})^{\times} \times (\mathbf{Z}/q_2 \mathbf{Z})^{\times}$$

when q_1 and q_2 are coprime integers. We have $\varphi(p) = p - 1$ if p is prime, and more generally $\varphi(p^k) = p^k - p^{k-1}$ for p prime and $k \geqslant 1$ (since an element x of $\mathbf{Z}/p^k\mathbf{Z}$ is invertible if and only if its unique lift in $\{0, \ldots, p^k - 1\}$ is not divisible by p). Hence, by factorization, we obtain the product expansion

$$\varphi(n) = \prod_{p|n} (p^{v_p(n)} - p^{v_p(n)-1}) = n \prod_{p|n} \left(1 - \frac{1}{p}\right),$$

where $v_p(n)$ is the power p-adic valuation of n, that is, the exponent of the power of p dividing exactly n.

We deduce from Lemma C.1.4 the expression

$$\sum_{n \geqslant 1} \varphi(n) n^{-s} = \prod_p \left(1 + (p-1)p^{-s} + (p^2 - p)p^{-2s} + \cdots + (p^k - p^{k-1})p^{-ks} + \cdots\right)$$
$$= \frac{\zeta(s-1)}{\zeta(s)},$$

again valid for $\mathrm{Re}(s) > 1$. This may also be deduced from the formula

$$\varphi(n) = \sum_{d|n} \mu(d) \frac{n}{d},$$

that is, $\varphi = \mu \star \mathrm{Id}$, where Id is the identity arithmetic function.

C.2 Additive Functions

We also often encounter *additive functions* (although they are not so important in this book), which are complex-valued functions g defined for integers $n \geqslant 1$ such that $g(nm) = g(n) + g(m)$ for all pairs of coprime integers n and m. In particular, we have then $g(1) = 0$.

If g is an additive function, then we can write

$$g(n) = \sum_p g\left(p^{v_p(n)}\right)$$

for any $n \geqslant 1$, where v_p is the p-adic valuation (which is zero for all but finitely many p). As for multiplicative functions, an additive function is therefore determined uniquely by its values at prime powers.

Some standard examples are given by $g(n) = \log n$, or more generally $g(n) = \log f(n)$, where f is a multiplicative function that is always positive. The arithmetic function $\omega(n)$ that counts the number of prime factors of an integer $n \geqslant 1$ (without multiplicity) is also additive; it is of course the subject of the Erdős–Kac Theorem.

Conversely, if g is an additive function, then for any complex number $s \in \mathbf{C}$, the function $n \mapsto e^{sg(n)}$ is a multiplicative function.

C.3 Primes and Their Distribution

For any real number $x \geqslant 1$, we denote by $\pi(x)$ the prime counting function, that is, the number of prime numbers $p \leqslant x$. This is of course one of the key functions of interest in multiplicative number theory. Except in the most elementary cases, interesting statements require some information on the size of $\pi(x)$.

The first nontrivial quantitative bounds are due to Chebychev, giving the correct order of magnitude of $\pi(x)$, and were elaborated by Mertens to obtain other very useful estimates for quantities involving primes.

Proposition C.3.1 (Chebychev and Mertens estimates) (1) *There exist positive constants c_1 and c_2 such that*

$$c_1 \frac{x}{\log x} \leqslant \pi(x) \leqslant c_2 \frac{x}{\log x}$$

for all $x \geqslant 2$.

(2) *For any $x \geqslant 3$, we have*

$$\sum_{p \leqslant x} \frac{1}{p} = \log \log x + O(1).$$

(3) *For any $x \geqslant 3$, we have*

$$\sum_{p \leqslant x} \frac{\log p}{p} = \log x + O(1).$$

See, e.g, [59, §2.2] or [52, Th. 7, Th. 414] (resp. [59, (2.15)] or [52, Th. 427]; [59, (2.14)] or [52, Th. 425]) for a proof of the first (resp. second, third) estimate.

Exercise C.3.2 (1) Show that the first estimate implies that the nth prime is of size about $n \log n$ (up to multiplicative constants) and also implies the bounds

$$\log \log x \ll \sum_{p \leqslant x} \frac{1}{p} \ll \log \log x$$

for $x \geqslant 3$.

(2) Let $\pi_2(x)$ be the numbers of integers $n \leqslant x$ such that n is the product of at most two primes (possibly equal). Prove that there exist positive constants c_3 and c_4 such that

$$c_3 \frac{x \log \log x}{\log x} \leqslant \pi_2(x) \leqslant c_4 \frac{x \log \log x}{\log x}$$

for all $x \geqslant 3$.

The real key result in the study of primes is the Prime Number Theorem with a strong error term:

Theorem C.3.3 *Let* A > 0 *be an arbitrary real number. For* $x \geqslant 2$, *we have*

$$\pi(x) = \mathrm{li}(x) + \mathrm{O}\left(\frac{x}{(\log x)^{\mathrm{A}}}\right), \tag{C.3}$$

where $\mathrm{li}(x)$ *is the logarithmic integral*

$$\mathrm{li}(x) = \int_2^x \frac{dt}{\log t}$$

and the implied constant depends only on A. *More generally, for* $\alpha \geqslant 0$ *fixed, we have*

$$\sum_{p \leqslant x} p^{\alpha} = \int_2^x t^{\alpha} \frac{dt}{\log t} + \mathrm{O}\left(\frac{x^{1+\alpha}}{(\log x)^{\mathrm{A}}}\right),$$

where the implied constant depends only on A.

For a proof, see, for instance, [59, §2.4 or Cor. 5.29]. By an elementary integration by parts, we have

$$\mathrm{li}(x) = \int_2^x \frac{dt}{\log t} = \frac{x}{\log x} + \mathrm{O}\left(\frac{x}{(\log x)^2}\right),$$

for $x \geqslant 2$, hence the "usual" simple asymptotic version of the Prime Number Theorem

$$\pi(x) \sim \frac{x}{\log x}, \qquad \text{as } x \to +\infty.$$

However, note that if one expresses the main term in the "simple" form $x/\log x$, the error term cannot be better than $x/(\log x)^2$.

The Prime Number Theorem easily implies a stronger form of the Mertens formula:

Corollary C.3.4 *There exists a constant* $\mathrm{C} \in \mathbf{R}$ *such that*

$$\sum_{p \leqslant x} \frac{1}{p} = \log \log x + \mathrm{C} + \mathrm{O}((\log x)^{-1}). \tag{C.4}$$

Exercise C.3.5 Show that (C.4) is in fact equivalent with the Prime Number Theorem in the form

$$\pi(x) \sim \frac{x}{\log x}$$

as $x \to +\infty$.

Another estimate that will be useful in Chapter 4 is the following:

Proposition C.3.6 *Let* A > 0 *be a fixed real number. For all* $x \geqslant 2$*, we have*

$$\prod_{p \leqslant x} \Bigl(1 + \frac{A}{p}\Bigr) \ll (\log x)^{A},$$
$$\prod_{p \leqslant x} \Bigl(1 - \frac{A}{p}\Bigr)^{-1} \ll (\log x)^{A},$$

where the implied constant depends only on A.

Proof In both cases, if we compute the logarithm, we obtain

$$\sum_{p \leqslant x} \Bigl(\frac{A}{p} + O\Bigl(\frac{1}{p^2}\Bigr)\Bigr),$$

where the implied constant depends on A and the result follows from the Mertens formula. □

In Chapter 5, we will also need the generalization of these basic statements to primes in arithmetic progressions. We recall that for $x \geqslant 1$, and any modulus $q \geqslant 1$ and integer $a \in \mathbf{Z}$, we define

$$\pi(x; q, a) = \sum_{\substack{p \leqslant x \\ p \equiv a \pmod{q}}} 1,$$

the number of primes $p \leqslant x$ that are congruent to a modulo q. If a is not coprime to q, then $\pi(x; q, a)$ is bounded as x varies; it was one of the first major achievements of analytic number theory when Dirichlet proved that, conversely, there are infinitely many primes $p \equiv a \pmod{q}$ if $(a, q) = 1$. This was done using the theory of Dirichlet characters and L-functions, which we will survey later (see Section C.5). Here we state the analogue of the Prime Number Theorem, which shows that, asymptotically, all residue classes modulo q are roughly equivalent.

Theorem C.3.7 *For any fixed* $q \geqslant 1$ *and* A $\geqslant 1$*, and for any* $x \geqslant 2$*, we have*

$$\begin{aligned} \pi(x; q, a) &= \frac{1}{\varphi(q)} \frac{x}{\log x} + O\Bigl(\frac{x}{(\log x)^{A}}\Bigr) \\ &\sim \frac{1}{\varphi(q)} \pi(x) \sim \frac{1}{\varphi(q)} \operatorname{li}(x). \end{aligned}$$

C.4 The Riemann Zeta Function

As recalled in Section 3.1, the Riemann zeta function is defined first for complex numbers s such that $\mathrm{Re}(s) > 1$ by means of the absolutely convergent series

$$\zeta(s) = \sum_{n\geqslant 1} \frac{1}{n^s}.$$

By Lemma C.1.4, it has also the Euler product expansion

$$\zeta(s) = \prod_p (1 - p^{-s})^{-1}$$

in this region. Using this expression, we can compute the logarithmic derivative of the zeta function, always for $\mathrm{Re}(s) > 1$. We obtain the Dirichlet series expansion

$$-\frac{\zeta'}{\zeta}(s) = \sum_p \frac{(\log p)p^{-s}}{1 - p^{-s}} = \sum_{n\geqslant 1} \frac{\Lambda(n)}{n^s} \tag{C.5}$$

(using a geometric series expansion), where the function Λ is called the *von Mangoldt function*, defined by

$$\Lambda(n) = \begin{cases} \log p & \text{if } n = p^k \text{ for some prime } p \text{ and some } k \geqslant 1, \\ 0 & \text{otherwise.} \end{cases} \tag{C.6}$$

In other words, up to the "thin" set of powers of primes with exponent $k \geqslant 2$, the function Λ is the logarithm restricted to prime numbers.

Beyond the region of absolute convergence, it is known that the zeta function extends to a meromorphic function on all of $\mathbf{C}$, with a unique pole located at $s = 1$, which is a simple pole with residue 1 (see the argument in Section 3.1 for a simple proof of analytic continuation to $\mathrm{Re}(s) > 0$). More precisely, let

$$\xi(s) = \pi^{-s/2}\Gamma\left(\frac{s}{2}\right)\zeta(s)$$

for $\mathrm{Re}(s) > 1$. Then ξ extends to a meromorphic function on $\mathbf{C}$ with simple poles at $s = 0$ and $s = 1$, which satisfies the functional equation

$$\xi(1 - s) = \xi(s).$$

Because the Gamma function has poles at integers $-k$ for $k \geqslant 0$, it follows that $\zeta(-2k) = 0$ for $k \geqslant 1$ (the case $k = 0$ is special because of the pole at $s = 1$). The negative even integers are called the *trivial zeros* of $\zeta(s)$. Hadamard and de la Vallée Poussin proved (independently) that $\zeta(s) \neq 0$ for $\mathrm{Re}(s) = 1$, and it follows that the nontrivial zeros of $\zeta(s)$ are located in the *critical strip* $0 < \mathrm{Re}(s) < 1$.

Proposition C.4.1 (1) *For* $1/2 < \sigma < 1$, *we have*

$$\frac{1}{2\mathrm{T}}\int_{-\mathrm{T}}^{\mathrm{T}} |\zeta(\sigma+it)|^2 dt \longrightarrow \zeta(2\sigma)$$

as $\mathrm{T} \to +\infty$.

(2) *We have*

$$\frac{1}{2\mathrm{T}}\int_{-\mathrm{T}}^{\mathrm{T}} |\zeta(\tfrac{1}{2}+it)|^2 dt \sim \mathrm{T}(\log \mathrm{T})$$

for $\mathrm{T} \to +\infty$.

See [117, Th. 7.2] for the proof of the first formula and [117, Th. 7.3] for the second (which is due to Hardy and Littlewood).

Exercise C.4.2 This exercise explains the proof of the first formula (which is easier than the second one).

(1) Prove that for $\frac{1}{2} \leqslant \sigma \leqslant \sigma' < 1$ and for $\mathrm{T} \geqslant 2$, we have

$$\sum_{1\leqslant m<n\leqslant \mathrm{T}} \frac{1}{(mn)^{\sigma}}\frac{1}{\log(n/m)} \ll \mathrm{T}^{2-2\sigma}(\log \mathrm{T}).$$

(Consider separately the sum where $m < \frac{1}{2}n$ and the remainder.)

(2) Prove that

$$\frac{1}{2\mathrm{T}}\int_{-\mathrm{T}}^{\mathrm{T}} \Big|\sum_{n\leqslant |t|} n^{-\sigma-it}\Big|^2 dt \to \zeta(2\sigma)$$

as $\mathrm{T} \to +\infty$. (Expand the square and integrate using (1).)

(3) Conclude using Proposition C.4.5 below.

For much more information concerning the analytic properties of the Riemann zeta function, see [117]. Note however that the deeper *arithmetic* aspects are best understood in the larger framework of L-functions, from Dirichlet L-functions (which are discussed below in Section C.5) to automorphic L-functions (see, e.g, [59, Ch. 5]).

We will also use the *Hadamard factorization* of the Riemann zeta function. This is an analogue of the factorization of polynomials in terms of their zeros, which holds for meromorphic functions on **C** with restricted growth.

Proposition C.4.3 *The zeros* ϱ *of* $\xi(s)$ *all satisfy* $0 < \mathrm{Re}(\varrho) < 1$, *and there exists constants* α *and* $\beta \in \mathbf{C}$ *such that*

$$s(s-1)\xi(s) = e^{\alpha+\beta s}\prod_{\varrho}\Big(1-\frac{s}{\varrho}\Big)e^{-s/\varrho}$$

for any $s \in \mathbf{C}$, where the product runs over the zeros of $\xi(s)$, counted with multiplicity, and converges uniformly on compact subsets of $\mathbf{C}$. *In fact, we have*

$$\sum_{\varrho} \frac{1}{|\varrho|^2} < +\infty.$$

Given that $s \mapsto s(s-1)\xi(s)$ is an entire function of finite order, this follows from the general theory of such functions (see, e.g, [116, Th. 8.24] for Hadamard's factorization theorem). What is most important for us is the following corollary, which is an analogue of partial fraction expansion for the logarithmic derivative of a polynomial – except that it is most convenient here to truncate the infinite sum.

Proposition C.4.4 *Let $s = \sigma + it \in \mathbf{C}$ be such that $\frac{1}{2} \leqslant \sigma \leqslant 1$ and $\zeta(s) \neq 0$. Then there are $\ll \log(2 + |t|)$ zeros ϱ of ξ such that $|s - \varrho| \leqslant 1$, and we have*

$$-\frac{\zeta'(s)}{\zeta(s)} = \frac{1}{s} + \frac{1}{s-1} - \sum_{|s-\varrho|<1} \frac{1}{s-\varrho} + O(\log(2 + |t|)),$$

where the sum is over zeros ϱ of $\zeta(s)$ such that $|s - \varrho| < 1$, counted with multiplicity.

Sketch of proof We first claim that the constant β in Proposition C.4.3 satisfies

$$\operatorname{Re}(\beta) = -\sum_{\varrho} \operatorname{Re}(\varrho^{-1}), \tag{C.7}$$

where ϱ runs over all the zeros of $\xi(s)$ with multiplicity. Indeed, applying the Hadamard product expansion to both sides of the functional equation $\xi(1 - s) = \xi(s)$ and taking logarithms, we obtain

$$2\operatorname{Re}(\beta) = \beta + \bar{\beta} = -\sum_{\varrho} \left(\frac{1}{s-\varrho} + \frac{1}{1-s-\bar{\varrho}} + \frac{1}{\varrho} + \frac{1}{\bar{\varrho}} \right).$$

For any fixed s that is not a zero of $\xi(s)$, we have $(s-\varrho)^{-1} - (1-s-\bar{\varrho})^{-1} \ll |\varrho|^{-2}$, where the implied constant depends on s. Similarly, $\varrho^{-1} + \bar{\varrho}^{-1} \ll |\varrho|^{-2}$, so the series

$$\sum_{\varrho} \left(\frac{1}{s-\varrho} + \frac{1}{1-s-\bar{\varrho}} \right) \quad \text{and} \quad \sum_{\varrho} \left(\frac{1}{\varrho} + \frac{1}{\bar{\varrho}} \right)$$

are absolutely convergent. So we can separate them; the first one vanishes, because the terms cancel out (both ϱ and $1 - \bar{\varrho}$ are zeros of $\zeta(s)$), and we obtain (C.7).

Now let $\mathrm{T} \geqslant 2$ and $s = 3 + i\mathrm{T}$. Using the expansion

$$-\frac{\zeta'(s)}{\zeta(s)} = \sum_{k\geqslant 0}\sum_{p} (\log p)p^{-ks},$$

we get the trivial estimate

$$\left|\frac{\zeta'}{\zeta}(s)\right| \leqslant \zeta'(3).$$

By Stirling's formula (Proposition A.3.3), we have $\frac{\Gamma'}{\Gamma}(s/2) \ll \log(2+\mathrm{T})$, and for any zero $\varrho = \beta + i\gamma$ of $\xi(s)$, we have

$$\frac{2}{9 + (\mathrm{T}-\gamma)^2} < \mathrm{Re}\left(\frac{1}{s-\varrho}\right) < \frac{3}{4 + (\mathrm{T}-\gamma)^2}.$$

If we compute the real part of the formula

$$-\frac{\zeta'}{\zeta}(s) = \frac{\Gamma'}{\Gamma}(s/2) - \beta + \frac{1}{s} + \frac{1}{s-1} - \sum_{\varrho}\left(\frac{1}{s-\varrho} + \frac{1}{\varrho}\right)$$

and rearrange the resulting absolutely convergent series (using (C.7)), we get

$$\sum_{\varrho} \frac{1}{1 + (\mathrm{T}-\gamma)^2} \ll \log(2+\mathrm{T}). \tag{C.8}$$

This convenient estimate implies, as claimed, that there are $\ll \log(2+\mathrm{T})$ zeros ϱ such that $|\mathrm{Im}(\varrho - \mathrm{T})| \leqslant 1$.

Now, finally, let $s = \sigma + it$ such that $\frac{1}{2} \leqslant \sigma \leqslant 1$ and $\xi(s) \neq 0$. We have

$$-\frac{\zeta'}{\zeta}(s) = -\frac{\zeta'}{\zeta}(s) + \frac{\zeta'}{\zeta}(3+it) + \mathrm{O}(\log(2+|t|)),$$

by the previous elementary estimate. Hence (by the Stirling formula again) we have

$$-\frac{\zeta'}{\zeta}(s) = \frac{1}{s} + \frac{1}{s-1} - \sum_{\varrho}\left(\frac{1}{s-\varrho} - \frac{1}{3+it-\varrho}\right) + \mathrm{O}(\log(2+|t|)).$$

In the series, we keep the zeros with $|s-\varrho| < 1$, and we estimate the contribution of the others by

$$\sum_{|s-\varrho|>1}\left|\frac{1}{s-\varrho} - \frac{1}{3+it-\varrho}\right| \leqslant \sum_{|s-\varrho|>1} \frac{3}{1+(\mathrm{T}-\gamma)^2} \ll \log(2+|t|)$$

by (C.8). ☐

We will use an elementary approximation for $\zeta(s)$ in the strip $\frac{1}{2} < \mathrm{Re}(s) < 1$.

Proposition C.4.5 *Let* $\mathrm{T} \geqslant 1$. *For* $\sigma > 1/2$, *and for any* $s = \sigma + it$ *with* $1/2 \leqslant \sigma < 3/4$ *and* $|t| \leqslant \mathrm{T}$, *we have*

$$\zeta(s) = \sum_{1 \leqslant n \leqslant \mathrm{T}} n^{-s} + \mathrm{O}\Big(\frac{\mathrm{T}^{1-\sigma}}{|t|+1} + \mathrm{T}^{-1/2}\Big).$$

Proof This follows from [117, Th. 4.11] (a result first proved by Hardy and Littlewood) which states that for any $\sigma_0 > 0$, we have

$$\zeta(s) = \sum_{1 \leqslant n \leqslant \mathrm{T}} n^{-s} - \frac{\mathrm{T}^{1-\sigma}}{1-s} + \mathrm{O}(\mathrm{T}^{-1/2})$$

for $\sigma \geqslant \sigma_0$, since $1/(1-s) \ll 1/(|t|+1)$ if $1/2 \leqslant \sigma < 3/4$. □

The last (and most subtle) result concerning the zeta function that we need is an important refinement of (2) in Proposition C.4.1.

Proposition C.4.6 *Let* $\mathrm{T} \geqslant 1$ *be a real number, and let* m, n *be integers such that* $1 \leqslant m, n \leqslant \mathrm{T}$. *Let* σ *be a real number with* $\frac{1}{2} \leqslant \sigma \leqslant 1$. *We have*

$$\begin{aligned}\frac{1}{2\mathrm{T}} \int_{-\mathrm{T}}^{\mathrm{T}} \Big(\frac{m}{n}\Big)^{it} |\zeta(\sigma+it)|^2 dt &= \zeta(2\sigma)\Big(\frac{(m,n)^2}{mn}\Big)^{\sigma} \\ &\quad + \frac{1}{2\mathrm{T}}\zeta(2-2\sigma)\Big(\frac{(m,n)^2}{mn}\Big)^{1-\sigma} \int_{-\mathrm{T}}^{\mathrm{T}} \Big(\frac{|t|}{2\pi}\Big)^{1-2\sigma} dt \\ &\quad + \mathrm{O}(\min(m,n)\mathrm{T}^{-\sigma+\varepsilon}).\end{aligned}$$

This is essentially due to Selberg [110, Lemma 6], and a proof is given by Radziwiłł and Soundararajan [95, §6].

C.5 Dirichlet L-Functions

Let $q \geqslant 1$ be an integer. The Dirichlet L-functions modulo q are Dirichlet series attached to characters of the group of invertible residue classes modulo q. More precisely, for any such character $\chi : (\mathbf{Z}/q\mathbf{Z})^\times \to \mathbf{C}^\times$, we extend it to $\mathbf{Z}/q\mathbf{Z}$ by sending noninvertible classes to 0, and then we view it as a q-periodic function on $\mathbf{Z}$. The resulting function on $\mathbf{Z}$ is called a *Dirichlet character modulo* q. (See Example B.6.2 (3) for the definition and basic properties of characters of finite abelian groups; an excellent elementary account can also be found in Serre's book [112, §VI.1].)

We denote by 1_q the trivial character modulo q (which is identically 1 on all invertible residue classes modulo q and 0 elsewhere). A character χ such that $\chi(n) \in \{\pm 1\}$ for all n coprime to q is called a *real character*. This condition is equivalent to having χ real-valued, or to having $\chi^2 = 1_q$.

By the duality theorem for finite abelian groups (see Example B.6.2, (3)), the set of Dirichlet characters modulo q is a group under pointwise multiplication with 1_q as the identity element, and it is isomorphic to $(\mathbf{Z}/q\mathbf{Z})^\times$; in particular, the number of Dirichlet characters modulo q is $\varphi(q)$. Moreover, the Dirichlet characters modulo q form an orthonormal basis of the space of complex-valued functions on $(\mathbf{Z}/q\mathbf{Z})^\times$.

Let χ be a Dirichlet character modulo q. By construction, the function χ is multiplicative on $\mathbf{Z}$, in the strong sense that $\chi(nm) = \chi(n)\chi(m)$ for all integers n and m (even if they are not coprime).

The orthonormality property of characters of a finite group implies the following fundamental relation:

Proposition C.5.1 *Let $q \geqslant 1$ be an integer. For any x and y in $\mathbf{Z}$, we have*

$$\frac{1}{\varphi(q)} \sum_{\chi \,(\mathrm{mod}\, q)} \chi(x)\overline{\chi(y)} = \begin{cases} 1 & \text{if } x \equiv y \pmod q \text{ and } x,\ y \text{ are coprime with } q, \\ 0 & \text{otherwise,} \end{cases}$$

where the sum is over all Dirichlet characters modulo q.

Proof If x or y is not coprime with q, then the formula is valid because both sides are zero. Otherwise, this is a special case of the general decomposition formula

$$\frac{1}{|\mathrm{G}|} \sum_{\chi \in \widehat{\mathrm{G}}} \chi(x)\overline{\chi(y)} = \begin{cases} 1 & \text{if } x = y, \\ 0 & \text{if } x \neq y \end{cases} \tag{C.9}$$

for any finite abelian group G and elements x and y of G. Indeed, if we view y as fixed and x as a variable, this is simply the decomposition of the characteristic function f_y of the element $y \in \mathrm{G}$ in the orthonormal basis of characters: this decomposition is

$$f_y = \sum_{\chi \in \widehat{\mathrm{G}}} \langle f_y, \chi \rangle \chi,$$

which becomes

$$f_y = \sum_{\chi \in \widehat{\mathrm{G}}} \overline{\chi(y)} \chi,$$

from which in turn (C.9) follows by evaluating at x. ☐

Let $q \geqslant 1$ be an integer and χ a Dirichlet character modulo q. One defines

$$\mathrm{L}(s, \chi) = \sum_{n \geqslant 1} \frac{\chi(n)}{n^s}$$

for all $s \in \mathbf{C}$ such that $\mathrm{Re}(s) > 1$; since $|\chi(n)| \leqslant 1$ for all $n \in \mathbf{Z}$, this series is absolutely convergent and defines a holomorphic function in this region, called the L-*function associated to* χ.

In the region where $\mathrm{Re}(s) > 1$, the multiplicativity of χ implies that we have the absolutely convergent Euler product expansion

$$\mathrm{L}(s,\chi) = \prod_p (1 - \chi(p)p^{-s})^{-1}$$

(by Lemma C.1.4 applied to $f(n) = \chi(n)n^{-s}$ for any $s \in \mathbf{C}$ such that $\mathrm{Re}(s) > 1$). In particular, we deduce that $\mathrm{L}(s,\chi) \neq 0$ if $\mathrm{Re}(s) > 1$. Moreover, computing the logarithmic derivative, we obtain the formula

$$-\frac{\mathrm{L}'}{\mathrm{L}}(s,\chi) = \sum_{n \geqslant 1} \Lambda(n)\chi(n)n^{-s}$$

for $\mathrm{Re}(s) > 1$.

For the trivial character 1_q modulo q, we have the formula

$$\mathrm{L}(s, 1_q) = \prod_{p \nmid q} (1 - p^{-s})^{-1} = \zeta(s) \prod_{p \mid q} (1 - p^{-s}).$$

Since the second factor is a finite product of quite simple form, we see that, when q is fixed, the analytic properties of this particular L-function are determined by those of the Riemann zeta function. In particular, it has meromorphic continuation with a simple pole at $s = 1$, where the residue is

$$\prod_{p \mid q} (1 - p^{-1}) = \frac{\varphi(q)}{q}.$$

For χ nontrivial, we have the following result (see, e.g., [59, §5.9]):

Theorem C.5.2 *Let χ be a nontrivial Dirichlet character modulo q. Define $\varepsilon_\chi = 0$ if $\chi(-1) = 1$ and $\varepsilon_\chi = 1$ if $\chi(-1) = -1$. Let*

$$\xi(s,\chi) = \pi^{-(s+\varepsilon_\chi)/2} q^{s/2} \Gamma\Big(\frac{s+\varepsilon_\chi}{2}\Big) \mathrm{L}(s,\chi)$$

for $\mathrm{Re}(s) > 1$. *Furthermore, let*

$$\tau(\chi) = \frac{1}{\sqrt{q}} \sum_{x \in (\mathbf{Z}/q\mathbf{Z})^\times} \chi(x) e\Big(\frac{x}{q}\Big).$$

Then $\xi(s,\chi)$ extends to an entire function on $\mathbf{C}$ which satisfies the functional equation

$$\xi(s,\chi) = \tau(\chi)\xi(1-s,\overline{\chi}).$$

In Chapter 5, we will require the basic information on the distribution of zeros of Dirichlet L-functions. We summarize it in the following proposition (see, e.g., [59, Th. 5.24]).

Proposition C.5.3 *Let χ be a Dirichlet character modulo q.*

(1) *For* $T \geqslant 1$, *the number* $N(T;\chi)$ *of zeros* ϱ *of* $L(s,\chi)$ *such that*

$$\mathrm{Re}(\varrho) > 0, \qquad |\mathrm{Im}(\varrho)| \leqslant T,$$

satisfies

$$N(T;\chi) = \frac{T}{\pi}\log\left(\frac{qT}{2\pi}\right) - \frac{T}{\pi} + O(\log q(T+1)), \tag{C.10}$$

where the implied constant is absolute.

(2) *For any* $\varepsilon > 0$, *the series*

$$\sum_{\varrho} |\varrho|^{-1-\varepsilon}$$

converges, where ϱ *runs over zeros of* $L(s,\chi)$ *such that* $\mathrm{Re}(\varrho) > 0$.

Remark C.5.4 These two statements are not independent, and in fact the first one implies the second by splitting the partial sum

$$\sum_{|\varrho|\leqslant T} \frac{1}{|\varrho|^{1+\varepsilon}}$$

for $T \geqslant 1$ in terms of zeros in intervals of length 1:

$$\sum_{|\varrho|\leqslant T} \frac{1}{|\varrho|^{1+\varepsilon}} \leqslant \sum_{1\leqslant N\leqslant T} \frac{1}{N^{1+\varepsilon}} \sum_{N-1\leqslant|\varrho|\leqslant N} 1 \ll \sum_{1\leqslant N\leqslant T} \frac{\log N}{N^{1+\varepsilon}}$$

by (1). Since this is uniformly bounded for all T, we obtain (2).

Corollary C.5.5 *Let χ be a Dirichlet character modulo q.*

(1) *We have*

$$\sum_{\substack{0<\gamma<T \\ L(\frac{1}{2}+i\gamma,\chi)=0}} \frac{1}{|\frac{1}{2}+i\gamma|} \gg (\log T)^2$$

for T *large enough.*

(2) *We have*

$$\sum_{\substack{\gamma>T \\ L(\frac{1}{2}+i\gamma,\chi)=0}} \frac{1}{|\frac{1}{2}+i\gamma|^2} \ll \frac{\log T}{T}$$

for $T \geqslant 1$.

Finally, we need a form of the *explicit formula* linking zeros of Dirichlet L-functions with the distribution of prime numbers.

Theorem C.5.6 *Let $q \geqslant 1$ be an integer, and let χ be a nontrivial Dirichlet character modulo q. For any $x \geqslant 2$ and any $\mathrm{X} \geqslant 2$ such that $2 \leqslant x \leqslant \mathrm{X}$, we have*

$$\sum_{n \leqslant x} \Lambda(n)\chi(n) = -\sum_{\substack{\mathrm{L}(\beta+i\gamma)=0 \\ |\gamma| \leqslant \mathrm{X}}} \frac{x^{\beta+i\gamma}}{\beta+i\gamma} + \mathrm{O}\Big(\frac{x(\log qx)^2}{\mathrm{X}}\Big),$$

where the sum is over nontrivial zeros of $\mathrm{L}(s,\chi)$*, counted with multiplicity, and the implied constant is absolute.*

Sketch of proof We refer to, for example, [59, Prop. 5.25] for this result. Here we wish to justify intuitively the existence of such a relation between sums (essentially) over primes and sums over zeros of the associated L-function.

Pick a function φ defined on $[0, +\infty[$ with compact support. Using the Mellin inversion formula (see Proposition A.3.1, (3)), we can write

$$\sum_{n \geqslant 1} \Lambda(n)\chi(n)\varphi\Big(\frac{n}{x}\Big) = \frac{1}{2i\pi}\int_{(2)} -\frac{\mathrm{L}'}{\mathrm{L}}(s,\chi)\widehat{\varphi}(s)x^s ds$$

for all $x \geqslant 1$. Assume (formally) that we can shift the integration line to the left, say, to the line where the real part is $1/4$, where the contribution would be $x^{1/4}$. The contour shift leads to poles located at all the zeros of $\mathrm{L}(s,\chi)$, with residue equal to the opposite of the multiplicity of the zero (since the L-function is entire, there is no contribution from poles). We can therefore expect that

$$\sum_{n \geqslant 1} \Lambda(n)\chi(n)\varphi\Big(\frac{n}{x}\Big) = -\sum_{\varrho} \widehat{\varphi}(\varrho)x^{\varrho} + \text{(small error)},$$

where ϱ runs over nontrivial zeros of $\mathrm{L}(s,\chi)$, counted with multiplicity.

If such a formula holds for the characteristic function φ of the interval $[0,1]$, then since

$$\widehat{\varphi}(s) = \int_0^1 x^{s-1} dx = \frac{1}{s},$$

we would obtain

$$\sum_{n \geqslant 1} \Lambda(n)\chi(n)\varphi\Big(\frac{n}{x}\Big) = -\sum_{\varrho} \frac{x^{\varrho}}{\varrho} + \text{(small error)}.$$

□

Remark C.5.7 There is nontrivial analytic work to do in order to justify the computations in this sketch, because of various convergence issues for instance (which also explains why the formula is most useful in a truncated form involving only finitely many zeros), but this formal outline certainly explains the *existence* of the explicit formula.

This explicit formula explains why the location of zeros of Dirichlet L-functions is so important in the study of prime numbers in arithmetic progressions. This motivates the *Generalized Riemann Hypothesis* modulo q:

Conjecture C.5.8 (Generalized Riemann Hypothesis) *For any integer $q \geqslant 1$ and for any Dirichlet character χ modulo q and any zero $\varrho = \beta + i\gamma$ of its* L*-function such that $0 < \beta \leqslant 1$, we have $\beta = \frac{1}{2}$.*

This is the most famous open problem of number theory. In practice, we will also speak of *Generalized Riemann Hypothesis modulo q* when considering only the fixed modulus q instead of all moduli. The case $q = 1$ corresponds to the original Riemann Hypothesis for the Riemann zeta function only.

By just applying orthogonality (Proposition C.5.1) and estimating trivially in the explicit formula with the help of Proposition C.5.3, we deduce:

Proposition C.5.9 *Let $q \geqslant 1$ be an integer. Assume that the Generalized Riemann Hypothesis modulo q holds. Then we have*

$$\sum_{\substack{n \leqslant x \\ n \equiv a \pmod{q}}} \Lambda(n) = \frac{x}{\varphi(q)} + \mathrm{O}(x^{1/2}(\log qx)^2).$$

Remark C.5.10 Compare the quality of the error term with the (essentially) best known unconditional result of Theorem C.3.7.

Another corollary of the explicit formula that will be helpful in Chapter 5 is the following:

Corollary C.5.11 *Let $q \geqslant 1$ be an integer and let χ be a nontrivial Dirichlet character modulo q. Assume that the Generalized Riemann Hypothesis holds for $\mathrm{L}(s,\chi)$, i.e., that all nontrivial zeros of $\mathrm{L}(s,\chi)$ have real part $1/2$. For any $x \geqslant 2$, we have*

$$\int_2^x \Big(\sum_{n \leqslant t} \Lambda(n)\chi(n) \Big) dt \ll x^{3/2},$$

where the implied constant depends on q.

Proof Pick $X = x$ in the explicit formula. Using the assumption on the zeros, we obtain by integration the expression

$$\int_2^x \Big(\sum_{n\leqslant t} \Lambda(n)\chi(n)\Big)\, dt$$

$$= \sum_{\substack{L(\frac{1}{2}+i\gamma)=0\\ |\gamma|\leqslant x}} \int_2^x \frac{t^{\frac{1}{2}+i\gamma}}{\frac{1}{2}+i\gamma} dt + O(x(\log qx)^2)$$

$$= \sum_{\substack{L(\frac{1}{2}+i\gamma)=0\\ |\gamma|\leqslant x}} \frac{x^{\frac{1}{2}+i\gamma+1} - 2^{\frac{1}{2}+i\gamma+1}}{(\frac{1}{2}+i\gamma)(\frac{1}{2}+i\gamma+1)} + O(x(\log qx)^2) \ll x^{3/2},$$

where the implied constant depends on q, since the series

$$\sum_{L(\frac{1}{2}+i\gamma)=0} \frac{1}{(\frac{1}{2}+i\gamma)(\frac{1}{2}+i\gamma+1)}$$

converges absolutely by Proposition C.5.3, (2). □

C.6 Exponential Sums

In Chapter 6, we studied some properties of exponential sums. Although we do not have the space to present a detailed treatment of such sums, we will give a few examples and try to explain some of the reasons why such sums are important and interesting. This should motivate the "curiosity driven" study of the shape of the partial sums. We refer to the notes [75] and to [59, Ch. 11] for more information, including proofs of the Weil bound (6.1) for Kloosterman sums.

In principle, any finite sum

$$S_N = \sum_{1\leqslant n\leqslant N} e(\alpha_n)$$

of complex numbers of modulus 1 counts as an exponential sum, and the goal is – given the *phases* $\alpha_n \in \mathbf{R}$ – to obtain a bound on S that improves as much as possible on the "trivial" bound $|S_N| \leqslant N$.

On probabilistic grounds, one can expect that for highly oscillating phases, the sum S_N is of size about $\sqrt{N}$. Indeed, if we consider α_n to be random variables that are independent and uniformly distributed in $\mathbf{R}/\mathbf{Z}$, then the

Central Limit Theorem shows that $S_N/\sqrt{N}$ is distributed approximately like a standard complex Gaussian random variable, so that the "typical" size of S_N is of order of magnitude $\sqrt{N}$. When this occurs also for deterministic sums (up to factors of smaller order of magnitude), one says that the sums have *square-root cancellation*; this usually only makes sense for an infinite sequence of sums where $N \to +\infty$.

Example C.6.1 For instance, the partial sums

$$M_N = \sum_{1 \leqslant n \leqslant N} \mu(n)$$

of the Möbius function can be seen in this light. Estimating M_N is vitally important in analytic number theory, because it is not very hard to check that the Prime Number Theorem, in the form (C.3), with error term $x/(\log x)^A$ for any $A > 0$, is *equivalent* with the estimate

$$M_N \ll \frac{N}{(\log N)^A}$$

for any $A > 0$, where the implied constant depends on A. Moreover, the best possible estimate is the square-root cancellation

$$M_N \ll N^{1/2+\varepsilon},$$

with an implied constant depending on $\varepsilon > 0$, and this is known to be equivalent to the Riemann Hypothesis for the Riemann zeta function.

The sums that appear in Chapter 6 are, however, of a fairly different nature. They are sums over finite fields (or subsets of finite fields), with summands $e(\alpha_n)$ of "algebraic nature." For a prime p and the finite field $\mathbf{F}_p$ with p elements,[2] the basic examples are of the following types:

Example C.6.2 (1) [Additive character sums] Fix a rational function $f \in \mathbf{F}_p(T)$. Then for $x \in \mathbf{F}_p$ that is not a pole of f, we can evaluate $f(x) \in \mathbf{F}_p$, and $e(f(x)/p)$ is a well-defined complex number. Then consider the sum

$$\sum_{\substack{x \in \mathbf{F}_p \\ f(x) \text{ defined}}} e(f(x)/p).$$

For fixed a and b in $\mathbf{F}_p^\times$, the example $f(T) = aT + bT^{-1}$ gives rise to the *Kloosterman sum* of Section 6.1. If $f(T) = T^2$, we obtain a *quadratic Gauss sum*, namely,

[2] For simplicity, we restrict to these particular finite fields, but the theory extends to all.

$$\sum_{x \in \mathbf{F}_p} e\left(\frac{x^2}{p}\right).$$

(2) [Multiplicative character sums] Let χ be a nontrivial character of the finite multiplicative group $\mathbf{F}_p^\times$; we define $\chi(0) = 0$ to extend it to $\mathbf{F}_p$. Let $f \in \mathbf{F}_p[\mathrm{T}]$ be a polynomial (or a rational function). The corresponding multiplicative character sum is

$$\sum_{x \in \mathbf{F}_p} \chi(f(x)).$$

One may also have finitely many polynomials and characters and sum their products. An important example of these is

$$\sum_{x \in \mathbf{F}_p} \chi_1(x)\chi_2(1-x),$$

for multiplicative characters χ_1 and χ_2, which is called a *Jacobi sum*.

(3) [Mixed sums] In fact, one can mix the two types, obtaining a family of sums that generalize both: fix rational functions f_1 and f_2 in $\mathbf{F}_p(\mathrm{T})$, and consider the sum

$$\sum_{x \in \mathbf{F}_p} \chi(f_1(x))e(f_2(x)/p),$$

where the summand is defined to be 0 if $f_2(x)$ is not defined, or if $f_1(x)$ is 0 or not defined.

Some of the key examples are obtained in this manner. Maybe the simplest interesting ones are the *Gauss sums* attached to χ, defined by

$$\sum_{x \in \mathbf{F}_p} \chi(x)e(ax/p),$$

where $a \in \mathbf{F}_p$ is a parameter. Others are the sums

$$\sum_{x \in \mathbf{F}_p^\times} \chi(x)e\left(\frac{ax + b\bar{x}}{p}\right)$$

for a, b in $\mathbf{F}_p$, which generalize the Kloosterman sums. When χ is a character of order 2 (i.e., $\chi(x)$ is either 1 or -1 for all $x \in \mathbf{F}_p$), this is called a *Salié sum*.

Remark C.6.3 We emphasize that the sums that we discuss range over the *whole* finite field (except for values of x where the summand is not defined). Sums over smaller subsets of $\mathbf{F}_p$ (e.g., over a segment $1 \leqslant x \leqslant \mathrm{N} < p$ of integers) are very interesting and important in applications (indeed, they are the topic of Chapter 6!), but behave very differently.

Except for a few special cases (some of which are discussed below in exercises), a simple "explicit" evaluation of exponential sums of the previous types is not feasible. Even deriving nontrivial bounds is far from obvious, and the most significant progress requires input from algebraic geometry. The key result, proved by A. Weil in the 1940s, takes the following form (in a simplified version that is actually rather weaker than the actual statement). It is a special case of the Riemann Hypothesis over finite fields.

Theorem C.6.4 (Weil) *Let χ be a nontrivial multiplicative character modulo q. Let f_1 and f_2 be rational functions in $\mathbf{F}_p[\mathrm{T}]$, and consider the sum*

$$\sum_{x\in\mathbf{F}_p} \chi(f_1(x))e(f_2(x)/p).$$

Assume that either *f_1 is not of the form g_1^d, where d is the order of χ and $g_1 \in \overline{\mathbf{F}}_p[\mathrm{T}]$,* or *$f_2$ has a pole of order not divisible by p, possibly at infinity.*

Then, there exists an integer β, depending only on the degrees of the numerator and denominator of f_1 and f_2, and for $1 \leqslant i \leqslant \beta$, there exist complex numbers α_i such that $|\alpha_i| \leqslant \sqrt{p}$, with the property that

$$\sum_{x\in\mathbf{F}_p} \chi(f_1(x))e(f_2(x)/p) = -\sum_{i=1}^{\beta} \alpha_i.$$

In particular, we have

$$\left|\sum_{x\in\mathbf{F}_p} \chi(f_1(x))e(f_2(x)/p)\right| \leqslant \beta\sqrt{p}.$$

In fact, one can provide formulas for the integer β that are quite explicit (in terms of the zeros and poles of the rational functions f_1 and f_2), and often one knows that $|\alpha_i| = \sqrt{q}$ for all i. For instance, if $f_1 = 1$ (so that the sum is an additive character sum) and f_2 is a polynomial such that $1 \leqslant \deg(f_2) < p$, then $\beta = \deg(f_2) - 1$, and $|\alpha_i| = \sqrt{p}$ for all p.

For more discussion and a proof in either the additive or multiplicative cases, we refer to [75].

The following exercises illustrate this general result in three important cases. Note however, that there is no completely elementary proof in the case of Kloosterman sums, where $\beta = 2$, leading to (6.1).

Exercise C.6.5 (Gauss sums) Let χ be a nontrivial multiplicative character of $\mathbf{F}_p^\times$ and $a \in \mathbf{F}_p^\times$. Denote

$$\tau(\chi, a) = \sum_{x \in \mathbf{F}_p} \chi(x) e(ax/p),$$

and put $\tau(\chi) = \tau(\chi, 1)$ (up to normalization, this is the same sum as occurs in the functional equation for the Dirichlet L-function $\mathrm{L}(s, \chi)$, see Theorem C.5.2).

(1) Prove that

$$|\tau(\chi, a)| = \sqrt{p}.$$

(This proves the corresponding special case of Theorem C.6.4 with $\beta = 1$ and $|\alpha_1| = \sqrt{p}$.) [**Hint**: Compute the modulus square, or apply the discrete Parseval identity.]

(2) Prove that for any automorphism σ of the field $\mathbf{C}$, we also have

$$|\sigma(\tau(\chi, a))| = \sqrt{p}.$$

(This additional property is also true for all α_i in Theorem C.6.4 in general; it means that each α_i is a so-called p-Weil number of weight 1.)

Exercise C.6.6 (Jacobi sums) Let χ_1 and χ_2 be nontrivial multiplicative characters of $\mathbf{F}_p^\times$. Denote

$$\mathrm{J}(\chi_1, \chi_2) = \sum_{x \in \mathbf{F}_p} \chi_1(x) \chi_2(1 - x).$$

(1) Prove that

$$\mathrm{J}(\chi_1, \chi_2) = \frac{\tau(\chi_1)\tau(\chi_2)}{\tau(\chi_1 \chi_2)},$$

and deduce that Theorem C.6.4 holds for the Jacobi sums $\mathrm{J}(\chi_1, \chi_2)$ with $\beta = 1$ and $|\alpha_1| = 1$. Moreover, show that α_1 satisfies the property of the second part of the previous exercise.

(3) Assume that $p \equiv 1 \pmod 4$. Prove that there exist integers a and b such that $a^2 + b^2 = p$ (a result of Fermat). [**Hint**: Show that there are characters χ_1 of order 2 and χ_2 of order 4 of $\mathbf{F}_p^\times$, and consider $\mathrm{J}(\chi_1, \chi_2)$.]

Exercise C.6.7 (Salié sums) Assume that $p \geqslant 3$.

(1) Check that there is a unique nontrivial real character χ_2 of $\mathbf{F}_p^\times$. Prove that for any $x \in \mathbf{F}_p$, the number of $y \in \mathbf{F}_p$ such that $y^2 = x$ is $1 + \chi_2(y)$.

(2) Prove that

$$\tau(\chi^2)\tau(\chi_2) = \chi(4)\tau(\chi)\tau(\chi\chi_2)$$

(Hasse–Davenport relation). [**Hint**: Use the formula for Jacobi sums, and compute $\mathrm{J}(\chi,\chi)$ in terms of the number of solutions of quadratic equations; express this number of solutions in terms of χ_2.]

For $(a,b) \in \mathbf{F}_p^\times$, define

$$\mathrm{S}(a,b) = \sum_{x \in \mathbf{F}_p^\times} \chi_2(x) e\left(\frac{ax + b\bar{x}}{p}\right).$$

(3) Show that for $b \in \mathbf{F}_p^\times$, we have

$$\mathrm{S}(a,b) = \sum_{\chi} s(\chi)\chi(a),$$

where

$$s(\chi) = \frac{\chi(b)\chi_2(b)\tau(\bar{\chi})\tau(\bar{\chi}\chi_2)}{q-1}.$$

[**Hint**: Use a discrete multiplicative Fourier expansion.]

(4) Show that

$$s(\chi) = \frac{\chi_2(b)\tau(\chi_2)}{q-1}\chi(4b)\tau(\bar{\chi}^2).$$

(5) Deduce that

$$\mathrm{S}(a,b) = \tau(\chi_2) \sum_{ay^2=4b} e\left(\frac{y}{p}\right).$$

(6) Deduce that Theorem C.6.4 holds for $\mathrm{S}(a,b)$ with either $\beta = 0$ or $\beta = 2$, in which case, $|\alpha_1| = |\alpha_2| = \sqrt{p}$.

References

[1] L.-P. Arguin, D. Belius, P. Bourgade, M. Radziwiłł, and K. Soundararajan: *Maximum of the Riemann zeta function on a short interval of the critical line*, Commun. Pure Appl. Math. 72 (2017), 500–535.

[2] R. Arratia, A. D. Barbour, and S. Tavaré: *Logarithmic combinatorial structures: a probabilistic approach*, EMS Monographs, 2003.

[3] P. Autissier, D. Bonolis, and Y. Lamzouri: *The distribution of the maximum of partial sums of Kloosterman sums and other trace functions*, preprint arXiv:1909.03266.

[4] B. Bagchi: *Statistical behaviour and universality properties of the Riemann zeta function and other allied Dirichlet series*, PhD thesis, Indian Statistical Institute, Kolkata, 1981; available at http://library.isical.ac.in:8080/jspui/bitstream/10263/4256/1/TH47.CV01.pdf.

[5] A. Barbour, E. Kowalski, and A. Nikeghbali: *Mod-discrete expansions*, Probability Theory and Related Fields, 2013; doi:10.1007/s00440-013-0498-8.

[6] P. Billingsley: *On the distribution of large prime factors*, Period. Math. Hungar. 2 (1972), 283–289.

[7] P. Billingsley: *Prime numbers and Brownian motion*, Am. Math. Monthly 80 (1973), 1099–1115.

[8] P. Billingsley: *The probability theory of additive arithmetic functions*, Ann. Probability 2 (1974), 749–791.

[9] P. Billingsley: *Probability and measure*, 3rd ed., Wiley, 1995.

[10] P. Billingsley: *Convergence of probability measures*, 2nd ed., Wiley, 1999.

[11] V. Blomer, É. Fouvry, E. Kowalski, Ph. Michel, D. Milićević, and W. Sawin: *The second moment theory of families of* L*-functions: the case of twisted Hecke* L*-functions*, Mem. AMS, to appear; arXiv:1804.01450.

[12] J. Bober, E. Kowalski, and W. Sawin: *On the support of the Kloosterman paths*, preprint (2019).

[13] E. Bombieri and J. Bourgain: *On Kahane's ultraflat polynomials*, J. Eur. Math. Soc. 11 (2009), 627–703.

[14] D. Bonolis: *On the size of the maximum of incomplete Kloosterman sums*, preprint arXiv:1811.10563.

[15] N. Bourbaki: *Éléments de mathématique*, *Topologie générale*, Springer, 2007.

[16] N. Bourbaki: *Éléments de mathématique, Fonctions d'une variable réelle*, Springer, 2007.

[17] N. Bourbaki: *Éléments de mathématique, Intégration*, Springer, 2007.

[18] N. Bourbaki: *Éléments de mathématique, Algèbres et groupes de Lie*, Springer, 2007.

[19] N. Bourbaki: *Éléments de mathématique, Théories spectrales*, Springer, 2019.

[20] L. Breiman: *Probability*, Classic in Applied Mathematics, Vol. 7, SIAM, 1992.

[21] E. Breuillard and H. Oh (eds): *Thin groups and super-strong approximation*, MSRI Publications Vol. 61, Cambridge University Press, 2014.

[22] F. Cellarosi and J. Marklof: *Quadratic Weyl sums, automorphic functions and invariance principles*, Proc. Lond. Math. Soc. (3) 113 (2016), 775–828.

[23] B. Cha, D. Fiorilli, and F. Jouve: *Prime number races for elliptic curves over function fields*, Ann. Sci. Éc. Norm. Supér. (4) 49 (2016), 1239–1277.

[24] P. L. Chebyshev: *Lettre de M. le Professeur Tchébychev à M. Fuss sur un nouveau théorème relatif aux nombres premiers contenus dans les formes* $4n+1$ *et* $4n+3$, in *Oeuvres de P.L. Tchebychef*, vol. I, Chelsea, 1962, pp. 697–698; also available at https://archive.org/details/117744684_001/page/n709.

[25] H. Cohen and H. W. Lenstra Jr.: *Heuristics on class groups of number fields*, in *Number theory (Noordwijkerhout, 1983)*, LNM, Vol. 1068, Springer, 1984, pp. 33–62.

[26] L. Devin: *Chebyshev's bias for analytic* L*-functions*, Math. Proc. Cambridge Philos. Soc. 169 (2020), 103–140.

[27] P. Donnelly and G. Grimmett: *On the asymptotic distribution of large prime factors*, J. London Math. Soc. 47 (1993), 395–404.

[28] J. J. Duistermaat and J. A. C. Kolk: *Distributions: theory and applications*, Birkhaüser, 2010.

[29] W. Duke, J. Friedlander, and H. Iwaniec: *Equidistribution of roots of a quadratic congruence to prime moduli*, Ann. Math. 141 (1995), 423–441.

[30] M. Einsiedler and T. Ward: *Ergodic theory: with a view towards number theory*, Graduate Texts in Mathematics, Vol. 259, Springer, 2011.

[31] N. D. Elkies and C. T. McMullen: *Gaps in* $\sqrt{n}$ *mod* 1 *and ergodic theory*, Duke Math. J. 123 (2004), 95–139.

[32] J. S. Ellenberg, A. Venkatesh, and C. Westerland: *Homological stability for Hurwitz spaces and the Cohen-Lenstra conjecture over function fields*, Ann. Math. 183 (2016), 729–786.

[33] P. Erdős: *On the density of some sequences of numbers, III*, J. LMS 13 (1938), 119–127.

[34] P. Erdős: *On the smoothness of the asymptotic distribution of additive arithmetical functions*, Am. J. Math. 61 (1939), 722–725.

[35] P. Erdős and M. Kac: *The Gaussian law of errors in the theory of additive number theoretic functions*, Am. J. Math. 62 (1940), 738–742.

[36] P. Erdős and A. Wintner: *Additive arithmetical functions and statistical independence*, Am. J. Math. 61 (1939), 713–721.

[37] W. Feller: *An introduction to probability theory and its applications*, Vol. 2, Wiley, 1966.

[38] D. Fiorilli and F. Jouve: *Distribution of Frobenius elements in families of Galois extensions*, preprint, https://hal.inria.fr/hal-02464349.

[39] Y. V. Fyodorov, G. A. Hiary, and J. P. Keating: *Freezing transition, characteristic polynomials of random matrices, and the Riemann zeta-function*, Phys. Rev. Lett. 108 (2012), 170601.

[40] G. B. Folland: *Real analysis*, Wiley, 1984.

[41] K. Ford: *The distribution of integers with a divisor in a given interval*, Ann. Math. 168 (2008), 367–433.

[42] É. Fouvry, E. Kowalski, and Ph. Michel: *An inverse theorem for Gowers norms of trace functions over* $\mathbf{F}_p$, Math. Proc. Cambridge Philos. Soc. 155 (2013), 277–295.

[43] J. Friedlander and H. Iwaniec: *Opera de cribro*, Colloquium Publ. 57, AMS, 2010.

[44] P. X. Gallagher: *The large sieve and probabilistic Galois theory*, in *Proc. Symp. Pure Math., Vol. XXIV*, AMS, 1973, pp. 91–101.

[45] P. X. Gallagher: *On the distribution of primes in short intervals*, Mathematika 23 (1976), 4–9.

[46] M. Gerspach: *On the pseudomoments of the Riemann zeta function*, PhD thesis, ETH Zürich, 2020.

[47] É. Ghys: *Dynamique des flots unipotents sur les espaces homogènes*, Séminaire N. Bourbaki 1991–92, exposé 747; http://numdam.org/item?id=SB_1991-1992__34__93_0.

[48] A. Granville: *The anatomy of integers and permutations*, preprint (2008), www.dms.umontreal.ca/~andrew/PDF/Anatomy.pdf.

[49] A. Granville and J. Granville: *Prime suspects*, illustrated by R. J. Lewis, Princeton University Press, 2019.

[50] A. Granville and G. Martin: *Prime number races*, Am. Math. Monthly 113 (2006), 1–33.

[51] A. Granville and K. Soundararajan: *Sieving and the Erdös-Kac theorem*, in *Equidistribution in number theory, an introduction*, Springer, 2007, pp. 15–27.

[52] G. H. Hardy and E. M. Wright: *An introduction to the theory of numbers*, 5th ed., Oxford University Press, 1979.

[53] A. Harper: *Two new proofs of the Erdős-Kac Theorem, with bound on the rate of convergence, by Stein's method for distributional approximations*, Math. Proc. Camb. Philos. Soc. 147 (2009), 95–114.

[54] A. Harper: *The Riemann zeta function in short intervals*, Séminaire N. Bourbaki, 71ème année, exposé 1159, March 2019.

[55] A. Harper: *On the partition function of the Riemann zeta function, and the Fyodorov–Hiary–Keating conjecture*, preprint (2019); arXiv:1906.05783.

[56] A. Harper and Y. Lamzouri: *Orderings of weakly correlated random variables, and prime number races with many contestants*, Probab. Theory Related Fields 170 (2018), 961–1010.

[57] C. Hooley: *On the distribution of the roots of polynomial congruences*, Mathematika 11 (1964), 39–49.

[58] K. Ireland and M. Rosen: *A classical introduction to modern number theory*, 2nd ed., GTM 84, Springer, 1990.

[59] H. Iwaniec and E. Kowalski: *Analytic number theory*, Colloquium Publ. 53, AMS, 2004.

[60] O. Kallenberg: *Foundations of modern probability theory*, Probability and Its Applications, Springer, 1997.

[61] N. M. Katz: *Gauss sums, Kloosterman sums and monodromy groups*, Annals of Mathematics Studies, Vol. 116, Princeton University Press, 1988.

[62] N. M. Katz and P. Sarnak: *Zeroes of zeta functions and symmetry*, Bull. Am. Math. Soc. 36 (1999), 1–26.

[63] J. P. Keating and N. C. Snaith: *Random matrix theory and* $\zeta(1/2+it)$, Commun. Math. Phys. 214 (2000), 57–89.

[64] D. Koukoulopoulos: *Localized factorizations of integers*, Proc. London Math. Soc. 101 (2010), 392–426.

[65] E. Kowalski: *The large sieve and its applications*, Cambridge Tracts Math., Vol. 175, Cambridge University Press, 2008.

[66] E. Kowalski: *Poincaré and analytic number theory*, in *The scientific legacy of Poincaré*, edited by É. Charpentier, É. Ghys, and A. Lesne, AMS, 2010.

[67] E. Kowalski: *Sieve in expansion*, Séminaire Bourbaki, Exposé 1028 (November 2010), in Astérisque 348, Soc. Math. France (2012), 17–64.

[68] E. Kowalski: *The large sieve, monodromy, and zeta functions of algebraic curves, II Independence of the zeros*, Art. ID rnn 091, IMRN (2008).

[69] E. Kowalski: *Families of cusp forms*, Pub. Math. Besançon 2013 (2013), 5–40.

[70] E. Kowalski: *An introduction to the representation theory of groups*, Graduatte Studies in Mathematics, Vol. 155, AMS, 2014.

[71] E. Kowalski: *The Kloostermania page*, http://blogs.ethz.ch/kowalski/the-kloostermania-page/.

[72] E. Kowalski: *Bagchi's Theorem for families of automorphic forms*, in *Exploring the Riemann Zeta function*, Springer, 2017, pp. 180–199.

[73] E. Kowalski: *Averages of Euler products, distribution of singular series and the ubiquity of Poisson distribution*, Acta Arithmetica 148.2 (2011), 153–187.

[74] E. Kowalski: *Expander graphs and their applications*, Cours Spécialisés 26, Soc. Math. France, 2019.

[75] E. Kowalski: *Exponential sums over finite fields, I: elementary methods*, www.math.ethz.ch/~kowalski/exp-sums.pdf.

[76] E. Kowalski and Ph. Michel: *The analytic rank of* $\mathrm{J}_0(q)$ *and zeros of automorphic L-functions*, Duke Math. J. 100 (1999), 503–542.

[77] E. Kowalski and A. Nikeghbali: *Mod-Poisson convergence in probability and number theory*, International Mathematics Research Notices 2010; doi:10.1093/imrn/rnq019.

[78] E. Kowalski and A. Nikeghbali: *Mod-Gaussian convergence and the value distribution of* $\zeta(1/2+it)$ *and related quantities*, J. LMS 86 (2012), 291–319.

[79] E. Kowalski and W. Sawin: *Kloosterman paths and the shape of exponential sums*, Compositio Math. 152 (2016), 1489–1516.

[80] E. Kowalski and K. Soundararajan: *Equidistribution from the Chinese Remainder Theorem*, preprint (2020).

[81] N. V. Krylov: *Introduction to the theory of random processes*, Graduate Studies in Mathematics, Vol. 43, AMS, 2002.

[82] Y. Lamzouri: *On the distribution of the maximum of cubic exponential sums*, J. Inst. Math. Jussieu 19 (2020), 1259–1286.

[83] D. Li and H. Queffélec: *Introduction à l'étude des espaces de Banach; Analyse et probabilités*, Cours Spécialisés, Vol. 12, SMF, 2004.

[84] A. Lubotzky, R. Phillips, and P. Sarnak: *Ramanuajan graphs*, Combinatorica 8 (1988), 261–277.

[85] W. Bosma, J. Cannon, and C. Playoust: *The Magma algebra system, I. The user language*, J. Symbolic Comput. 24 (1997), 235–265; also http://magma.maths.usyd.edu.au/magma/.

[86] J. Markov and A. Strömbergsson: *The three gap theorem and the space of lattices*, Am. Math. Monthly 124 (2017), 741–745.

[87] D. Milićević and S. Zhang: *Distribution of Kloosterman paths to high prime power moduli*, preprint, arXiv:2005.08865.

[88] S. J. Montgomery-Smith: *The distribution of Rademacher sums*, Proc. AMS 109 (1990), 517–522.

[89] D. W. Morris: *Ratner's theorems on unipotent flows*, Chicago Lectures in Mathematics, University of Chicago Press, 2005.

[90] J. Najnudel: *On the extreme values of the Riemann zeta function on random intervals of the critical line*, Probab. Theory Related Fields 172 (2018), 387–452.

[91] J. Neukirch: *Algebraic number theory*, Springer, 1999.

[92] C. Perret-Gentil: *Some recent interactions of probability and number theory*, Newsl. Eur. Math. Soc. (March 2019).

[93] PARI/GP, version `2.6.0`, Bordeaux, 2011, http://pari.math.u-bordeaux.fr/.

[94] J. Pintz: *Cramér vs. Cramér: on Cramér's probabilistic model for primes*, Functiones Approximatio 37 (2007), 361–376.

[95] M. Radziwiłł and K. Soundararajan: *Selberg's central limit theorem for* $\log|\zeta(1/2+it)|$, L'enseignement Math. 63 (2017), 1–19.

[96] M. Radziwiłł and K. Soundararajan: *Moments and distribution of central* L*-values of quadratic twists of elliptic curves*, Invent. Math. 202 (2015), 1029–1068.

[97] M. Radziwiłł and K. Soundararajan: *Value distribution of* L*-functions*, Oberwolfach report 40/2017, to appear.

[98] O. Randal-Williams: *Homology of Hurwitz spaces and the Cohen–Lenstra heuristic for function fields (after Ellenberg, Venkatesh, and Westerland)*, Seminaire N. Bourbaki 2019, exposé 1162; arXiv:1906.07447; in Astérisque, to appear.

[99] M. Ratner: *On Raghunathan's measure conjecture*, Ann. Math (2) 134 (1991), 545–607.

[100] A. Rényi and P. Turán: *On a theorem of Erdős and Kac*, Acta Arith. 4 (1958), 71–84.

[101] G. Ricotta and E. Royer: *Kloosterman paths of prime powers moduli*, Comment. Math. Helv. 93 (2018), 493–532.

[102] G. Ricotta, E. Royer, and I. Shparlinski: *Kloosterman paths of prime powers moduli, II*, Bull. SMF, to appear.

[103] J. Rotman: *Advanced modern algebra*, part I, 3rd ed., Graduate Studies in Mathematics, Vol. 165, AMS, 2015.

[104] D. Revuz and M. Yor: *Continuous martingales and Brownian motion*, 3rd ed., Springer, 1999.
[105] M. Rubinstein and P. Sarnak: *Chebyshev's bias*, Exp. Math. 3 (1994), 173–197.
[106] W. Rudin: *Real and complex analysis*, McGraw-Hill, 1970.
[107] P. Sarnak: letter to B. Mazur on the Chebychev bias for $\tau(p)$, https://publications.ias.edu/sites/default/files/MazurLtrMay08.PDF.
[108] I. Schoenberg: *Über die asymptotische Verteilung reeller Zahlen mod* 1, Math. Z. 28 (1928), 171–199.
[109] I. Schoenberg: *On asymptotic distributions of arithmetical functions*, Trans. AMS 39 (1936), 315–330.
[110] A. Selberg: *Contributions to the theory of the Riemann zeta function*, Arch. Math. Naturvid. 48 (1946), 89–155; or in *Collected works*, I.
[111] J.-P. Serre: *Linear representations of finite groups*, Graduate Texts in Mathematics, Vol. 42, Springer, 1977.
[112] J.-P. Serre: *A course in arithmetic*, Graduate Texts in Mathematcis, Vol. 7, Springer, 1973.
[113] V. T. Sós: *On the theory of diophantine approximations*, Acta Math. Acad. Sci. Hungar. 8 (1957), 461–472.
[114] M. Talagrand: *Concentration of measure and isoperimetric inequalities in product spaces*, Publ. Math. IHÉS 81 (1995), 73–205.
[115] G. Tenenbaum: *Introduction to analytic and probabilistic number theory*, Cambridge Studies in Advanced Mathematics, Vol. 46, Cambridge University Press, 1995.
[116] E. C. Titchmarsh: *The theory of functions*, 2nd ed., Oxford University Press, 1939.
[117] E. C. Titchmarsh: *The theory of the Riemann zeta function*, 2nd ed., Oxford University Press, 1986.
[118] A. Tóth: *Roots of quadratic congruences*, Int. Math. Res. Notices 2000 (2000), 719–739.
[119] S. M. Voronin: *Theorem on the universality of the Riemann zeta function*, Izv. Akad. Nauk SSSR, Ser. Matem. 39 (1975), 475–486; English translation in Math. USSR Izv. 9 (1975), 443–445.
[120] H. Weyl: *Über die Gleichverteilung von Zahlen mod. Eins*, Math. Ann. 77 (1914), 313–352.
[121] A. Zygmund: *Trigonometric sums*, 3rd ed., Cambridge Math Library, Cambridge, University Press 2002.

Index

Printed in the United States
by Baker & Taylor Publisher Services